Electron Transport in Nanostructures and Mesoscopic Devices

Electron Transport in Nanostructures and Mesoscopic Devices

An Introduction

Thierry Ouisse

Series Editor
Mireille Mouis

First published in Great Britain and the United States in 2008 by ISTE Ltd and John Wiley & Sons, Inc.

ISTE Ltd
6 Fitzroy Square
London W1T 5DX
UK

John Wiley & Sons, Inc.
111 River Street
Hoboken, NJ 07030
USA

www.iste.co.uk

www.wiley.com

© ISTE Ltd, 2008

Library of Congress Cataloging-in-Publication Data

Ouisse, Thierry.
  Electron transport in nanostructures and mesoscopic devices / Thierry Ouisse.
    p. cm.
  Includes bibliographical references and index.
  ISBN 978-1-84821-050-9
  1. Electron transport. 2. Nanostructured materials--Electric properties. 3. Nanostructures--Electric properties. 4. Mesoscopic phenomena (Physics) I. Title.
  QC176.8.E4O95 2008
  530.4'1--dc22

                                                                                    2008008768

British Library Cataloguing-in-Publication Data
A CIP record for this book is available from the British Library
ISBN: 978-1-84821-050-9

Table of Contents

Chapter 1

Introduction

1.1. Introduction and preliminary warning

Matter stability and the way in which rigid crystalline or amorphous arrays of atoms can be formed are ruled by two pillars of physics: electromagnetism and quantum mechanics; nothing else, provided that we admit the existence of elementary constituents such as atom nuclei without having to derive their internal structure from the first principles (then we need to add nuclear forces to our bunch of tools). The postulates and basic equations of these two theories can be written on a couple of pages, and everything can be derived from them[1]. If the world was ruled by classical mechanics, it would simply be impossible to obtain stable atoms[2] or stable chemical bonding to ensure the existence of matter as we all experience it in our everyday life. Thus, it is something of a misnomer to say that we are going to study quantum devices as opposed to devices which would not be quantum. Everything is ruled by quantum mechanics, from the insulating or conducting character to the color of any piece of matter or object that you can see inside the room where you are now reading this introduction (see also Figure 1.1). To understand our macroscopic world, we often feel that once we admit the existence of stable matter, we can content ourselves with using the second Newton's law of motion and classical gravitational forces. An aeronautics engineer does not put too much quantum mechanics in his calculations, but this is certainly no longer the case

---

1 Of course with a substantial amount of hard work and mathematics, and adding some thermodynamics. Note also that if quantum mechanical predictions can be verified with an astonishingly high precision, their interpretation was (and is) the source of thousands of scientific articles and books.
2 Classical electrons accelerated over orbits radiate electromagnetic waves and thus lose energy. Thus, bound electrons would collapse onto the atoms.

if we want to justify the way in which electrons and therefore the electrical current behaves in a bulk semiconductor. Without a periodic atomic lattice and quantum mechanics, we could not find free electrons able to carry a current in a *p-n* junction, or in the channel of transistors which form the integrated circuits inside our computers. Thus, the reason why the devices under study in this book are called quantum is that we can straightforwardly apply to them the basic quantum effects that students are accustomed to calculating in an introductory quantum mechanics course.

Figure 1.1. *The ubiquitous character of quantum mechanics*

In nanostructures, electrons can be confined in potential wells narrow enough to obtain energy quantization along the confining direction. Their dimension is small enough for probing the dual wave-particle nature of the electron in a straightforward manner, because the electron wave function phase can be kept coherent over the whole device length. Thus, it becomes possible to observe wave interference effects just by measuring the average current which can be passed through such components, and particle-like properties from current noise data. As once stated by the physicist Esaki, this looks like some kind of "do-it-yourself" quantum mechanics: you are not required to become a specialist in group theory and irreducible representations, or of field-theoretic methods to get in touch with the essence of the topic (see also Figure 1.2). In addition, other specific effects, although not quantum-mechanical, are also due to reduced dimensions: if you can inject a few electrons into a nanostructure and if the capacitance between this nanostructure and the rest of the world is very small, we can probe effects which are due to charge granularity (we cannot divide the electron charge), and which are known as Coulomb blockade. Such effects are the subject of intensive research in R&D laboratories, because many people hope to put them to good use to produce new types of memories and devices that are smaller, faster and require a smaller amount of operating power. The aim of this book is to give an introduction to the basic concepts which govern the conduction mechanisms taking place in such small devices.

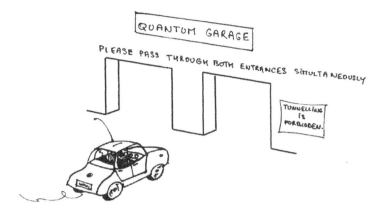

Figure 1.2. *The quantum garage*

Many (not to say most) of the phenomena described in this book usually take place at quite low temperatures, or in devices not yet (and for some of them never) used in the industry. The physics described here is not useful for understanding how industrial semiconductor devices behave in most applications right now, with the notable exception of resonant tunneling. Nevertheless, "today's" silicon (*Si*) Metal-Oxide-Semiconductor Field-Effect Transistors (MOSFETs) definitely exhibit non-stationary and ballistic transport effects. Explaining these effects requires us to use some of the concepts developed in this book, even the high electric fields involved in MOSFET operation make the application of such concepts much more complicated than what is described in this introduction. At room temperature, the electron mean free path in silicon is in the 5-10 nm range, not far from the 45 nm channel length of the current CMOS technology, and integrated chips using a 32 nm process technology have already been demonstrated by the INTEL corporation in 2007. Figure 1.3 shows the picture of a 20 nm channel length prototype MOSFET produced in 2006 by LETI-CEA. Thus, even at room temperature some commercial electronic devices are close to the ballistic regime. Those *industrial* MOSFET's are fabricated with an incredibly high reproducibility in order to form extremely complex integrated circuits (and as a side note such precision and reproducibility are actually far from being achieved in most research laboratories working in the realm of mesoscopic physics and nanostructures, or with semiconductors more exotic and physically more appealing than silicon). Device-modeling based on ballistic properties has thus become an active research field, even in the case of silicon devices (see, e.g., [NAT 94] for one of the pioneering Si papers).

In addition, mesoscopic effects are important in four respects:

(i) they are often of great physical significance, and give a deep and straightforward insight into some of the most striking implications of quantum mechanics (for instance, they provide unambiguous and clear demonstrations of the dual electron nature, particle and wave);

(ii) although often obtained at low temperatures or high magnetic fields they are very useful for extracting physical parameters dealing with (nano)structures actually used in applications;

(iii) some of the effects are already used in (e.g. resonant tunneling) or potentially useful for (e.g. Coulomb blockade) applications;

(iv) although still difficult to engineer, devices made from graphene or carbon nanotubes exhibit truly ballistic and *quantum-coherent* effects even at room temperature. Thus, it is quite possible that not only ballistic, but also quantum-coherent effects may be present in electronic applications in the near future.

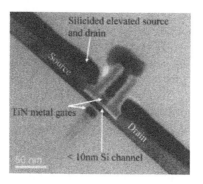

Figure 1.3. *A transmission electron microscope view of a planar double-gate MOSFET fabricated by LETI-CEA with a 20 nm channel length; reproduced by permission after J. Widiez et al., IEEE Transactions on nanotechnology, vol. 5, p. 643 (2006), copyright ©2006IEEE ([WID 06])*

As a consequence, in most of the largest semiconductor companies, and in a very large number of university labs, intensive research work is devoted to such structures. Scarcely applied though it may seem at first sight, this field of activity is in fact the leading edge of semiconductor research.

This book is designed to be accessible to the independent reader, and to students not having a strong background in solid-state physics (e.g. issued from engineering disciplines). As a matter of fact, this book is an attempt to answer the following question: what must be taught to students starting from scratch to make them understand the bases of electron transport in mesoscopic devices? A professor placed in such a situation soon realizes that a good deal of solid-state physics and quantum mechanics is required. This explains the incorporation of chapters which are usually absent from the more specialized, already-existing books, and marks the difference between them and this. In addition, to follow the classification once given by J.M. Ziman, this book does not fall into the category of a "treatise" but into that of a "textbook", with the purpose of introducing and explaining concepts. The text has been written with the aim of being as self-contained as possible, and is based on an oral course delivered at an international European master's degree course involving three technical universities (GrenobleINP, EPFLausanne and Polit'oTorino). It is a deliberate choice of the author to keep in the book the spirit of the oral course, and this is the reason why the reader should not be surprised to be sometimes interpellated or hailed in a somewhat familiar way[3].

Assimilating the quantum-mechanical rules summarized at the very beginning of the book suffices to derive any subsequent result, but should by no means be considered as enough to master quantum mechanics itself. Hence, and despite the fact that the text remains at an introductory level, a complete understanding of the course probably requires a minimum prior knowledge and self-maturation of the basic quantum-mechanical concepts. A reader not acquainted with this field will certainly feel the need to consult more authoritative manuals, due to the innumerable number of questions, either technical or fundamental, that a concise and *incomplete* presentation of quantum mechanics must arouse in any normally constituted mind. Some knowledge of solid-state and semiconductor physics certainly help as well, but all concepts useful for understanding the book can in principle be found in the book itself, and since this book is an introduction dedicated to a broad audience, maybe some of you are probably already acquainted with the required solid-state physics notions. For those who are experienced in solid-state physics it is possible to simply skip most of the reminders which make up Chapter 2. Besides, many of those reminders are not always quite rigorously demonstrated. All undemonstrated or heuristically-derived quantum-mechanical formulae can be found and are rigorously derived in a self-contained, encyclopedic textbook: [COH 77]. Solid-state physics has its self-contained book too: [ASH 76]. For *bulk* semiconductor physics and transport, an advanced and quite remarkable and complete textbook was written by [RID 82], but it is not essential for understanding this book. Eventually, we can find books specifically devoted to mesoscopic electron transport, which can be of great support for a better understanding or for gaining more information (the list below is not exhaustive): [BEE 91], [KEL 95], [DAT 95] and [FER 97]. The book which is the closest in spirit to this course is the one by Datta. It includes many exercises and also contains more advanced formalisms (e.g. Green's functions) and discussions,

---

3 As you may have already noticed, the familiar way of addressing the reader began in the very first lines of this introduction.

which are not necessarily required at this introductory level. The book by Kelly presents a very large amount of data and also deals with aspects which are either more related to technological aspects or closer to the applications.

This book is an introduction and as such a number of important aspects have been omitted, mainly those which imply the use of mathematical concepts too involved to be developed in front of an audience new to the field. In particular, the reader will not find here a rigorous description of Green's function formalism, which is necessary to include electron-electron interactions in transport modeling. A general discussion and study of many-body effects is also absent, which would be mandatory to understand a physical phenomenon such as the fractional quantum Hall effect, metal-based mesoscopic devices, carbon nanotubes operating in the 1D form of a Luttinger liquid and many others. Justice has not been done to the electron spin and its possible applications. This book could thus be given a second title: how far can we go using only independent electrons and the exclusion Pauli principle (see also Figure 1.4)? Surprising though it may seem, a good deal of nanostructure physics can still be grasped that way, but the reader will not find in this book a wealth of phenomena associated with electron-electron interactions. If they are not discouraged by this introductory text their study should constitute the next step, to be achieved by studying more specialized treatises and articles. Thus, if after studying the various chapters the student decides to read further and deeper, the main objective of this book will have been fulfilled. In the same spirit, we shall skip some difficult demonstrations which would be required for a rigorous derivation of some important solid-state physics results[4]. However, even if difficult theoretical techniques have been deliberately banished from the text, "the language of physics is mathematics", and none of the chapters escape from the rule.

Figure 1.4. *The quantum society and Pauli's exclusion principle*

---

4 Whenever this occurs, the unsatisfied reader will always be left with the possibility of consulting the more advanced textbooks or specialized articles mentioned in the bibliography.

Most exercises proposed at the end of each chapter are easy and their purpose is to provide the reader with a means of checking that they have correctly assimilated the chapter content and concepts. However, some of them require more time, and have been inserted to complete points not detailed in the main text.

Not all the sections were dealt with during the original oral course. I have put indicators at the beginning of each section:

◑ This section is a reminder. Thus it can be skipped if the reader is already familiar with the corresponding field.

● This section is essential to the book (and, quite accessorily, it may be helpful to prepare an exam). Some reminders belong to this category.

◍ This section is not a reminder, but is not considered as essential to understand the other parts.

○ This section can be skipped at first reading.

1.2. Bibliography

[ASH 76] ASHCROFT N.W. and MERMIN N.D., *Solid State Physics*, Wiley, New York, 1976.

[BEE 91] BEENAKER C.W.J. and VAN HOUTEN H., *Quantum Transport in Semiconductor Nanostructures*, Solid State Physics 44, Academic Press, 1991.

[COH 77] COHEN-TANNOUDJI C., DIU B. and LALOË F., *Quantum Mechanics*, Wiley, New York, 1977.

[DAT 95] DATTA S., *Electronic Transport in Mesoscopic Systems*, Cambridge University Press, 1995.

[FER 97] FERRY D.K. and GOODNICK S.M., *Transport in Nanostructures*, Cambridge University Press, 1997.

[KEL 95] KELLY M.J., *Low-dimensional Semiconductors*, Oxford University Press, 1995.

[NAT 94] NATORI K., "Ballistic metal-oxide-semiconductor field effect transistor", *Journal of Applied Physics*, vol. 76, no. 8, 1994, p. 4879-4890.

[RID 82] RIDLEY, B.K. *Quantum Processes in Semiconductors*, Clarendon Press, Oxford, 1982.

[WID 06] WIDIEZ J., POIROUX T., VINET M., MOUIS M., DELEONIBUS S., "Experimental comparison between sub-0.1 μm ultrathin SOI single- and double-gate MOSFETs: performance and mobility", *IEEE Transactions on Nanotechnology*, vol. 5, no. 6, 2006, p. 643-648.

Chapter 2

Some Useful Concepts and Reminders

2.1. Quantum mechanics and the Schrödinger equation ◎

2.1.1. *A more than brief introduction*

The following is only a summary which includes the basic quantum-mechanical (QM) equations required for understanding the book. It is by no means a rigorous introduction to the topics, and if you want to go further, a wise thing to do would be to immerse yourself, e.g., in the introductory textbook by R.P. Feynman [FEY 65], and then in the book by Cohen-Tannoudji *et al.* [COH 77] for a while[1]. Besides, several formulations can be used to describe quantum mechanics, and here we shall not really make the effort of differentiating them from one another. A concise description of those different formulations can be found in [STY 02].

In classical mechanics the elementary constituents of matter are massive point particles whose movement is controlled by electromagnetic or gravitational forces. At any instant we can precisely define the particle position and, provided that at a time t we are given the position and velocity of all the system particles, we can calculate everything at any other time, and obtain well-defined trajectories (with a powerful enough computer if the particles are numerous, etc.), even if the system remains isolated. Thus, the whole picture is *in principle* perfectly deterministic. In quantum mechanics the situation is far more subtle. Experimentally, it appears that if we let a system evolve isolated for a while, the maximum information concerning

---

1 Of course these are not the only useful introductory QM textbooks, and the reader can also consult, e.g., [DIR 58], [SCH 68], [MES 62], [BOH 54], [MER 70], [LAN 65], among which many present a historical interest in addition to their scientific value, and there are many others.

this system that is physically accessible to human knowledge does not allow us to predict in a deterministic and unique way the result which will be obtained once we act on this system to measure some of its properties.

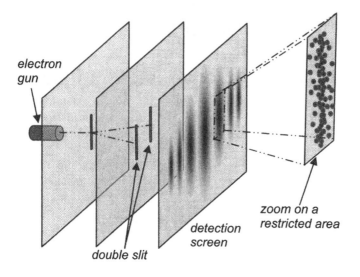

Figure 2.1. *Quantum-mechanical interference experiment illustrating the dual wave-particle nature of the electron*

The celebrated double-slit interference experiment is probably one of the more striking and meaningful illustrations of the quantum nature of matter. This effect figures in due place in almost any introduction chapter on quantum mechanics, and we shall respect this very justified habit. Interference experiments such as that illustrated by Figure 2.1 reveal that it is no longer possible to consider an entity such as an electron or a proton as a particle, and that it is not possible to consider it as a pure wave either [FEY 65]. "Identically prepared" electrons propagating through double slits exhibit interference patterns like waves [JON 61], but if we put a screen behind the plane of those slits we always obtain localized spots, as for particles [TON 89]. It is the statistical collection of a large number of such individual events which forms the interference pattern. Thus, in quantum mechanics (and in the real world) we have to assign a dual nature to electrons, whose behavior can be modeled only as a combination of both a particle and a wave. Suppress one slit and we lose the interference pattern. The wave really passes through both slits. Try to detect the electron at one of the two slits and we also lose the pattern, because the particle-like detection at one slit instantaneously reduces the extended propagating wave.

In some textbooks it is stated that quantum mechanics does not allow us, *even in principle*, to calculate any trajectory, and that it is a probabilistic theory *in essence*. This is not a correct statement, because one interpretation, known as the de Broglie-Bohm theory, gives a perfectly deterministic picture of quantum mechanics (at least for massive particles). In such an interpretation both a wave and a particle co-exist. The wave guides the particle, and in Bohm's version the *only* guiding rule states that the particle momentum is equal to the phase gradient of the complex wave obeying the Schrödinger equation multiplied by $\hbar$, a physical quantity known as the Planck constant [HOL 93]. In such a picture we can calculate well-defined trajectories (which are quite weird compared to classical ones, due to the action of the guiding wave). The unknown parameters, or "hidden variables", which make experiments exhibit a statistical aspect are nothing but the initial particle coordinates with respect to the wave. Thus, it is not possible to say from quantum mechanics that the basic facts of nature are undeterministic *in principle*. However, since this interpretation does not provide any new prediction with respect to the usual quantum rules, and exhibits the moral drawback of being obviously non-local, it has not attracted the favor of most physicists[2].

The "orthodox" interpretation of the quantum-mechanical formalism is that in between measurements we cannot precisely define a thing such as a particle; the experimental indeterminacy obtained when repeating the same experiment a large number of times with identically prepared systems results from the indeterminacy of nature itself, and not from a difference in system preparation which would be unknown to the observer. This was quite an incredible statement when quantum mechanics emerged, but it has now become a common "philosophical" view among scientists. In this book we shall not enter into those considerations any longer. We shall just use the quantum-mechanical rules, which up to now have always been experimentally validated with a numerical precision unprecedented by any other physical theory.

---

2 Non-locality, or the possibility of instantaneous action-at-a-distance, does not agree with another pillar of modern physics, special relativity, so physicists are reluctant to accept it in other theories. In Bohm's interpretation experiments conducted on space-separated but correlated particles make this non-locality quite explicit. Note that in private most scientists would admit that quantum mechanics is deeply non-local, whatever the interpretation, but in public they see it as an unforgiveable flaw, as soon as this non-locality is no longer buried in the mysteries of the "Copenhagen's school" or "orthodox" interpretation, and is given a straightforward meaning. Note that Bohm's interpretation also has its "weaknesses". For instance, if at some time t the particle distribution is given by the square modulus of the wave function, this implies that it will be true forever; however, there is nothing which rigorously justifies when and how it began to be so.

2.1.2. *The postulates of quantum mechanics*

Here we reproduce the postulates as they are expressed in most quantum mechanics textbooks (see, e.g., [COH 77]). Maybe some of you are already acquainted with them, but for others it is perhaps not completely useless to give a reminder. If you already followed a good course in quantum mechanics just skip this part; you will learn nothing from it. If you are more inexperienced and require further explanation consult any quantum mechanics textbook.

First postulate: *At a given time t, the state of a physical system is described by an abstract state vector $|\psi(t)\rangle$, also called a ket, which belongs to the state space.*

In practice, we shall essentially appeal to the "quantum mechanics of the poor man", most often contenting ourselves with identifying those states with the wave functions $\psi(\vec{r},t)$ obtained by solving the Schrödinger equation inside our nanostructures. Before a position measurement, the nature of the electron wave prevents us from assigning a uniquely defined space-time position to the particle. These wave functions are mathematical devices associated with a given electron and *link to each point of space a complex number*. As with any wave, they can propagate or lead to stationary phenomena. The scalar product of two kets $|\varphi\rangle$ and $|\psi\rangle$ is defined as

$$\langle \varphi | \psi \rangle = \int \varphi^*(\vec{r}) \psi(\vec{r}) d^3r \tag{2.1}$$

where φ^* is the complex conjugate of φ (the state space is a mathematical, complex Hilbert space; it is formed by *complex functions* operating on *real space*, and not using simple real vectors). The notation $\langle \varphi |$ allows us to manipulate easily scalar products such as in equation (2.1). This was proposed by Dirac and is called a bra. The wave nature of an electron forbids its precise localization as long as it is not subject to a position measurement, during which it reveals its particle nature. Therefore, it is clear that in a quantum system, classical measurable quantities which depended on position, or on position and velocity (such as, for instance, the energy), cannot be assigned a precise and unique value unless they are specifically measured. Their description thus requires an operation which can act on the whole wave field, and this leads us to the formulation of the second postulate.

Second postulate: *Any measurable physical quantity $\mathcal{M}$ (such as, e.g., a position, or a momentum, or an energy) is described by an operator M which acts on the state vector $|\psi(t)\rangle$. This operator is called an observable.*

We can apply to a state vector the operator B, and then another operator A. The operator corresponding to those two successive actions is noted AB, but this product is not always commutative. AB is not necessarily equal to BA, and when these two operator products differ we say that A and B do not commute. A famous example of non-commuting operators is the couple formed by position and momentum. Non-commutativity of some operators is indeed at the heart of many strange consequences of quantum mechanics. The quantity $AB-BA$ is called the commutator of A and B, and is denoted $[AB]$.

Third postulate: *The result of a measurement of a physical quantity $\mathcal{M}$ is always an eigenvalue of the corresponding observable M.*

Consider for instance the position operator. Its eigenvalues are formed by the ensemble of all three-dimensional real vectors $\vec{r}_0$. The corresponding eigenstates are Dirac peaks centered at $\vec{r}_0$, which are written under the form $|\vec{r}_0\rangle = \delta(\vec{r} - \vec{r}_0)$. For the next postulate we shall limit ourselves to the case of a non-degenerate spectrum (i.e. we assume that to one eigenvalue corresponds one and only one eigenstate).

Fourth postulate: *If the spectrum of the observable $\mathcal{M}$ is discrete, and if the state vector is normalized, the probability $P(m_n)$ of obtaining the eigenvalue m_n as a measurement result is equal to $P(m_n)=|\langle u_n | \psi \rangle|^2$, where $|u_n\rangle$ is the normalized eigenvector of M associated with the eigenvalue m_n. If the spectrum is continuous, the probability $dP(\mu)$ of obtaining a result between μ and $\mu+d\mu$ is equal to $dP(\mu)=|\langle u_\mu|\psi\rangle|^2$, where $|u_\mu\rangle$ is the eigenvector of the eigenvalue μ.*

An example of a continuous spectrum is the position, and an example of a discrete spectrum is the energy inside a quantum well. From this postulate we can give a probabilistic interpretation of the wave function (initially proposed by M. Born): if we make a position measurement, from the fourth postulate the probability of finding the particle at a position $\vec{r}_0$ is given by $P(\vec{r}_0)=|\langle \vec{r}_0 |\psi\rangle|^2=\int\delta(\vec{r} - \vec{r}_0)\psi(\vec{r})d^3r=|\psi(\vec{r}_0)|^2$. Thus, from the fourth postulate $|\psi(\vec{r})|^2$ is nothing but the probability density of finding the particle at a position $\vec{r}$. However, be careful: in the "orthodox" interpretation this is *not* the probability of the particle being at $\vec{r}$, and we cannot say that we do not know its position before the measurement simply because we are missing some information. Before a measurement the entity "electron" exists, but your particle is literally "nowhere". Otherwise we could not obtain any wave interference effect. No matter how strange nature may seem, it is really the "measurement process" which turns an entity such as an electron into a corpuscle. In between measurements (or energy exchange with the macroscopic external world) the only relevant physical entity to prescribe the system evolution is the electron wave and nothing else.

Fifth postulate (also called the measurement postulate): *If the measurement result of the physical quantity $\mathcal{M}$ gives the result m_n, the state vector of the system immediately after the measurement is the normalized projection*

$$\frac{P_n|\psi\rangle}{\sqrt{\langle\psi|P_n|\psi\rangle}}$$

of $|\psi\rangle$ onto the sub-eigenspace associated with m_n (P_n is the projection operator).

If the spectrum is non-degenerate this means that just after a measurement the state vector is necessarily equal to the normalized eigenstate corresponding to the obtained eigenvalue m_n. This postulate has caused more ink to flow than all the newspaper issues devoted to Princess Diana and Madonna put together. Nevertheless, in this book we shall not even try to discuss the subtle issues which are attached to it.

Sixth (and final) postulate: *The time evolution of the state vector obeys Schrödinger's equation*

$$H(t)|\psi(t)\rangle = i\hbar\frac{d|\psi\rangle}{dt} \quad , \tag{2.2}$$

in which the Hamiltonian H(t) is the operator associated with the energy of the system.

2.1.3. Essential properties of observables

An operator A is *Hermitian* or *self-adjoint* if it verifies the property

$$\int\varphi * A\psi\, d^3r = \int(A\varphi) * \psi\, d^3r\,. \tag{2.3}$$

As a matter of fact, in quantum mechanics *all observables are both linear and Hermitian*. Linearity ensures the validity of the superposition principle of our wave functions, which derives from the form of the Schrödinger equation. Hermiticity is required because if we measure something, we always obtain a real number, not a complex one, even if the scalar products are calculated using complex waves. Now from equation (2.3) and the fifth postulate, it is easy to demonstrate that this is achieved with Hermitian operators, because we have

$$m_n = \langle u_n | Mu_n \rangle = \int u_n^* Mu_n \, d^3r \tag{2.4}$$
$$= \int (Mu_n)^* u_n \, d^3r = (\int u_n * Mu_n d^3r)^* = m_n^* .$$

Since the eigenvalue is equal to its complex conjugate, it must be real.

It is worth noting that the eigenstates of an observable always form an orthogonal basis of the state space, even if it is of infinite dimension (in fact this should be taken as the definition of an observable). Thus, it is possible to develop any state vector as a linear combination of the observable eigenstates $|u_n\rangle$:

$$|\psi\rangle = \sum \langle u_n | \psi \rangle | u_n \rangle \tag{2.5}$$

A useful relation can be easily derived from that property. The unity operator I (i.e. the operator which leaves any state vector unchanged) can be expressed as

$$I = \sum |u_n\rangle\langle u_n| \tag{2.6}$$

where the sum runs over all eigenstates (this is immediately demonstrated by making this operator act on a state vector, since this gives its expansion in terms of the eigenfunctions).

Assume that M is an observable and that the state vector is $|\psi\rangle$. By expressing the ket $|\psi\rangle$ in the eigenstate basis $|u_n\rangle$ the action of M upon $|\psi\rangle$ can be written as

$$M|\psi\rangle = M \sum \langle u_n | \psi \rangle | u_n \rangle = \sum m_n \langle \psi | u_n \rangle | u_n \rangle \tag{2.7}$$

Using the fourth postulate and equation (2.7), the quantity $\langle \psi | M\psi \rangle$ can thus be transformed as

$$\langle \psi | M\psi \rangle = \langle \psi | M | \psi \rangle = \sum m_n |\langle \psi | u_n \rangle|^2 = \langle M \rangle \tag{2.8}$$

Thus, $\langle \psi | M | \psi \rangle$ is nothing but the expectation value of M (i.e. the average value which would be approached after carrying out many measurements with identically prepared systems, all described by the state $|\psi\rangle$).

2.1.4. *Momentum operator*

The relation between momentum and wavevector $\vec{p} = \hbar\vec{k}$ proposed by de Broglie is one of the key ideas which paved the way for the advent of a rigorous version of quantum mechanics. His proposal was to associate both a wave and a particle to describe an entity such as an electron, and that a plane wave function of the form $\exp(i\vec{k}\vec{r})$ carries a momentum $\vec{p} = \hbar\vec{k}$. A *heuristic* way[3] to find again the expression of the momentum operator

$$p = -i\hbar\nabla$$

$$(2.9)$$

is to apply operator p to a plane wave $\exp(i\vec{k}\vec{r})$, and to state that we must find the eigenvalue $\hbar\vec{k}$. It immediately appears that the momentum operator must have the form above. An important relation links the momentum operator to the position operator, that we can also easily find again using a plane wave. Apply the operator rp to a plane wave, and then apply pr. The difference between both results is not equal to zero and we will easily find the non-commutation relation

$$[r, p] = rp - pr = i\hbar ,$$

$$(2.10)$$

which can be shown to lead in turn to the famous Heisenberg's uncertainty relation (see section 10.1 for a demonstration based on simple mathematics):

$$\Delta p_x \Delta x \geq \hbar$$

$$(2.11)$$

Equation (2.11) means that it is not possible to measure with an arbitrary precision both the momentum and position. This can also be viewed another way: from the de Broglie relation the momentum is proportional to the wavevector, and thus happens to be the Fourier transform of the position, within a constant proportionality factor $\hbar$. If the reader has followed a course in signal processing they will already know that the more something is bounded (e.g. in time), the more its Fourier transform spreads (e.g. in frequency), and reciprocally. Thus, we cannot restrict one without extending the uncertainty of the other. The momentum and position are not the only non-commuting observables. For instance, two orthogonal

---

3 This note will not really help us to understand, but this relation can indeed be *derived* from general arguments, which show that the Hamiltonian of a particle is expressed in a unique way as a function of the generator of the space translations $-i\hbar\nabla$ and of the position, because on quite general grounds this Hamiltonian must satisfy the commutation relations of the Galilei transformation group (see [JOR 75]; note that this reference is not an article that can be read quickly).

components of an orbital angular momentum do not commute. From our comment on the Fourier transform we can also understand that time and energy, whose unit appears as a *frequency* multiplied by a constant action $\hbar$ in the Schrödinger equation, cannot be measured simultaneously with an arbitrary precision. Even if time is not an operator, we can also write

$$\Delta E \Delta t \geq \hbar .$$
(2.12)

2.1.5. *Stationary states*

Assume that the potential energy $V(\vec{r})$ does not depend on time. The Hamiltonian is $H = p^2/2m + V(\vec{r})$, and along with equation (2.9) the Schrödinger equation can be written as

$$-\frac{\hbar^2}{2m} \nabla^2 \psi (\vec{r}, t) + V(\vec{r}) \psi (\vec{r}, t) = i\hbar \frac{\partial \psi (\vec{r}, t)}{\partial t} .$$
(2.13)

Here we are going to show that it is possible to separate the space and time dependence. If we look for solutions of the form

$$\psi(\vec{r}, t) = \varphi(\vec{r}) \zeta(t) ,$$
(2.14)

after a few manipulations we can obtain from equation (2.13) that

$$i\hbar \frac{1}{\zeta(t)} \frac{d\zeta(t)}{dt} = -\frac{\hbar^2}{2m\varphi(\vec{r})} \nabla^2 \varphi(\vec{r}) + V(\vec{r}) .$$
(2.15)

The left-hand side is a function of t only, and the right-hand side is a function of $\vec{r}$. Thus, to be equal for any value of t and $\vec{r}$ these two quantities must be a constant, which we shall note $\hbar\omega$. Then we can integrate the left-hand side to obtain

$$\zeta(t) = Ce^{-i\omega t}$$
(2.16)

and the right-hand side leads to

$$-\frac{\hbar^2}{2m} \nabla^2 \varphi(\vec{r}) + V(\vec{r}) \varphi(\vec{r}) = \hbar\omega \varphi(\vec{r}) \quad ,$$
(2.17)

which can also be written under the form

$$H\varphi\left(\vec{r}\right) = E\varphi\left(\vec{r}\right) \qquad\qquad (2.18)$$

defining $E=\hbar\omega$. We can incorporate the C factor in function φ, because if we do so φ will still be a solution of equation (2.15), and we find that

$$\psi\left(\vec{r},t\right) = \varphi\left(\vec{r}\right) e^{-i\omega t} \qquad\qquad (2.19)$$

is a solution to the stationary Schrödinger's equation. This type of solution is called a stationary solution, because with a form such as equation (2.19) it is clear that the probability density $|\psi(\vec{r})|^2$ does not depend on time and is just a function of position. Since the function $\varphi\left(\vec{r}\right)$ satisfies equation (2.18) it is an eigenstate of the Hamiltonian operator, and $E=\hbar\omega$ is an energy eigenvalue.

Equation (2.18) is nothing but the eigenvalue equation of the Hamiltonian operator. From the fifth postulate, after an energy measurement on such a system we can only obtain one of the eigenvalues of H. This is the famous energy quantization phenomenon, which is an essential ingredient of semiconductor nanostructures and devices such as semiconductor quantum wells. Be careful: the fact that we can only obtain one energy eigenvalue does not mean that before the measurement the electron is in the corresponding eigenstate, even though the scalar product between the actual state vector and the eigenstate must be different from zero (see the fifth postulate). The stationary eigenstates do form a basis onto which can be expanded any other state, but if the actual state vector is a linear combination of several stationary states its evolution becomes time-dependent, because if we add several complex functions such as equation (2.19) the time does not only appear in the phase factor.

2.1.6. *Probability current*

In the general case the wave function depends on time and thus the probability density to find the particle somewhere also depends on time. This means that there is a probability density flow $\vec{J}$, and to calculate an average electron current we must be able to express this flow as a function of the wave function. Since probability should be conserved we expect to find a relation such as the charge conservation equation established from Maxwell's equations. In this section we limit ourselves to the case of a scalar potential.

To find $\vec{J}$ write the wave function under the form $\psi = Rexp(iS/\hbar)$, where $R = (\psi\psi^*)^{1/2}$ is the modulus of the wave function and S its phase multiplied by $\hbar$. Introduce this form into the Schrödinger equation and separate the real and imaginary parts. This healthy calculation exercise will lead us to two new equations, and the one corresponding to the imaginary part of Schrödinger equation is

$$\frac{\partial P}{\partial t} + \nabla\left(P\frac{\nabla S}{m}\right) = 0 \quad , \tag{2.20}$$

where $P = R^2 = \Psi\Psi^*$ is the probability density. This should remind us of the charge conservation equation obtained from Maxwell's equations of electromagnetism [JAC 98]:

$$\frac{\partial \rho}{\partial t} + \nabla\vec{j} = 0 \tag{2.21}$$

(in equation (2.21) ρ is the charge density and $\vec{j}$ the current density). By comparing equations (2.20) and (2.21) we should easily be convinced that equation (2.20) expresses nothing but the probability conservation, and from equation (2.20) the probability current density is obviously given by

$$\vec{J} = \psi\psi^* \frac{\nabla S}{m} \quad . \tag{2.22}$$

A more convenient and common way to write equation (2.22) is

$$\vec{J} = -\frac{i\hbar}{2m}\left(\psi^*\nabla\psi - \psi\nabla\psi^*\right) \tag{2.23}$$

We can easily check that equation (2.22) and equation (2.23) are identical by replacing Ψ by $Rexp(iS/\hbar)$ in equation (2.22) and by developing the resulting expression. If we calculate the probability current carried by a plane wave $exp(i(kx-\omega t))$, by applying equation (2.23) we immediately find that it is equal to $\hbar k/m$. Fortunately enough this is also the velocity expected from the de Broglie relationship $p = \hbar k$. Note that with a plane wave, the probability density does not change with time, but nevertheless we do have a probability flow. Electrons can be found everywhere but only move one way.

2.1.7. *Electrons in vacuum and group velocity*

With a scalar potential vanishing everywhere $V(\vec{r}) = 0$ the Schrödinger equation becomes

$$-\frac{\hbar^2}{2m}\nabla^2\varphi = -\frac{\hbar^2}{2m}\left(\frac{\partial^2\varphi}{\partial x^2} + \frac{\partial^2\varphi}{\partial y^2} + \frac{\partial^2\varphi}{\partial z^2}\right) = E\varphi \quad . \tag{2.24}$$

If we seek solutions of the form $\varphi(x,y,z) = \varphi_x(x)\varphi_y(y)\varphi_z(z)$ we can separate the space variables and for each space coordinate we only have to deal with an ordinary second order linear differential equation. Any (positive) energy value is allowed and the eigenstates are plane waves of the form

$$\psi(\vec{r},t) = e^{i(\vec{k}\,\vec{r} - \omega t)} \tag{2.25}$$

where the wavevector $\vec{k}$ and pulsation ω are related to the energy E by the relationship

$$E = \frac{\hbar^2 k^2}{2m} = \hbar\omega \tag{2.26}$$

A plane wave spreading over the whole space cannot be normalized, so that it is clear that a physically meaningful wave must be a linear combination of plane waves in order to form a normalizable wavepacket (note that if the plane waves are confined to a given volume V they can be normalized just by dividing them by $V^{1/2}$). The velocity operator $\vec{v} = \vec{p}/m$ of a plane wave is given by

$$\vec{v} = \frac{\hbar\,\vec{k}}{m} \tag{2.27}$$

A localized electron wave function must be represented by a wavepacket, i.e. by a linear combination of plane waves of the form

$$\psi(\vec{r},t) = \int_{-\infty}^{+\infty} g(\vec{k})\, e^{i(\vec{k}\,\vec{r} - \omega t)}\, d\vec{k} \tag{2.28}$$

The spreading in k values is required by the uncertainty principle to obtain a finite spreading of the wave function in real space. To simplify the notations we restrict the following analysis to one dimension, but it is straightforwardly generalizable to three. First remember that ω is a function of k. If the wavevector spreading remains limited around a mean value k_0 so that $k=k_0+\Delta k$ we can use the expansion [SMI 69]

$$\omega = \omega_0 + \Delta k \frac{\partial \omega}{\partial k} \qquad (2.29)$$

and obtain from equation (2.28)

$$\Psi(x,t) = \int\limits_{-\infty}^{+\infty} g(k_0 + \Delta k)\exp\left(i(k_0 x - \omega_0 t)+i\Delta k\left(x - \frac{\partial \omega}{\partial k}t\right)\right)dk \quad . \qquad (2.30)$$

If we define the function $h(\Delta k)=g(k_0+\Delta k)$ it takes appreciable values just for small Δk values, and changing the integration variable we can express the wave function as

$$\Psi(x,t) = \exp(i(k_0 x - \omega_0 t))\int\limits_{-\infty}^{+\infty} h(\Delta k)\exp\left(i\Delta k\left(x - \frac{\partial \omega}{\partial k}t\right)\right)d(\Delta k). \qquad (2.31)$$

For $t=0$ equation (2.31) reads

$$\Psi(x,0) = \exp(ik_0 x)\int\limits_{-\infty}^{+\infty} h(\Delta k)\exp(i\Delta kx)d(\Delta k) = f(x)\exp(ik_0 x). \qquad (2.32)$$

$f(x)$ is the inverse Fourier transform of h, whose variable is Δk, and since h takes appreciable values only over a small interval Δk_{max}, $f(x)$ is restricted to an interval of order $x_{max}=1/\Delta k_{max}$. Thus, we see that the wave function is normalizable. We can re-write equation (2.31) as

$$\Psi(x,t) = \exp(i(k_0 x - \omega_0 t))f\left(x - \frac{\partial \omega}{\partial k}t\right). \qquad (2.33)$$

From this equation the phase velocity is still equal to ω_0/k_0, as for a single plane wave, but the velocity which characterizes the wavepacket envelope f is obviously given by

$$v_G = \frac{\partial \omega}{\partial k} = \frac{1}{\hbar} \frac{\partial E}{\partial k} \qquad (2.34)$$

and is called the group velocity. It is the average velocity which would be found by measuring the velocity of electrons prepared in a state described by the wave packet considered above.

2.2. Energy band structure in a periodic lattice ◎

This section is a reminder. For those interested in rigorous derivations and not afraid of indexes and Fourier transforms or group theory they can consult the book by Ashcroft and Mermin [ASH 76]. All semiconductors of interest for our book are formed by the crystalline assembly of atoms, forming a three-dimensional periodic lattice and thus a periodic, three-dimensional potential landscape which is seen by the electrons of the materials under consideration. As we will explain in more detail in section 2.6, the lattice periodicity renders possible the existence of allowed and forbidden energy bands, electrons being free to move in the former. To make the whole picture clearer, hereafter we restrict ourselves to the 1D case. If $V(x)$ is the potential energy seen by the electrons, with lattice periodicity a, we have to solve Schrödinger's equation in the stationary case:

$$\left[\frac{(-i\hbar\nabla)^2}{2m} + V(x) \right] \Psi(x) = E\Psi(x) \qquad (2.35)$$

where $\Psi(x)$ is the electron wave function and E is the energy eigenvalue. If we remember that $|\Psi(x)|^2$ represents the probability of finding the electron at abscissa x if we make a position measurement, it is quite reasonable to assume that for extended states, if there is any, the probability density should follow the lattice periodicity (in fact this can be rigorously demonstrated, see [ASH 76]). Thus, we can write

$$\left|\psi(x+a)\right|^2 = \left|\psi(x)\right|^2 \qquad (2.36)$$

From equation (2.36) it immediately follows that $\Psi(x)$ and $\Psi(x+a)$ differ only by a phase factor, hence

$$\psi(x+a) = e^{i\phi}\,\psi(x) \qquad (2.37)$$

where Φ is a real number. If we now define the real number k as $k=\Phi/a$, we easily obtain

$$\psi(x+na)=e^{inka}\psi(x) \tag{2.38}$$

Defining the function

$$u_k(x)=e^{-ikx}\psi(x) \tag{2.39}$$

so that we can re-express the wave function as $\Psi(x)=e^{ikx}u_k(x)$, we straightforwardly arrive at

$$u_k(x+na)=u_k(x) \tag{2.40}$$

by using equations (2.38) and (2.39). Therefore, u_k has the lattice periodicity. As a matter of fact, this is generalizable to three dimensions and is known as the Bloch theorem, which we state below.

The stationary electron wave functions in a periodic crystal are Bloch waves of the form $\psi(\vec{r})=u_k(\vec{r})\exp(i\vec{k}\vec{r})$ *where* $u_k(\vec{r})=u_k(\vec{r}+n\vec{a})$ *has the lattice periodicity and* $\vec{k}$ *is a real wavevector.*

An equivalent formulation is the statement that a 3D Bloch wave verifies the property

$$\psi_k(\vec{r}+\vec{R})=e^{i\vec{k}\vec{R}}\psi(\vec{r}) \tag{2.41}$$

where $\vec{R}$ is any lattice vector[4]. Note that equation (2.41) is nothing but the 3D extension of equation (2.38).

If the lattice were perfectly periodic, the electrons could really propagate freely! Their wave function is a plane wave modulated by a Bloch function. All k values are allowed if the medium is infinite, and subject to some boundary conditions if the sample is of finite size. However, if the reader has already followed an introductory course in quantum mechanics they will have certainly studied the 1D Kronig-Penney model (see, e.g., [COH 77]) and will know that not all energy values are authorized,

---

4 In this section if the reader finds terms or concepts which are new to them and not explained in the text, they can have a look at section 10.2, which is a summary of the basic notions and definitions required to describe a crystalline lattice (base vectors, first Brillouin zone, etc.).

but the propagating states are restricted to allowed energy bands separated by forbidden bands (the existence of those bands and the way in which they are formed is discussed in more detail in section 2.6). In fact, this is a general feature: whatever the three-dimensional periodic lattice, the periodic potential leads to the formation of a band structure, which defines continuous and allowed energy intervals with extended states described by Bloch waves, separated by forbidden energy gaps. This band structure is fully defined by the relationship which exists between energy E and wavevector k.

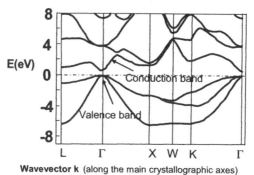

Figure 2.2. *Band structure of gallium arsenide*

In practice, once we know the periodic lattice, it is a matter of (sometimes quite hard) calculation to recover the band structure of the considered materials (a way to do this in a quite approximate but simple way is exposed in section 2.7). Of course in the case of commercial semiconductors such as silicon (Si) and gallium arsenide (GaAs) this was done a long time ago and in great detail. Thus, if we want to understand how these semiconductors behave we just have to examine the E versus k curves that are available in the literature, such as the one above (Figure 2.2). It must be noted that the energy value does not depend only on the magnitude of the wavevector, but also on its orientation. In general, the E vs k curves are given for the main crystallographic orientations in the first Brillouin zone (the latter is defined in section 10.2).

A quite interesting feature is that at a local extremum in the E *versus* k curve in a band (i.e. at the bottom or at the top of an allowed energy band, see Figure 2.2) the first derivative is equal to zero and we can of course develop E versus k up to second order:

$$E(k) = E_C + \frac{(k-k_0)^2}{2}\frac{\partial^2 E}{\partial k^2} = E_C + \frac{\hbar^2(k-k_0)^2}{2m_{eff}} \tag{2.42}$$

where k_0 is the extremum position in reciprocal space, and where we consider for the sake of simplicity that the extremum is isotropic. Equation (2.42) defines the effective mass, and is indeed extremely useful: take as the origin of energy the bottom (or the top) of the band E_C. From equation (2.42) we see that close to this threshold we obtain the electron kinetic energy in exactly the same form as for free electrons in vacuum, but with a change in apparent mass. In addition, suppose that our electron has a definite momentum. It can be readily demonstrated that its velocity will be given as in the vacuum case by the usual group velocity expression equation (2.34).

These are quite amazing results, because we discovered that we really can get rid of all this complicated part, the Bloch function, which is very difficult to calculate from the knowledge of the lattice and must be calculated numerically. However, there is still more than this: it turns out that all what we have to do, if we want to study electron dynamics in a band and their response to an external field or (long-range) potential, is to consider that they behave just like electrons in vacuum, but with a different mass (i.e. we just have to consider the wave function envelope). Since the E vs k relation depends on orientation, this mass may also depend on angle (this is the case for Si). However, we can use this extraordinary (and convenient) result that the periodic lattice is fully taken into account by just a mass renormalization, which allows us to drop the Bloch function term of the electron wave function and just keep the free wave envelope (this is further demonstrated in section 2.8). Nevertheless, do not forget that if the electron kinetic energy further increases (e.g. by applying a high electric field), then they can reach a region far from the bottom of the band, in which the second order term is not necessarily the prevailing one. However, even in such a case we can use an equation which keeps just the wave function envelope, even though it may become slightly more complicated than simply using an effective mass. What you must retain is that in a semiconductor, at the bottom (or at the top) of an energy band, the energy is proportional to k^2 and then we have an effective mass along a given axis which is given by

$$m_{\text{eff } x} = \frac{\hbar^2}{\dfrac{\partial^2 E}{\partial k_x^2}} \tag{2.43}$$

2.3. Semi-classical approximation ◎

Even if the effective mass approximation is not valid, from the Bloch theorem we can describe an electron wave function with a function owning the periodicity of the lattice, i.e. with short-range, atomic scale variations, modulated by a plane wave with wavevector $\vec{k}$. In the semi-classical approximation we consider that any perturbing potential (e.g. due to an externally applied electric field) is long range with respect to the Bloch wave variations and to the spread in the electron wavepacket. We can thus assume that the wave packet extends over a small distance, over which the potential is a well-defined constant (see Figure 2.3), and that the packet itself is considered as long-range with respect to the atomic lattice.

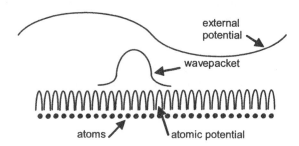

Figure 2.3. *Relative size of the elements considered in the semi-classical approximation*

As a consequence, if we are concerned with an electrostatic potential $\Phi(\vec{r})$, the energy conservation requires that the sum $E(\vec{k}) - e\Phi(\vec{r})$, which includes both the kinetic electron energy and the electron electrostatic energy, remains constant (the potential energy included in $E(\vec{k})$ is of course a constant). The derivative of this sum with respect to the time must be equal to zero and can be straightforwardly turned into

$$\frac{\partial E}{\partial \vec{k}}\frac{d\vec{k}}{dt} - e\nabla\Phi\frac{d\vec{r}}{dt} = \vec{v}_G \times \hbar\frac{\partial\vec{k}}{\partial t} - e\nabla\Phi \times \vec{v}_G = 0. \tag{2.44}$$

From equation (2.44) we immediately obtain

$$\hbar\frac{d\vec{k}}{dt} = -e\vec{\mathcal{E}} \tag{2.45}$$

where $\vec{\varepsilon}$ is the electric field. Addition of the magnetic field-induced Lorentz force leads to the more complete equation[5]

---

5 Let us say honestly that although equation (2.46) seems very reasonable it is in fact quite difficult to demonstrate the full equality including the Lorentz force, and this is not discussed in this book because a rigorous derivation would involve a lot of space and more than simple mathematics.

$$\hbar \frac{d\vec{k}}{dt} = -e(\vec{\mathcal{E}} + \vec{v}_G \wedge \vec{B}).$$ (2.46)

Equation (2.34), together with equation (2.46), form the semi-classical equations of motion. In these equations the dynamical aspects due to the forces exerted on the electron by the periodic lattice are fully taken into account by the knowledge of the dispersion relation.

2.4. Electrons and holes ◉

In this section we shall restrict the discussion to one-dimensional systems in order to simplify the concepts. We consider a lattice cell of size a. As explained in section 10.2, in a band the allowed energies are a periodic function of wavevector (see also Figure 2.4), and $\vec{k}$ is defined modulo $2\pi/a$. Therefore a unit cell of the reciprocal lattice, which is defined as the interval $[-\pi/a, \pi/a]$, contains all $\vec{k}$ values necessary to recover the full dispersion relationship and all eigenfunctions. It is convenient (and usual) to consider only the first Brillouin zone, since any value of $\vec{k}$ can be reduced to another wavevector value comprised inside it. All allowed states being described by a wavevector of the first Brillouin zone, we can enumerate all of them just by counting the number of allowed wavevectors in this zone, and below we will see that this consideration still applies if we submit our electrons to a force. From a dispersion relation such as in Figure 2.4 a number of interesting points can be deduced and are worth being mentioned.

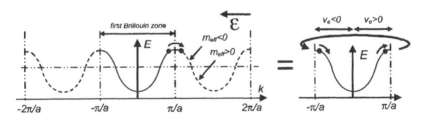

Figure 2.4. *An electron crossing the first Brillouin zone under the action of an electric field is equivalent to an electron re-entering into the opposite side of the same zone, because two wavevectors which differ by a reciprocal lattice vector describe the same wave function*

First, from equation (2.43) we see that at the band bottom the effective mass is positive, but at the top of a band it is *negative* (see Figure 2.4). Therefore, in the latter case, the electron sees a force which exhibits the same sign as that exerted by the external electric field (and which is of course due to the combined action of the field and the lattice)! Put just one electron in a band, and assume that there is no scattering. Under the action of a negative electric field and from equation (2.45) the wavevector value increases linearly with time, and when k crosses the first Brillouin zone boundary, due to the periodicity of the dispersion relationship, its is exactly equivalent to consider that the wavevector goes on increasing after having passed the boundary (left part of Figure 2.4), or that the electron crossing the first Brillouin zone boundary re-enters through the opposite side of the zone (see the right part of Figure 2.4). Hence, the movement is described by a sawtooth oscillation of the wavevector, and from equation (2.34) the velocity follows periodic oscillations known as Bloch oscillations (Figure 2.5a). In such a picture an electron cannot be indefinitely accelerated by the field, but is decelerated whenever it enters a part of the zone where the effective mass is negative[6].

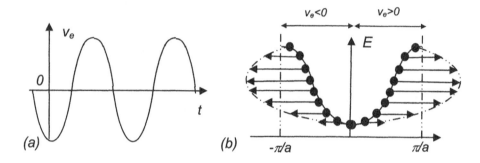

Figure 2.5. *(a) Bloch oscillations of the velocity and (b) a filled band cannot conduct electricity because all electron velocities compensate one another*

Now assume that a band is totally filled. All states can be described by wavevectors inside the first Brillouin zone, and in the absence of scattering, all electrons follow the same process as before. However, since there are as many electrons going one way as the opposite, and since the electric field cannot lead to any overall change (see Figure 2.5b) in the state occupation, the electrical current is equal to zero. *A filled band does not conduct electricity.* Thus, in practice, in a semiconductor the existence of an electrical current is achieved by partially filling an empty conduction band, or by partially emptying an initially full valence band.

In the almost empty band case, there is no conceptual difficulty because the electrons lie in the lowest energy states and have a positive effective mass, so that their velocity increases in the direction opposite to that of the field, as usual. In addition, in practical devices, they are in fact never accelerated up to the top of a

6 Although there have been many claims to the actual observation of Bloch oscillations in semiconductors, this specific topic is still subject to a certain amount of scientific controversy.

band by the electric field, as they suffer from inelastic collisions which prevent the observation of Bloch oscillations. However, in the almost filled band case, it would seem that the state of affairs is more difficult, as we now have to consider the movement of many electrons with a negative mass, which in addition acquire a velocity in the electric field direction. However, a great simplification is obtained by appealing to the concept of holes.

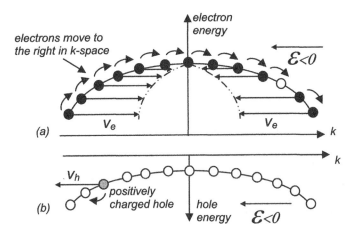

Figure 2.6. *An electron vacancy in an almost full band as in (a) can be replaced by a negative electron and a positively charge particle, both following the same movement as the vacancy. The added electron and the other electrons form a filled band which does not conduct, and the positive particle, whose velocity increases with time in the direction of the field, can be replaced by a positively charged "hole" moving as in the band of figure (b); this hole carries the same current as the almost filled band, exhibits a positive effective mass and follows the equation* $\hbar dk/dt = +e\varepsilon$

Have a look at Figure 2.6. All states are filled but one, and under the action of a negative electric field, the wavevector of all electrons continuously increases, until it crosses the Brillouin zone and re-enters at the right side through an instantaneous negative jump. The "absent electron" wavevector obviously follows the same evolution. This situation is thus completely equivalent to a picture in which we replace this "electron vacancy" by a pair formed by a fictitious electron of charge $-e$ and a fictitious, positively charged particle following the same movement as the vacancy. Charge and current are obviously conserved, and the real electrons plus the fictitious electron form a filled band, not carrying any current, and you can just forget them. We are thus left with the positive particle, which can advantageously be replaced by a positively charged "hole" with an effective mass m_h and a wavevector opposite to that of the electron. The hole velocity is the same as that of the fictitiously introduced electron, and it acquires a velocity in the direction of the electric field. In an almost filled band the electrons fill the lower energy states and therefore holes are located at the bottom of a hole band. The hole dispersion relation is usually plotted upside down so as to keep the same appearance as the corresponding electron band. We can thus model electrical transport just by considering positively charged holes, with a positive effective mass equal to $m_h = -m_e$, and follow our classical intuition[7].

2.5. Semiconductor heterostructure ◉

A semiconductor heterostructure is a stack of different semi-conducting materials, the lattices of which are well matched so as to form acceptable interfaces. These stacks are most often obtained by an advanced growth technique known as Molecular Beam Epitaxy (MBE), in which the successive atomic layers are deposited one by one in an ultra-high vacuum chamber. If we consider such a device, we have a misalignment between the energy bands of the different materials we put together, and in the conduction band (or valence band for the holes) we can obtain something which is very close to the potential wells that the reader may have studied in an introductory quantum mechanics course, and which leads to quantized energy values corresponding to a kinetic energy along the confining direction (Figure 2.7).

Generally, if we add a particular potential U which is long-range with respect to the periodic lattice, since we dropped the Bloch term, we now have to solve Schrödinger's equation but with an effective mass:

$$\left[E_C + \frac{(i\hbar\nabla)^2}{2m_{eff}} + U(\vec{r}) \right] \Psi(\vec{r}) = E\ \Psi(\vec{r}). \tag{2.47}$$

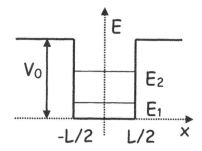

Figure 2.7. *Conduction band energy diagram of a finite quantum well*

---

7 For a more rigorous demonstration than this simple description, see [ASH 76].

This is the approach that we will take in many cases. It is known as the Effective Mass Approximation (EMA). It is valid as long as the variations of the potential U remain long-range considering that of the periodic lattice, but it is very often extended to cases where the potential variation is abrupt, because its use enormously simplifies the physical discussion. This relation is demonstrated in section 2.8, and also discussed in section 2.9. We shall often drop the energy E_C by taking it as the origin. Note that the effective mass is in general not the same in the two materials which form the heterostructure, a point which renders realistic calculations somewhat more subtle than the demonstrations presented in this book, but which does not radically alter the conclusions of our physical discussions.

2.6. Quantum well ●

2.6.1. *1D case*

Take two materials with a nice interface between them (e.g. GaAs and InGaAs) and make a sandwich with the materials with the lowest bandgap in between (a "heterostructure" is formed). For the conduction band we obtain something similar to Figure 2.7. To simplify we assume that the effective mass is the same in both materials, and then we have to solve the Schrödinger equation which can be straightforwardly put in the form below:

$$-\frac{\hbar^2}{2m}\frac{\partial^2\psi}{\partial x^2} = (E - V_0)\psi \quad if \quad |x| \geq L/2 ,$$ (2.48)

$$-\frac{\hbar^2}{2m}\frac{\partial^2\psi}{\partial x^2} = E\psi \quad if \quad |x| \leq L/2 .$$

Here we note that for a 1D system the solutions of the Schrödinger equation are non-degenerate (i.e. there is only one type of independent wave function for one energy level). Suppose that the potential $V(x)$ is even ($V(x)=V(-x)$), and that $\Psi(x)$ is a solution. Replacing x by $-x$ in the Schrödinger equation immediately yields that $\Psi(-x)$ is also a solution with the same energy. Since the levels are non-degenerate and the wavefuctions are normalized we thus have

$$\psi(-x) = e^{i\varphi}\psi(x)$$ (2.49)

where φ is an arbitrary phase factor. Applying the transformation $x \rightarrow -x$ twice gives

$$\psi(x) = e^{2i\varphi}\psi(x) \quad \Rightarrow \quad (e^{i\varphi})^2 = 1 .$$ (2.50)

Thus, we find that either $\Psi(x) = \Psi(-x)$, or $\Psi(x) = -\Psi(-x)$. *If the 1D potential is even then the solutions are either odd or even.* Let us apply this result to our potential.

If we are interested in the bound states with energy E lower than V_0, defining

$$k = \frac{\sqrt{2mE}}{\hbar} \qquad (2.51)$$

and

$$\beta = \frac{\sqrt{2m(V_0 - E)}}{\hbar} \qquad (2.52)$$

the previous equations (equation (2.48)) simplify to

$$\frac{\partial^2 \psi}{\partial x^2} - \beta^2 \psi = 0 \quad if \quad |x| \geq L/2$$

$$\frac{\partial^2 \psi}{\partial x^2} + k^2 \psi = 0 \quad if \quad |x| \leq L/2 \qquad (2.53)$$

These are simple second order differential equations, whose solutions are

$$\psi(x) = A\sin(kx) + B\cos(kx) \quad if \quad |x| \leq L/2$$

$$\psi(x) = Ce^{-\beta x} \quad if \quad x \geq L/2$$

$$\psi(x) = De^{\beta x} \quad if \quad x \leq -L/2 \qquad (2.54)$$

In the last two expressions we only keep the vanishing exponential terms, because the physical solutions cannot grow exponentially when going to infinity. Then, we impose the continuity of Ψ and the continuity of $d\Psi/dx$ for $x = \pm L/2$. These two properties might be demonstrated directly from the Schrödinger equation, as long as the potential energy remains finite (you can find the demonstration in [COH 77]; it is not reproduced here because understanding the demonstration is not essential for our discussion). We obtain for the even solutions

$$D\exp\left(-\frac{\beta L}{2}\right) - B\cos\left(\frac{kL}{2}\right) = 0$$

$$\beta D\exp\left(-\frac{\beta L}{2}\right) - Bk\sin\left(\frac{kL}{2}\right) = 0 \quad . \qquad (2.55)$$

To obtain solutions other than zero for B and D the determinant of this system must be equal to zero:

$$\begin{vmatrix} \exp\left(-\dfrac{\beta L}{2}\right) & -\cos\left(\dfrac{kL}{2}\right) \\ \beta \exp\left(-\dfrac{\beta L}{2}\right) & -k\sin\left(\dfrac{kL}{2}\right) \end{vmatrix} = 0 \quad . \tag{2.56}$$

Therefore, equation (2.55) leads to

$$\tan\left(\frac{kL}{2}\right) = \frac{\beta}{k} \tag{2.57}$$

for the even solutions, and we would obtain from a similar reasoning that

$$\tan\left(\frac{kL}{2}\right) = -\frac{k}{\beta} \tag{2.58}$$

for the odd solutions. These implicit equations give the wavevector and thus the quantized energy levels as a function of the well width L. In general we consider a given well, thus L is fixed, and in textbooks it is often proposed to find graphical solutions to these equations (see [COH 77]). However, if we want to plot analytically the variation of these levels as a function of L, a trick for finding analytical solutions is to express L *versus* E, instead of E *versus* L [FOU 94]. Using equation (2.57) (respectively equation (2.58)), we can write L as a function of k and β, and then replace the latter quantities by their expression as a function of energy (equation (2.51) and (2.52)). From the definition of the *Arc*tan function the solutions are defined *modulo* π, so that they can be indexed as

$$L_{2p} = \frac{2\hbar}{\sqrt{2mE}}\left(p\pi + Arc\,\tan\left(\sqrt{\frac{V_0 - E}{E}} \right) \right) \text{ (even levels)} \tag{2.59}$$

and

$$L_{2p+1} = \frac{2\hbar}{\sqrt{2mE}}\left((p+1)\pi - Arc\,\tan\left(\sqrt{\frac{E}{V_0 - E}} \right) \right) \text{ (odd levels).} \tag{2.60}$$

In the case of an infinite square well, we obtain the well known result

$$E_n = n^2 \frac{\pi^2 \hbar^2}{2mL^2} \tag{2.61}$$

and the wave functions are given by

$$\psi(x) = \sqrt{\frac{2}{L}} \sin\left(\frac{n\pi}{L}\left(x + \frac{L}{2}\right)\right) \quad . \tag{2.62}$$

In Figure 2.8 the quantum levels and the first three wave functions of a finite quantum well have been plotted as a function of position. Their shape resembles that of the infinite square well, but due to the finite barrier height the wave functions now spread in the classically forbidden area around the well (determination of the exact analytical form and of the normalization factors is proposed as an exercise in section 2.19.2). The higher the energy, the deeper the wave penetration into the classically forbidden region.

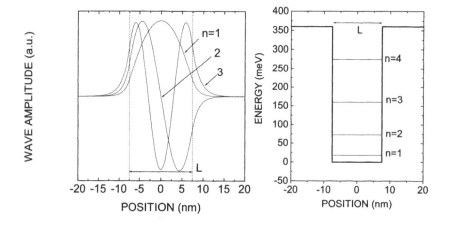

Figure 2.8. *First three wave functions (left) and energy levels (right) of a finite quantum well; the effective mass is that of GaAs (m=0.067 m$_0$), the energy barrier is equal to 0.36 eV and the well width is L=15 nm*

Equations (2.59) and (2.60) lead to the analytical curves of Figure 2.9, calculated for two common semiconductors. In Figure 2.9 we have also plotted the calculation for an infinite square well. We can conclude that for very small dimensions it is no longer a good approximation. In both cases, reducing the size over which our electrons are confined considerably increases the quantized energy levels.

Since this phenomenon occurs both in the conduction and valence bands, it can be used to adjust at will a quantum well laser wavelength, which will be fixed by the difference between the lower electron and hole quantized levels, between which recombination occurs with emission of a photon. The engineer "only" has to adjust the semiconductor growth process carefully so as to obtain the correct quantum well width. Thus, from a three-dimensional, unconfined semiconductor medium, by reducing adequately the dimension along z, we obtain a quantum well which is 2D (electrons can still freely move along x and y in the layer with a smaller bandgap), as in Figure 2.10. However, if we etch our semiconductor slab along, e.g., the y axis we now have two quantized directions, and the movement is 1D (Figure 2.11).

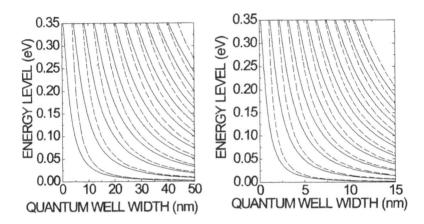

Figure 2.9. *Variation of the quantized energy levels of a finite quantum well (solid lines) and of an infinite quantum well (dashed lines) as a function of the well width. Left figure: $Al_{0.45}Ga_{0.55}As/GaAs/Al_{0.45}Ga_{0.55}As$ ($V_0=0.36$ eV and $m=0.067$ m_0) and right figure: $SiO_2/Si(100)/SiO_2$ quantum well ($V_0=3.2$ eV and $m=0.92$ m_0)*

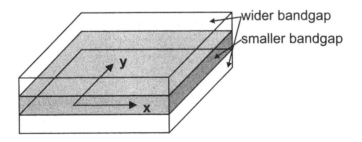

Figure 2.10. *Scheme of a semiconductor heterostructure as it can be grown by molecular beam epitaxy (MBE) on a substrate*

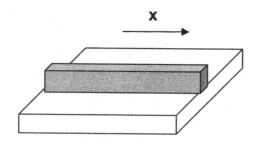

Figure 2.11. *Quantum wire that can be obtained by selectively etching a heterostructure as in the previous figure*

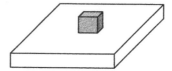

Figure 2.12. *Quantum box*

Quantizing the momentum along the last axis gives us a quantum box or dot (Figure 2.12). For a given structure, a finite temperature will impose thermal fluctuations to the distance between atoms (they vibrate) and to the electron energies. Since the thermal energy corresponding to a degree of freedom is $k_BT/2$, where k_B is the Boltzmann's constant, these random fluctuations will tend to broaden the initially discrete levels and eventually wipe them out. With a given confinement, it is interesting to assess roughly the temperature below which we can really expect quantum effects. Equating the thermal energy to the first level of an infinite quantum well gives us the *order of magnitude* of the device reduction required to quantumly confine a structure along a given axis. From equation (2.61) this readily gives us a length equal to $L=h/(8m_{eff}k_BT)^{1/2}$. For reasons linked to developments in statistical mechanics[8], we usually prefer to calculate a quantity quite close to the former, known as the de Broglie thermal length (Figure 2.13):

$$\lambda_{DB}^{th} = \frac{h}{\sqrt{2\pi m_{eff} k_B T}} \cdot \qquad (2.63)$$

---

8 Roughly speaking, the thermal de Broglie length represents the thermodynamic uncertainty in the localization of a particle possessing the average thermal momentum of the particle gas.

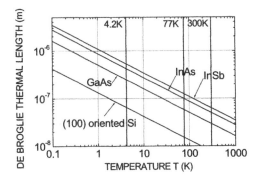

Figure 2.13. *Variation of the thermal de Broglie length as a function of temperature in the conduction band of various semi-conducting materials*

The de Broglie thermal length gives us a good indication of the critical size needed to observe quantum confinement, and allows us to compare various semi-conducting materials, as has been done in Figure 2.13.

2.6.2. *Coupled quantum wells*

Consider two identical 1D quantum wells, spatially separated from one another (Figure 2.14). If the wells are very far apart they do not interact and all levels are identical in both wells, so that each energy level is doubly-degenerate. As we move the two wells closer to one another there will be a growing overlap between the wave function of one well and the potential of the other. It is no longer possible to consider the two wells as independent and here we show how the interaction between both wells lifts the initial degeneracy of the levels. This case is also important if we consider a large number of identically spaced quantum wells because it illustrates how initially degenerate states can lead to the formation of allowed and forbidden energy bands. This is thus a good introduction to the calculations of the next section (section 2.7, tight-binding approximation).

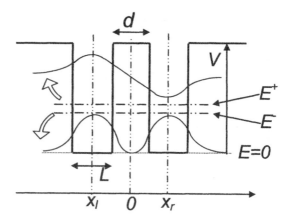

Figure 2.14. *Two coupled, 1D quantum wells*

As demonstrated in section 10.4.2, the degenerate ground state is split into two different states which to the first order of perturbation theory are given by a linear combination of the initial degenerate states (which we design by $|\psi_l\rangle$, where index l stands for "left", and $|\psi_r\rangle$, where index r stands for "right"):

$$|\psi\rangle = c_1|\varphi_l\rangle + c_2|\varphi_r\rangle \tag{2.64}$$

Coefficients c_1 and c_2 and energies E^+ and E^- of the "new" states can be determined by solving the secular equation (see section 10.4.2 for a demonstration)

$$\begin{pmatrix} W_r & W_{rl} \\ W_{lr} & W_l \end{pmatrix}\begin{pmatrix} c_1 \\ c_2 \end{pmatrix} = \Delta E \begin{pmatrix} c_1 \\ c_2 \end{pmatrix} \quad , \tag{2.65}$$

where the matrix coefficients can be calculated from the knowledge of the states of the two isolated systems:

$$\begin{aligned}
W_r &= \langle\varphi_r|V_r|\varphi_r\rangle & W_l &= \langle\varphi_l|V_l|\varphi_l\rangle \\
W_{rl} &= \langle\varphi_r|V_r|\varphi_l\rangle & W_{lr} &= \langle\varphi_l|V_l|\varphi_r\rangle
\end{aligned} \quad . \tag{2.66}$$

V_r is the potential induced by the left well and seen as a perturbation by the right well, and V_l represents the equivalent quantity for the left well (the full Hamiltonian can be written either as $H = H_r + V_r$ or as $H = H_l + V_l$, see section 10.4.2). Since $|\psi_l\rangle$ does not take appreciable values over the part where V_r does not vanish, W_r (and W_l)

can be neglected in front of W_{rl}, and if the wells are identical we also have $W_{rl} = W_{lr}$. Thus, as described in the last case of section 10.4.2 the levels are found by solving the system

$$
\begin{pmatrix} 0 & W_{rl} \\ W_{rl} & 0 \end{pmatrix} \begin{pmatrix} c_1 \\ c_2 \end{pmatrix} = \Delta E \begin{pmatrix} c_1 \\ c_2 \end{pmatrix} \quad ,
$$

(2.67)

where ΔE is the difference between the perturbed energy level and that of the initially degenerate system. This can be readily solved to give the two energies

$$
E^+ = E_1 - W_{rl} \qquad E^- = E_1 + W_{rl}
$$

(2.68)

(to first order) and the corresponding approached eigenstates

$$
|+\rangle = \frac{1}{\sqrt{2}} \big(|\varphi_l\rangle - |\varphi_r\rangle \big) \qquad |-\rangle = \frac{1}{\sqrt{2}} \big(|\varphi_l\rangle + |\varphi_r\rangle \big) \quad .
$$

(2.69)

Note that W_{rl} is negative, so that E^- is smaller than E^+. The lower state is symmetric with respect to the exchange of the two initial wave functions, and is called the bonding state (as the mechanism at the origin of the chemical bond is completely similar). The next level is anti-symmetric and is called the anti-bonding state. Let us calculate the degeneracy lift in the case of two square wells of width L with energy barriers equal to V and centered at positions x_r and x_l, respectively. The matrix element is given by

$$
W_{rl} = \int \varphi_l(x) V_l \varphi_r(x) dx
$$
$$
= -VBD \int_{x_r - L/2}^{x_r + L/2} \exp\big(- \beta(x - x_l)\big) \cos\big(k(x - x_r)\big) dx \quad ,
$$

(2.70)

where B and D are the coefficients in front of the wave function which are defined in equation (2.54), and k and β are defined by equations (2.51) and (2.52), respectively. The calculation of D and B is proposed as an exercise at the end of this chapter (section 2.19.2), and here we just quote the result:

$$
B = \sqrt{\frac{2}{L + 2/\beta}} \qquad D = \exp\left(\frac{\beta L}{2}\right) \sqrt{\frac{2}{\left(L + \frac{2}{\beta}\right)\left(1 + \frac{\beta^2}{k^2}\right)}}
$$

(2.71)

The integral appearing in the bottom line of equation (2.70) can be straightforwardly calculated and taking into account both equation (2.57) and the fact that $\cos(kL/2)$ is equal to $(1+\tan^2(kL/2))^{-1}$, so as to replace the cosine by an algebraic function of k and β, a few lines of calculations and trigonometric simplifications lead to a splitting equal to

$$E^+ - E^- = 2W_{rl} = \frac{8\beta^2 k^2}{(2+\beta L)(k^2+\beta^2)^2} V \exp(-\beta d) \quad , \tag{2.72}$$

where $d=x_r-x_l-L$ is defined as the separation between the two wells. The splitting decreases exponentially with the separation and Figure 2.15 shows calculated data in the case of two coupled GaAs quantum wells.

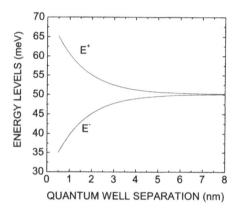

Figure 2.15. *Energy splitting of the ground state in coupled GaAs quantum wells, as a function of well separation; the energy barrier is V=0.36 eV and the well width is d=8 nm*

2.6.3. Quantum-confined Stark effect

When calculating quantum well properties, a last important effect which must be accounted for is the level variation induced by the action of an electric field. For instance, the operation of resonant-tunneling devices[9] often requires using high electric fields, which can exert a strong impact on the actual energy level values. Let us examine a situation as illustrated by Figure 2.16.

---

9 Their operation principle is described in Chapter 5.

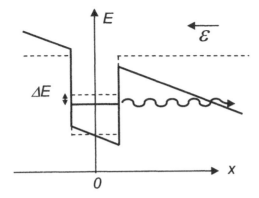

Figure 2.16. *The electric field modifies the energy levels of a quantum well, and may also induce electron escape by quantum tunneling*

The electric field $\mathcal{E}_x = \mathcal{E}$ adds a contribution to the potential energy equal to $-ex\mathcal{E}$, and to the first order of perturbation theory (section 10.4.1) the energy level modification should be given by

$$\Delta E = -e\mathcal{E}\left\langle \varphi_n^0 \left| x \right| \varphi_n^0 \right\rangle \quad . \tag{2.73}$$

However, since the perturbing potential is odd and the square of the wave function is even, the first order term is equal to zero, and to calculate the effect of the electric field one must go to the second order. This is in general easily calculable if we keep only the terms which correspond to the two first levels:

$$E_{1,2} = E_{1,2}^0 + e^2\mathcal{E}^2 \frac{\left|\left\langle \varphi_{2,1}^0 \left| x \right| \varphi_{1,2}^0 \right\rangle\right|^2}{E_{2,1}^0 - E_{1,2}^0} \tag{2.74}$$

In the infinite square well case and using equation (2.62), if we turn the sine/cosine product appearing in $\langle \varphi_1^0 | x | \varphi_2^0 \rangle$ into a sum of sine functions, we can integrate by parts and straightforwardly obtain the dipole matrix element

$$\left\langle \varphi_1^0 \left| x \right| \varphi_2^0 \right\rangle = \frac{16L}{9\pi^2} \quad . \tag{2.75}$$

Then, from equation (2.74) we have

$$E_1 = -\frac{512e^2mL^4}{243\hbar^2\pi^6}\mathcal{E}^2 = -\frac{256}{243\pi^4}\frac{(e\mathcal{E}L)^2}{E_1^0} .$$

(2.76)

From the equation above we can see that the energy change is substantial as soon as the electrostatic potential drop inside the well is in the same range as the initial quantum level value. From equation (2.74) the change in the energy difference between the two first levels is equal to twice the value above.

As the electric field increases the approximation above is often not accurate enough, and higher-order perturbation, a variational approach or brute force numerical calculations may be preferred. Analytical solutions expressed in terms of Airy functions may also appear quite useful [VAL 04] (see exercise 2.19.3). It is worth noting that if the well is not symmetric a first-order term must also be included. In addition, before the application of the field, the states are bound, but with the field there is always some possibility that the electrons will escape from the well by tunneling. This results in a dilution of the discrete state in the continuum under the form of a resonance with a finite width (this phenomenon is detailed in section 5.7).

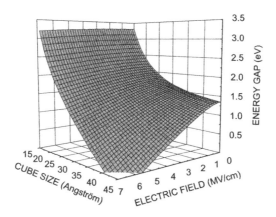

Figure 2.17. *Modification of the bandgap by the electric field calculated for cubic silicon dots embedded in Si0₂; after [OUI 00]*

Note that the Stark effect applies to both electrons and holes, so that if the quantum confinement resulting from the formation of a 3D dot can considerably increase the bandgap with respect to the bulk value, the Stark effect can strongly affect and reduce the impact of this increase, as illustrated by Figure 2.17 in the case of 3D silicon dots. The Stark effect also induces a spatial separation of the electrons and holes in a dot, and thus considerably increases the radiation lifetime, which is not necessarily a good point for light emission. However, this effect may also be put to good use in III-V semiconductor quantum wells, because the gap may be rendered smaller or higher than a given photon energy just by changing the electric field, so that it is possible to fabricate devices in which the optical absorption can be switched electrically.

2.7. Tight-binding approximation ○

In this section we will present a method devised a long time ago to calculate the band structure of solids[10]. Although not adequate to give a precise and quantitative picture of a real band structure, it will allow us to understand in a simple way how separate energy bands can be formed by assembling an atomic periodic lattice and bringing together the initially degenerate electron states of identical and separated atoms. In addition, we shall use this method in Chapter 9 to derive the main qualitative properties of the graphene band structure (graphene is a quite exciting and emerging material for fabricating quantum-coherent devices, to which a full chapter of this book is devoted). Although not necessary for obtaining a good idea of what coherent electron transport consists of, understanding this section will enable us to accept the most striking property of graphene, which exhibits a linear relationship between energy and wavevector, just as for photons, and in contrast with the effective mass approximation presented in section 2.2.

The general idea lying behind the tight binding approximation consists of using adequate linear combinations of atomic orbitals (LCAO) to find the dependence of energy on the Bloch function wavevector. A first assumption is to assume that close to the atomic core, the wave function is well represented by the initial wave function of an isolated atom. Thus, a first step is to form a Bloch wave from an adequate linear combination of atomic orbitals, each one belonging to one of the N atoms forming the lattice. If only one type of orbital per atom is to be considered and if there is only one atom per primitive unit cell we will see that only one such linear combination ψ is of the Bloch type for a given wavevector $\vec{k}$. Then the corresponding energy should be given by the usual expectation value $\langle \psi|H|\psi \rangle$ where H is the Hamiltonian of the full lattice. When there is more than one atom per primitive unit cell, we will show that as many Bloch waves as different atomic orbitals are involved in one unit cell for a given $\vec{k}$, but then use of the variational principle detailed in section 10.6 still allows us to determine an approximate band structure.

---

10 An introduction to the tight-binding approximation and other alternative methods can be found in [ASH 76] or [ZIM 65].

Remember the coupled quantum wells discussed in the previous section (section 2.6.2): as the two wells get closer to one another the initially degenerate levels are progressively split. If we take N atoms and follow the same procedure we expect the same type of behavior; as an electron occupying a given orbital in a single atom begins to feel the potential exerted by the other atoms, the N-fold degenerate orbital states of one atomic level split into N states and form an energy band, as in Figure 2.18. As the atoms are kept far apart and atomic spacing remains large one band results only from the interaction between atomic wave functions of one kind (s, p, d, etc.). When atoms get very close to one another of course there will also be some mixing between the bands, resulting in a more complicated picture.

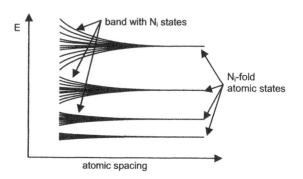

Figure 2.18. *Apparition of energy bands as the atom lattice spacing becomes smaller; for an infinite lattice each band is a continuum*

Let us first assume that one band is formed by the combination of one type of atomic state. Design by $\Phi_{at}(\vec{r})$ one single atomic wave function. If the perturbation brought to one atomic state by the other is not too large, we expect to find an acceptable solution of the Bloch type by combining together N atomic wave functions of the N atoms to form the ansatz

$$\psi_k(\vec{r}) = \sum_{\vec{R}} \alpha_{\vec{R}} \phi_{at}(\vec{r} - \vec{R}) \,, \tag{2.77}$$

where $\vec{R}$ is a vector of the lattice and is summed over all lattice sites. As N tends to infinity and with a periodic lattice we know that the wave function must be of the Bloch type, and the way to produce this with a wave function as in equation (2.77) is to write the wave function as

$$\psi_k(\vec{r}) = \sum_{\vec{R}} e^{i\vec{k}\vec{R}} \phi_{at}(\vec{r} - \vec{R}). \tag{2.78}$$

This can be demonstrated as follows, taking into account the fact that the sum is over an infinite lattice [ASH 76]:

$$
\begin{aligned}
\psi_k(\vec{r} + \vec{R}) &= \sum_{\vec{R}'} e^{ik\vec{R}'} \phi_{at}(\vec{r} + \vec{R} - \vec{R}') \\
&= e^{ik\vec{R}} \left(\sum_{\vec{R}'} e^{ik(\vec{R}' - \vec{R})} \phi_{at}\left(\vec{r} - (\vec{R}' - \vec{R})\right) \right) \\
&= e^{ik\vec{R}} \left(\sum_{\vec{R}''} e^{ik\vec{R}''} \phi_{at}(\vec{r} - \vec{R}'') \right) = e^{ik\vec{R}} \psi_k(\vec{r}) .
\end{aligned}
\tag{2.79}
$$

It is easily seen that equation (2.79) expresses nothing but the Bloch wave condition defined by equation (2.41).

There is one wave function for each wavevector $\vec{k}$ and the corresponding energy can now be assessed by calculating the expectation value

$$
E = \frac{\langle \psi_k | H | \psi_k \rangle}{\langle \psi_k | \psi_k \rangle} ,
\tag{2.80}
$$

where we "only" have to calculate integrals using the presumably known atomic wave function and H is the Hamiltonian of the full lattice. In practice, a considerable simplification is obtained by keeping only the interaction between nearest neighbors (the corresponding terms are called hopping integrals) and neglecting some overlap integrals, so that for each atom there is only a low number of integrals to calculate[11]. It is somewhat troubling that no attempt at energy minimization is allowed by the method. However, rough and simple-minded though it may seem, the tight-binding approach is often extremely useful to derive some important qualitative properties of a given lattice band structure.

---

11 Of course we do not expect high precision, and a way to improve it is to replace a single atomic wavefunction at each lattice site by a linear combination of atomic wavefunctions corresponding to a set of different atomic orbitals:

$$
\psi_k(\vec{r}) = \sum_{\vec{R}, j} e^{ik\vec{R}} \alpha_j \phi_{at}^j (\vec{r} - \vec{R})'
$$

Then we determine the coefficients $\alpha_j s$ by using the variational principle. This is known as the *LCAO* method.

Maybe one of the simplest examples we can work out is that of a three-dimensional cubic lattice of identical atoms (see Figure 2.19) [ZIM 65].

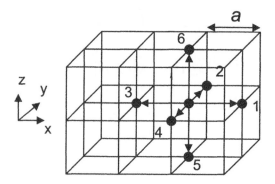

Figure 2.19. *A simple cubic lattice with an atom singled out along with its 6 nearest neighbors*

Equation (2.80) can be re-written as

$$E = \frac{\sum_{\vec{R},\vec{R}'} e^{i\vec{k}(\vec{R} - \vec{R}')} \langle \phi_{at}(\vec{r} - \vec{R}) | H | \phi_{at}(\vec{r} - \vec{R}') \rangle}{\sum_{\vec{R},\vec{R}'} \langle \phi_{at}(\vec{r} - \vec{R}) | \phi_{at}(\vec{r} - \vec{R}') \rangle} \qquad (2.81)$$

We neglect the overlap integrals of the $\langle \phi_{at}(\vec{r} - \vec{R}) | \phi_{at}(\vec{r} - \vec{R}') \rangle$ type if $\vec{R}$ is different from $\vec{R}'$ in the denominator. We single out the terms with $\vec{R} = \vec{R}'$ in the numerator. For $\vec{R}$ different from $\vec{R}'$ in the numerator we keep in the sum over $\vec{R}'$ only those terms which correspond to interaction with the nearest neighbors as defined in Figure 2.19. This leads to

$$E = \frac{N \langle \phi_{at}(\vec{r}) | H | \phi_{at}(\vec{r}) \rangle + N \sum_{i=1}^{i=6} e^{i\vec{k}\vec{r}_i} \langle \phi_{at}(\vec{r}) | H | \phi_{at}(\vec{r} - \vec{r}_i) \rangle}{N} . \qquad (2.82)$$

in which the first term of the right-hand side is calculated numerically and does not depend on $\vec{k}$. We define it as E_0. The transfer integrals involving the nearest neighbors are obviously all equal to the same value, which we design by γ_0, so that we can write

$$E = E_0 + \gamma_0 \sum_{i=1}^{6} e^{i\vec{k}\vec{r_i}}.$$

(2.83)

The imaginary terms cancel each other out because any of the six vectors has a counterpart exactly opposed to it (see Figure 2.19), and simple trigonometric simplifications lead to the dispersion relationship

$$E = E_0 + 2\gamma_0 \left(\cos(k_x a) + \cos(k_y a) + \cos(k_z a) \right).$$

(2.84)

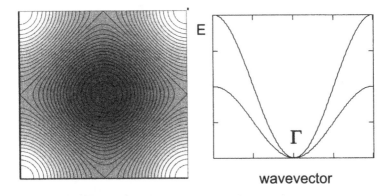

Figure 2.20. *Left: energy contours of the cubic lattice in the first Brillouin zone and in the plane $k_z=0$. Right: energy variation in some particular crystallographic directions with $k_z=0$ (upper curve: along a diagonal and lower curve: along k_x)*

In the first Brillouin zone (a quantity defined in section 10.2) the energy contours are represented as in Figure 2.20. Also shown in Figure 2.20 is the dispersion relationship along two particular orientations. Note that developing the *cosine* functions close to $k=0$ gives you a k^2 variation of energy with wavevector, which defines the effective mass at the bottom of the band.

A further complication arises when applying the method to lattices which include more than one atom per primitive lattice cell (see, e.g., the hexagonal 2D lattice of Figure 2.21, in which two inequivalent atoms A and B populate each primitive cell). Then it becomes necessary to use a linear combination of Bloch wave functions, each one belonging to one of the different sets of inequivalent lattice sites. In the simple case where there are only two inequivalent lattice sites A and B, the wave function in equation (2.78) must be replaced by a linear combination of two Bloch functions:

$$\psi_k(\vec{r}) = \alpha_A \psi_k^A(\vec{r}) + \alpha_B \psi_k^B(\vec{r})$$

$$= \alpha_A \sum_{\vec{R}_A} e^{i\vec{k}\vec{R}_A} \phi_{at}^A(\vec{r} - \vec{R}_A) + \alpha_B \sum_{\vec{R}_B} e^{i\vec{k}\vec{R}_B} \phi_{at}^B(\vec{r} - \vec{R}_B) \quad . \tag{2.85}$$

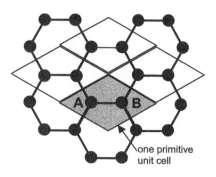

Figure 2.21. *A 2D hexagonal lattice with two inequivalent atoms per unit cell. Two observers positioned at atomic sites A and B and looking in the same direction do not see the same landscape*

Then finding the best coefficients and minimizing the energy expectation value is achieved by applying the variational principle to a linear combination of two tight-binding wave functions as demonstrated in section 10.6, from which we find that to each k value we can assign two energy levels, the full set of values thus forming two energy bands:

$$E(k) = H_{11} \pm |H_{12}| \tag{2.86}$$

with

$$H_{11} = \left\langle \psi_k^A | H | \psi_k^A \right\rangle$$

$$H_{12} = \left\langle \psi_k^A | H | \psi_k^B \right\rangle \tag{2.87}$$

(in equation (2.87) the overlap integrals defined in section 10.6 have been neglected, and in practice H_{11} and H_{12} are calculated by considering only the terms corresponding to nearest neighbors, as we did for the simple cubic lattice). We shall put this method to good use to establish some important properties of graphene in Chapter 9.

2.8. Effective mass approximation ○

2.8.1. *Wannier functions*

In section 2.2 we stated that we can incorporate the action of an external potential just by using a Schrödinger-type equation, substituting for the electron mass an effective mass determined by the energy band curvature. Here we are going to demonstrate and extend this fundamental result. This demonstration can be skipped at first reading because you can understand most of the book content just by admitting this property: *within a parabolic band, we incorporate all effects of the lattice just by changing the electron mass*; thus, this section will not help you to better understand quantum-coherent transport. However, we provide it for two reasons: firstly to prepare the ground for a better understanding of the way electron dynamics must be treated in a material such as graphene (Chapter 9), for which the band is *not* parabolic, and secondly, to maintain the self-contained character of the book. This section is thus dedicated to the curious student, and to follow the demonstration in detail is a good exercise in itself. The effective mass result was first demonstrated long ago in this form by Wannier and Slater. An extension to the action of a magnetic field and to many-valley semiconductors was proposed by Kohn and Luttinger using a slightly different but even more powerful formalism [KOH 55]. Here we detail a condensed form of the demonstration initially proposed by J.M. Ziman [ZIM 64].

In the tight-binding approximation we use linear combinations of atomic, and therefore *localized* orbitals to form Bloch waves. Unfortunately the orbitals of different sites are not orthogonal, and this prevents us from using them to obtain in a rigorous fashion the effective mass result. What we seek are expansions over localized and orthogonal functions, which will allow us to find an equation of motion retaining just the envelope of the wave, and getting rid of the complicated, short-range part. This is achieved by using Wannier functions, named after their inventor. The Wannier function of a band of index n is defined as

$$a_n(\vec{r} - \vec{R}_A) = \frac{1}{\sqrt{N}} \sum_{\vec{k}} e^{-i\vec{k}\cdot\vec{R}_A} \psi_{n,\vec{k}}(\vec{r}) \qquad (2.88)$$

There is one function a_n per energy band, as there is one type of atomic orbital corresponding to a band, but as for the atomic orbitals it can be centered at any lattice site, and therefore we have N Wannier functions per band, all translated from one another. Let us show that any Bloch wave can be expressed as a linear combination of Wannier functions. First we need to demonstrate two useful relations:

$$\sum_k e^{i\vec{k}(\vec{R}_A - \vec{R}_B)} = N\delta_{A,B} \quad and \quad \sum_A e^{i(\vec{k}-\vec{k}').\vec{R}_A} = N\delta_{\vec{k},\vec{k}'} \tag{2.89}$$

where the sum is carried over the first Brillouin zone and the full lattice, respectively. δ is the Kronecker symbol, equal to 1 if A and B are identical and to zero otherwise. To demonstrate the right equality we can use a concise argument attributed to Peierls: since the sum is carried over an infinite lattice it does not change if we add a particular lattice vector to the $\vec{R}_a s$ inside the exponential. Thus we can write

$$\sum_A e^{i(\vec{k}-\vec{k}').\vec{R}_A} = \sum_A e^{i(\vec{k}-\vec{k}').\vec{R}_A} \times e^{i(\vec{k}-\vec{k}')\vec{b}} \tag{2.90}$$

which implies either that the sum is equal to zero or that $(\vec{k}-\vec{k}').\vec{b} = 2n\pi$. However, the latter case implies that $\vec{k}$ and $\vec{k}'$ differ only by a reciprocal lattice vector, so that they are equivalent, and if we restrict the values to the first Brillouin zone $\vec{k} = \vec{k}'$ (and then the sum is equal to N). This completes the proof. To demonstrate the left equality in equation (2.89) just change the sum over $\vec{k}$ into an integral and integrate over the first Brillouin zone (do it at least in 1D to convince yourself).

Multiply equation (2.88) by $\exp(i\vec{k}'.\vec{R}_A)$ and sum over all lattice sites. This gives

$$\sum_A e^{i\vec{k}'.\vec{R}_A} a_n(\vec{r} - \vec{R}_A) = \frac{1}{\sqrt{N}} \sum_A \sum_{\vec{k}} e^{i(\vec{k}'-\vec{k}).\vec{R}_A} \psi_{n,\vec{k}}(\vec{r})$$

$$= \frac{1}{\sqrt{N}} \sum_{\vec{k}} \left(\sum_A e^{i(\vec{k}'-\vec{k}).\vec{R}_A} \right) \psi_{n,\vec{k}}(\vec{r}) = \frac{1}{\sqrt{N}} N \psi_{n,\vec{k}}(\vec{r}), \tag{2.91}$$

where we inverted the order of the summation and then used the right expression in equation (2.89). From equation (2.91) we obtain any Bloch function as a linear combination of Wannier functions:

$$\psi_{n,\vec{k}}(\vec{r}) = \frac{1}{\sqrt{N}} \sum_A e^{i\vec{k}\vec{R}_A} a_n(\vec{r} - \vec{R}_A) \tag{2.92}$$

From the definition of the Wannier function, and accounting for the orthogonality of Bloch functions and equation (2.89) (left), it is readily proved that even inside the same band Wannier functions are orthogonal:

$$\int a_n^*(\vec{r} - \vec{R}_A)a_n(\vec{r} - \vec{R}_B)d\vec{r} = \frac{1}{N}\sum_{k,k'} e^{-i\vec{k}\cdot\vec{R}_A}e^{i\vec{k}'\cdot\vec{R}_B}\int \psi_{n,\vec{k}}^*(\vec{r})\psi_{n,\vec{k}'}(\vec{r})d\vec{r}$$

$$= \frac{1}{N}\sum_{\vec{k}} e^{i\vec{k}\cdot(\vec{R}_B - \vec{R}_A)} = \delta_{A,B} \quad . \tag{2.93}$$

Any wave function can thus be represented as a proper linear combination of Wannier functions. It can be proved that these functions are not only centered but also localized on a given site, much like atomic orbitals. Even if their actual determination may appear quite involved, we do not need to calculate them explicitly in order to demonstrate the effective mass equation, the derivation of which is achieved by putting to good use the orthogonality property. This is the purpose of the next section.

2.8.2. *Effective mass Schrödinger equation*

With a potential V added to that of the lattice, the Schrödinger equation reads

$$\left(H_0 + V(\vec{r})\right)\psi(\vec{r},t) = i\hbar\frac{\partial\psi(\vec{r},t)}{\partial t} \quad , \tag{2.94}$$

where H_0 is the pure lattice Hamiltonian. We wish to investigate the solutions as the product of an envelope by the sum of localized Wannier functions, and to find an equation which just keeps the wave envelope. Thus, we seek solutions of the form

$$\psi(\vec{r},t) = \sum_{n,A} F_n(\vec{R}_A,t)a_n(\vec{r} - \vec{R}_A) \tag{2.95}$$

where F_n is the envelope corresponding to one energy band. Substitute this form into the Schrödinger equation, multiply the latter by the Wannier function $a_m(\vec{r} - \vec{R}_A)$ and integrate over the whole lattice. This amounts to writing

$$\sum_n \sum_A \int a_m^*(\vec{r} - \vec{R}_B)(H_0 + V)\, a_n(\vec{r} - \vec{R}_A)F_n(\vec{R}_A,t)d\vec{r} =$$

$$i\hbar\sum_n \sum_A \frac{\partial F_n(\vec{R}_A,t)}{\partial t}\int a_m^*(\vec{r} - \vec{R}_B)a_n(\vec{r} - \vec{R}_A)d\vec{r} \quad . \tag{2.96}$$

The right-hand side is easily simplified into $i\hbar \partial F_m(\vec{R}_B,t)/\partial t$, using the fact that Wannier functions are orthogonal. In the left-hand side we first operate a transformation of the part depending on the lattice Hamiltonian:

$$H_0 a_n(\vec{r} - \vec{R}_A) = \frac{1}{\sqrt{N}} \sum_k e^{-i\vec{k}\cdot\vec{R}_A} H_0 \psi_{n,\vec{k}}(\vec{r})$$

$$= \frac{1}{\sqrt{N}} \sum_{\vec{k}} e^{-i\vec{k}\cdot\vec{R}_A} E_n(\vec{k}) \psi_{n,\vec{k}}(\vec{r}) \tag{2.97}$$

where $E_n(\vec{k})$ is the eigenvalue of the Bloch function of indexes $n, \vec{k}$. Substituting the Bloch function by its expansion equation (2.92) we obtain

$$H_0 a_n(\vec{r} - \vec{R}_A) = \frac{1}{\sqrt{N}} \sum_{\vec{k}} e^{-i\vec{k}\cdot\vec{R}_A} E_n(\vec{k}) \frac{1}{\sqrt{N}} \sum_C e^{i\vec{k}\cdot\vec{R}_C} a_n(\vec{r} - \vec{R}_C) \ . \tag{2.98}$$

We invert the summation order:

$$H_0 a_n(\vec{r} - \vec{R}_A) = \frac{1}{N} \sum_C \left(\sum_{\vec{k}} e^{-i\vec{k}\cdot(\vec{R}_A-\vec{R}_C)} E_n(\vec{k}) \right) a_n(\vec{r} - \vec{R}_C) \ . \tag{2.99}$$

Then, defining $\varepsilon_n(\vec{r})$ as the Fourier transform of $E_n(\vec{k})$:

$$\varepsilon_n(\vec{R}_A) = \frac{1}{N} \sum_{\vec{k}} e^{-i\vec{k}\cdot\vec{R}_A} E_n(\vec{k}) \ , \tag{2.100}$$

we obtain from equation (2.99)

$$H_0 a_n(\vec{r} - \vec{R}_A) = \sum_C \varepsilon_n(\vec{R}_A - \vec{R}_C) a_n(\vec{r} - \vec{R}_C) \ . \tag{2.101}$$

Replacing index C by index B and injecting the latter expression into equation (2.96) leads to

$$\sum_n \sum_A \left(\int a_m^*(\vec{r} - \vec{R}_B) \left(\sum_C \varepsilon_n(\vec{R}_A - \vec{R}_C) a_n(\vec{r} - \vec{R}_C) \right) d\vec{r} \right) F_n(\vec{R}_A,t)$$

$$+ \sum_n \sum_A \left(\int a_m^*(\vec{r} - \vec{R}_A) V(\vec{r}) a_n(\vec{r} - \vec{R}_B) d\vec{r} \right) F_n(\vec{R}_A,t)$$

$$= i\hbar \frac{\partial F_m(\vec{R}_B,t)}{\partial t} \ . \tag{2.102}$$

Using the orthogonality of different Wannier functions we see that the term of the first line is different from zero only if $n=m$ and $C=B$. The term in the second line is nothing but the sum of the matrix elements $V_{m,n,A,B}$ of the potential between two Wannier functions of bands m,n and sites A,B:

$$V_{m,n,A,B} = \int a_m^*(\vec{r} - \vec{R}_A)V(\vec{r})a_n(\vec{r} - \vec{R}_B)d\vec{r} \quad . \tag{2.103}$$

Therefore we can re-write equation (2.102) as

$$\sum_n \sum_A \left(\delta_{m,n}\varepsilon_n(\vec{R}_A - \vec{R}_B) + V_{m,n,A,B} \right) F_n(\vec{R}_A,t) = i\hbar \frac{\partial F_m(\vec{R}_B,t)}{\partial t} \tag{2.104}$$

Now we are going to replace the terms with the Fourier-transformed energies by a more compact form. For this we calculate the quantity $E_n(-i\nabla)F_n(\vec{r})$, assuming that the dispersion relation $E_n(\vec{k})$ is already known. It can be written as the inverse Fourier transform of $\varepsilon_n(\vec{r})$:

$$E_n(-i\nabla)F_n(\vec{r}) = \left(\sum_A \varepsilon_n(\vec{R}_A)e^{i\vec{R}_A.(-i\nabla)} \right) F_n(\vec{r}) \quad . \tag{2.105}$$

If we expand the exponential operator inside the parenthesis we obtain[12]

$$\exp(\vec{R}_A\nabla) = 1 + \vec{R}_A\nabla + \frac{\vec{R}_A\nabla}{2} + ... + \frac{(\vec{R}_A\nabla)^n}{n!} + ... \tag{2.106}$$

and reporting this into equation (2.105) we have

$$E_n(-i\nabla)F_n(\vec{r}) = \sum_A \varepsilon_n(\vec{R}_A)\left(F_n(\vec{r}) + \vec{R}_A\nabla F_n(\vec{r}) + ... + \frac{(\vec{R}_A\nabla)^n F_n(\vec{r})}{n!} + .. \right). \tag{2.107}$$

The term inside the brackets is nothing but the Taylor expansion of $F_n(\vec{r} + \vec{R}_A)$ around $\vec{r}$, so that we can turn equation (2.107) into

$$E_n(-i\nabla)F_n(\vec{r}) = \sum_A \varepsilon_n(\vec{R}_A)F_n(\vec{r} + \vec{R}_A). \tag{2.108}$$

---

12 The exponential of an operator is calculated by replacing the exponential by its Taylor expansion; see [COH 77] for a demonstration.

Note that the first term inside the first bracket in equation (2.104) is precisely of the same form as the right-hand side of equation (2.108). Just make the variable change $\vec{r} \to \vec{R}_A - \vec{R}_B$ in equation (2.108). This gives the expression

$$E_n(-i\nabla)F_n(\vec{R}_A - \vec{R}_B) = \sum_A \varepsilon_n(\vec{R}_A)F_n(\vec{R}_B) \tag{2.109}$$

which substituted in equation (2.104) leads to

$$E_m(-i\nabla)F_m(\vec{R}_B,t) + \sum_n \sum_A V_{m,n,A,B}F_n(\vec{R}_A,t) = i\hbar\frac{\partial F_m(\vec{R}_B,t)}{\partial t}. \tag{2.110}$$

This is just the set of (exact) coupled differential equations that we were looking for to describe the time evolution of the envelope functions. It is exact and operates on the lattice points, but if we approximate the envelope by passing to the continuous limit, we obtain the equation of motion

$$E_m(-i\nabla)F_m(\vec{r},t) + \sum_n \sum_A V_{m,n,\vec{R}_A,\vec{r}}F_n(\vec{R}_A,t) = i\hbar\frac{\partial F_m(\vec{r},t)}{\partial t}. \tag{2.111}$$

In equation (2.111) a term of the $V_{m,n,\vec{R}_A,\vec{r}}$ type can be defined, e.g., by replacing $\vec{r}$ by the lattice point closest to it. Depending on the potential V, we can now operate some further approximating steps. If it is not large enough or abrupt so as to induce interband transitions, it is reasonable to assume that we just have to take into account the Wannier functions in one energy band. In such a case in the sum of the left-hand side in equation (2.111) only the term with $m=n$ survives. We have to calculate terms of the form

$$V_{m,m,\vec{R}_A,\vec{r}} = \int a_m^*(\vec{r}'-\vec{R}_A)V(\vec{r}')a_n(\vec{r}'-\vec{r})d\vec{r}' \quad. \tag{2.112}$$

If $\vec{R}_A$ and $\vec{r}$ are far apart, we can neglect the corresponding overlap integrals. Then, if the potential $V(\vec{r})$ is long-range with respect to the lattice constant, for sites close to $\vec{r}$ it can be factorized out of the overlap integrals and the orthogonality of Wannier functions makes those terms vanish too. It is then obvious that we just have to keep the matrix element corresponding to the site closest to the point $\vec{r}$, and if V is long-range with respect to the lattice constant this is quite well approximated as the value of V at point $\vec{r}$. We thus obtain the desired equation of motion:

$$E_m(-i\nabla)F_m(\vec{r},t) + V(\vec{r})F_m(\vec{r},t) = i\hbar\frac{\partial F_m(\vec{r},t)}{\partial t}. \tag{2.113}$$

Note that the dispersion relation between energy and wavevector does not need to be parabolic. In the case of a parabolic band $E=\hbar^2 k^2/2m$, this equation immediately reduces to the effective mass equation

$$-\frac{\hbar^2}{2m}\nabla^2 F(\vec{r},t)+V(\vec{r})F(\vec{r},t)=i\hbar\frac{\partial F(\vec{r},t)}{\partial t}.$$

(2.114)

This relation is valid as long as the wavevector is not located in an interval where the dispersion relation exhibits a degeneracy or several bands are very close to one another. However, even in such a case it is possible to obtain similar equations of motion. In general, close to a degeneracy point as many envelope functions as degenerate bands must be kept to form a set of coupled differential equations. We shall skip this more complicated demonstration, with the hope that understanding the derivation of the simplest case will have prepared the reader. The general case is discussed at length in [KOH 55], where use of Bloch functions at the band bottoms is preferred over that of Wannier functions.

2.9. How good is the effective mass approximation in a confined structure? O

This is not an easy question, and we shall only discuss it qualitatively. In an infinite medium it is quite correct to use it. However, as we reduce the dimensions, we must take two new facts into account: on the one hand, the interface plays an increasingly important role in shaping the wave functions, and the interface potential variation may be too abrupt to ensure that its variation takes place over more than a lattice unit. On the other hand, if the confining structure does not include more than a small number of basic lattice cells, it is all but evident that we can still apply the effective mass approximation, since it was established thanks to the invariance of the Hamiltonian system using translations. How can we talk about periodicity with only three or four crystalline cells? In some cases the effective mass approximation is no longer quantitatively correct, and only gives us a first order estimation of the real energy levels and electron wave functions. Nevertheless, this is so convenient an approximation to discuss nanostructure physics that we shall keep it throughout the text. Although it may substantially depart from an accurate numerical result, we shall see below on a simple 1D example that it can still remain quite close to the correct solution. There is a second strong argument in favor of its use: even if the quantized energy levels resulting from confinement in one direction cannot be accurately estimated using the effective mass approximation, transport along the unconfined directions is still pretty well described by the effective mass model.

In Figure 2.22 we can see a one-dimensionally confined potential with just a few periodic cells. Just to the right of the potential is plotted the E versus k relationship, numerically calculated for the infinite periodic lattice (solid line); the points represent the quantized levels for a well containing eight periodic cells. From this diagram we can numerically calculate the effective masses in each bottom or upper part of an allowed energy band. In the right figure we can see the first quantized levels of the nanostructure, calculated either numerically with the full lattice, or in the effective mass approximation.

Strikingly enough, even for two basic crystalline cells, the effective mass approximation still gives results very close to the mark, as long as the effective mass of the considered band is not too small. In this simple numerical example, the interfaces are perfectly reflecting ("hard wall"), and this illustrates the fact that what makes the effective mass approximation inefficient is not necessarily the small number of basic cells, but most often *the increasing influence of the interface with respect to the dot volume.*

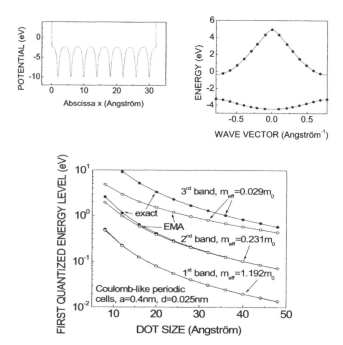

Figure 2.22. *Band structure and energy levels of a periodic potential bounded by hard walls; the upper left figure is the potential; the upper right figure is the dispersion relation for an infinite lattice onto which have been added the discrete levels corresponding to the bounded potential; the lower figure gives the exact energy levels as a function of the dot size and the energy levels obtained in the effective mass approximation*

2.10. Density of states ●

2.10.1. *3D case*

First we shall consider the three-dimensional case. Take a macroscopic slab of semiconductor, and, to simplify, a cube with side L. Of course in such a sample the energy spacing between the energy levels becomes infinitely small, and to count them it is convenient to rely upon a continuous quantity, *the density of available states per unit energy and unit volume D(E)*. Our assumption is that the bulk properties of this semiconductor are not affected by the boundaries. This can be rigorously derived, but the demonstration is largely outside the scope of this course, and it is obviously a quite reasonable hypothesis. In addition, once we have admitted this, we are given some more freedom in the choice of the boundary conditions, since they should not change the final result. Rigorously, we should impose the boundary condition $\Psi=0$ at the surface of our cube (and it is in fact easy to obtain the same result as the one we shall obtain in this way; this is a good exercise to check it). However, then our solutions would be stationary waves, and such solutions are not really convenient if we want to treat electrical transport in an open system because they do not carry current. It is much preferable to use running waves and to do so we shall use the Born-von Karmann (BVK) boundary conditions [ZIM 65]:

$$
\begin{aligned}
\psi(x + L, y, z) &= \psi(x, y, z)) \\
\psi(x, y + L, z) &= \psi(x, y, z)) \\
\psi(x, y, z + L) &= \psi(x, y, z)) \quad .
\end{aligned}
\tag{2.115}
$$

It is just as if an electron going in, e.g., the x direction was coming out of the semiconductor at $x=L$ and re-entering at $x=0$, the opposite side of the cube. This is topologically impossible in three dimensions, but in one dimension we can imagine a line closed so as to form a ring, and in two dimensions that the electrons are confined on a torus. Then if we look for solutions in the form of plane waves the BVK conditions immediately tell us that the wavevector components must be of the form

$$
k_x = n_x \frac{2\pi}{L} \qquad k_y = n_y \frac{2\pi}{L} \quad k_z = n_z \frac{2\pi}{L}
\tag{2.116}
$$

with the energy given by

$$
E = \frac{\hbar^2 k^2}{2m} = \frac{\hbar^2}{2m}\left(k_x^2 + k_y^2 + k_z^2\right).
\tag{2.117}
$$

From equation (2.116) it is clear that the allowed $\vec{k}$ values form a periodic cubic lattice in k-space, and each allowed state occupies an elementary volume $V=(2\pi/L)^3$. Thus, the total number of states *per unit volume* with a momentum lower than k is given by

$$N = 2\frac{4\pi k^3}{3V}\frac{1}{L^3} = \frac{k^3}{3\pi^2} .$$

(2.118)

In equation (2.118) we have multiplied by a factor of 2 in order to take into account the fact that each state can be occupied by two electrons with opposite spins, and not only one (this is called spin degeneracy). Then, using equations (2.117) and (2.118), the density of states per unit energy and volume is

$$D(E) = \frac{dN}{dE} = \frac{dN}{dk}\frac{dk}{dE} = \frac{1}{\pi^2\hbar^3}\sqrt{2m^3 E} .$$

(2.119)

This is a very important quantity, because the conductance of a sample depends on the number of available states which are present at the Fermi energy. We shall see in the following that this density of states depends on the system dimensionality. Thus, dimensionality crucially affects the electron transport properties.

2.10.2. *2D case*

Now we shall consider the pure 2D case: We set the bottom of the conduction band as the origin of energy. This means that we have confined all our electrons along one direction. Of course our electrons are still free to move along x and y, so that their overall energy is not simply E_0, but also depends on their momentum along x and y:

$$E = E_0 + \frac{\hbar^2}{2m}\left(k_x^2 + k_y^2\right)$$

(2.120)

However, as long as the Fermi energy remains well below E_1, they cannot occupy levels with a quantized energy along z different from E_0. In such a situation we say that all electrons lie in the first subband (if E_F becomes larger than E_1, two subbands are occupied). Applying the BVK conditions to k_x and k_y, each state now occupies an elementary area $S=(2\pi/L)^2$ in k-space (see Figure 2.23). The total number of states with momentum lower than k, taking spin degeneracy into account, is

$$N_{TOT} = 2 \times \frac{\pi k^2}{\left(\dfrac{2\pi}{L}\right)^2} \times \frac{1}{L^2} = \frac{k^2}{2\pi}. \tag{2.121}$$

The density of states is zero below E_0, and

$$D(E) = \frac{dN_{TOT}}{dk}\frac{dk}{dE} = \frac{m}{\pi\hbar^2} \tag{2.122}$$

above it. We see that in 2D this density has the important property of being a constant. Of course, if we consider an energy higher than E_1, we populate more than one subband, because the quantized energy corresponding to the transverse confinement may take at least two values, E_0 or E_1. As a matter of fact, the density of states increases by a factor of $m/\pi\hbar^2$ each time the energy becomes higher than the bottom of a new subband (E_1, E_2, E_3, etc.), so that it exhibits a staircase-like shape.

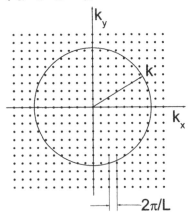

Figure 2.23. *Allowed k values in a 2D system with BVK boundary conditions*

2.10.3. *1D case*

Further reduce the number of degrees of freedom and constrain the electrons to move only along *y*. With a reasoning exactly similar to the previous ones, it is straightforward to find that the density of states is now given by

$$D(E) = \frac{1}{\pi\hbar}\sqrt{\frac{2m}{E - E_0}} \tag{2.123}$$

if E is larger than E_0, the bottom of the first subband, and equal to zero below E_0. Of course, as in the previous case, as the energy increases more than one subband can be populated, and each time E crosses a new quantized level corresponding to the confinement along x or z, we have to add a new component to equation (2.123), similar to it but replacing E_0 by E_1, E_2, etc.

2.10.4. *Summary*

We calculated the density of available electron states per unit energy and per unit volume (3D), surface (2D) or line (1D). This is a very important notion, to be intensively used in the remaining part of this book. $D(E)$ exhibits a shape which depends on dimensionality, as schematized in Figure 2.24.

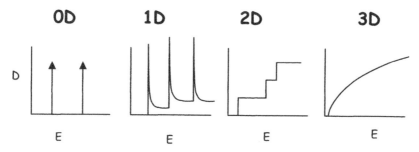

Figure 2.24. *Density of available quantum states as a function of energy for systems of various dimensionalities*

2.11. Fermi-Dirac statistics ●

Once we know the number of quantum eigenfunctions that we have in a given device, as well as the number of electrons, we still have to apply a rule to determine how the electrons are distributed among the various available quantum states, which are defined by their wavevector and spin. The statistics which governs this distribution is that of Fermi-Dirac, and it is driven by two underlying principles: first of all, electrons are fermions with half-integer spin, and although we shall not demonstrate it, this makes it impossible for two electrons to occupy the same quantum state. This is the exclusion principle, first stated by Pauli, who was also the man who *demonstrated* the connection existing between spin and statistics (and whose origin lies in special relativity) [PAU 40]. Without a magnetic field, for each available energy level, the electrons can be in two different quantum states, one with "spin-up" and one with "spin-down" (applying a magnetic field would lift this degeneracy by splitting these levels). Thus, each electron state with a given wavevector k and energy E can be occupied by at most two electrons. The second principle is that the most probable energy distribution is the one which can be realized using the highest number of different configurations (so as to maximize the entropy). Together with the exclusion principle this leads to the Fermi-Dirac occupancy function (demonstrated in Chapter 2 of [ASH 76]):

$$f_0(E) = \frac{1}{1 + \exp\left(\dfrac{E - E_F}{k_B T}\right)} \quad . \tag{2.124}$$

E_F is the Fermi energy or chemical potential of the system, and k_B is the Boltzmann constant. Two limiting cases can be found. In the "non-degenerate" or "Boltzmann" case the subband bottom energy is far above E_F, and thus

$$f_B(E) \cong \exp\left(-\frac{E - E_F}{k_B T}\right). \tag{2.125}$$

In the degenerate and low temperature case, E_F is well above E_S, and we can make the approximation

$$f_D(E) \cong H(E_F - E) \tag{2.126}$$

where H is the Heaviside (or step) function. These two cases are illustrated by Figure 2.22.

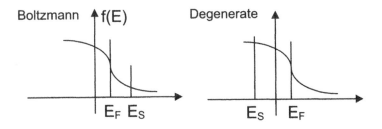

Figure 2.25. *Fermi-Dirac distribution in the Boltzmann and degenerate cases, respectively*

Now if we have a large number of available energy levels and if we know the density $D(E)$ of such levels per unit energy and volume (3D), area (2D) or length (1D), the total electron density n_S will of course be given by

$$n_S = \int D(E) f_0(E) dE \qquad (2.127)$$

At $T=0$, we thus have to fill all the quantum states with the available electrons, from the lowest energy state up to the Fermi energy. If the states of the system are running waves with wavevector $\bar{k}$ as in section 2.10, and if you remember that $E=\hbar^2 k^2/2m$, this means that in k-space and 3D you have to fill all the states included in the sphere of radius k_F, where $k_F=(2mE_F)^{1/2}/\hbar$ is the Fermi wavevector. This volume is called the Fermi sphere. In 2D you have to fill all the states in the disk enclosed in the circle of radius k_F, and in 1D you have to take all states between $-k_F$ and $+k_F$ (the "Fermi line"). This means that in the absence of excitation, the net current carried by the states is equal to zero. However, this zero current results from the fact that in the time-honored system with Born-von Karmann boundary conditions, as many electrons run in one direction as in the opposite one!

2.12. Examples of 2D systems ●

Here we shall briefly describe two archetypes of 2D systems currently used, either in applications or in research. The first one is the electronic device which has known the most tremendous industrial and technological success of the whole electronics history: the Metal Oxide Semiconductor Field Effect Transistor (MOSFET)[13]. It is based on a combination of silicon (Si) and silicon dioxide (SiO_2). Its incredible success is mostly due to the uppermost electronic interface quality between the semiconductor and its natural insulator SiO_2. In addition, the processing capabilities and function versatility of silicon make it a formidable competitor for almost any application except the optoelectronic market. In applications (CMOS digital circuits) MOSFETs are operated at high drain voltage and room temperature, but if we decrease both drain voltage and temperature, we can obtain a very good 2D system. As a matter of fact, many fundamental solid-state physics discoveries have been achieved with such a useful component (integer quantum Hall effect, universal conductance fluctuations, Coulomb blockade and many others).

As illustrated by Figure 2.23, the inversion layer electrons are maintained close to the interface by the transverse electric field, which creates a narrow, triangular-shaped potential well perpendicularly to the Si-SiO_2 interface. This confining potential leads to the creation of 2D subbands, and depending on the gate voltage (which determines the number of inversion electrons) and temperature, we can actually put all the electrons in only one 2D subband. This is still a very important device with respect to nanoscale devices, because silicon is the most convenient

---

13 For more information about MOSFET physics see [SZE 81] or [BRE 81], and see [AND 82] for low temperature and quantum aspects.

material for fabricating dots and single-electron transistors, which we shall introduce in the last part of this book. The bad point is that surface mobilities are not very high, remaining in the 10^3-$10^4 cm^2/V/s$ range at temperatures around 4 K, depending on the channel doping level and interface quality. The good point is that such a device can accommodate huge variations in 2D concentration, from 10^9 up to $10^{13} cm^{-2}$, only depending on gate voltage.

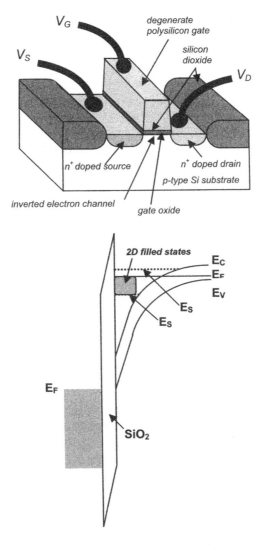

Figure 2.26. *Schematic representation of a MOSFET and corresponding energy band diagram in a direction perpendicular to the gate plane*

The second type of device is now often preferred by solid-state physicists, as they exhibit much higher mobilities and enhanced quantized energy spacing. Ballistic and quantum effects can be obtained with larger dimensions or higher temperatures. These are the III-V compound heterostructures. Most of them are processed from gallium arsenide (GaAs) and AlGaAs, but other III-V compounds seem very good as well (InAs, etc.). Many high frequency devices are based on these materials (such as the High Electron Mobility Transistors or HEMTs), and mobility can reach huge values, possibly larger than $10^7 cm^2/Vs$ at liquid helium temperature or several $10^4 cm^2/Vs$ at room temperature. The principle used to ensure high mobility is to separate spatially the doping impurities from the free carriers. The heterostructure is a stack of two well-matched materials (Figure 2.27). The wider bandgap material is delta-doped (i.e. the impurities are present only on a very thin layer of the material), close to the GaAs-AlGaAs interface. The electrons are transferred from the impurities into the intrinsic GaAs layer, whose conduction band bottom is lower than that of AlGaAs and the impurity levels. However, since the ionized impurities are positively charged, they create an electric field which prevents the electrons from leaving the interface. The electrons are thus trapped in a potential well as illustrated by Figure 2.28.

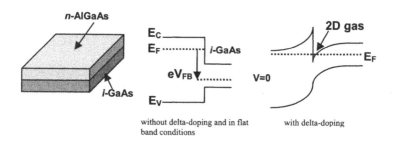

Figure 2.27. *III-V heterostructure and associated energy bands*

The potential well is narrow enough to induce a quantization of their transverse kinetic energy. In addition, the electrons do not suffer from a strong scattering by the ionized impurities, since they are located farther away and not at the interface, and only see the almost perfect MBE interface. Therefore, their mobility is close to that of the intrinsic materials. Unfortunately, in contrast to their Si MOSFET counterparts, such devices cannot be operated at as high an electron concentration, because the AlGaAs/GaAS conduction band offset is much smaller than that of the Si/SiO$_2$ system, and the carriers would escape as their concentration and the corresponding transverse electric field increase.

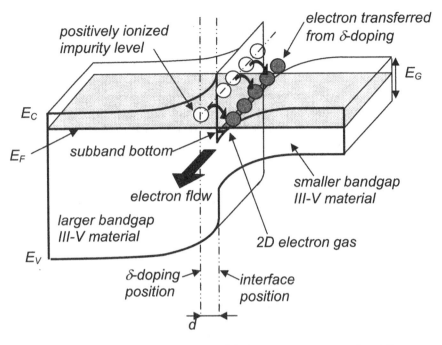

Figure 2.28. *Schematic representation of the heterostructure energy band diagram*

2.13. Characteristic lengths and mesoscopic nature of electron transport ●

In order to evaluate which physical model is appropriate to describe electron transport in a given device, it is important to compare some characteristic lengths, whose respective values are all essential in order to determine which physical effects prevail.

A first important length is the *Fermi wavelength* λ_F. At low temperature we have seen that available electron states are filled from the lowest one up to the Fermi energy. λ_F is the electron wavelength at the Fermi energy (and thus the smallest electron wavelength present in the system). This important quantity depends on the total electron number. For example, if there is a short-range potential fluctuation, i.e. an impurity, electrons will tend to accumulate or to leave the area over which such a potential is exerted. This is known as screening (accumulated electrons around a positively charged impurity tend to make the potential vanish with respect to electrons located farther away: they "screen" this potential). Since λ_F is the shortest wavelength, Fourier components of the potential higher than $1/\lambda_F$ cannot be screened.

A second important length is the *mean free path* λ_M. It is defined as the average distance that an electron travels before losing its initial momentum. Suppose that the physically relevant quantity is the average time τ_m elapsed before the electron loses its initial momentum, then for electrons at the Fermi energy the mean free path is given by $\lambda_m = v_F \tau_m$, where v_F is the Fermi velocity (note that a single collision is most often not sufficient for an electron to lose its momentum, and several are required; thus, the average time between two collisions may be shorter than τ_m). We shall deepen our understanding of the physical significance of λ_m in the section dealing with Drude's mobility model.

The third important length with respect to mesoscopic or ballistic transport is the *phase coherence length* λ_φ. It is defined as the average length over which an electron travels before loosing its initial phase. An electron can change its momentum without losing its phase coherence. Here we can imagine a simple case to illustrate this notion: suppose that a 1D electron, represented by a wavepacket, impinges on a hard wall (say from the left side). If we know the exact wavepacket shape before the rebound, we can calculate at any time the wave function shape after the collision, when the wavepacket motion is now directed towards the right side, just by solving Schrödinger's equation. Before and after the collision, the momentum distribution resulting from the wavepacket existence exhibits opposite average values, and during the collision we have interference effects between the reflected and incident parts of the packet. Since we can always calculate the phase of the electron wave function at any time and position if we know the initial wavepacket, we say that the electron keeps phase coherence. Now suppose that we displace the wall by a small and random amount at arbitrary times. It is no longer possible to calculate the phase of the electron wave function, as the wall motion is random, and averaging over many electrons will not give us interference effects. We can also understand this last point by imagining a two-slit experiment with electrons. If we assume the wall containing the slits is perfectly immobile, repeating the experiment produces an interference pattern on the screen placed behind. However, if we randomly change the slit position each time an electron is emitted from the source, repeating a large number of experiments will not give us an interference pattern *even if each electron wave function is coherent*. Another possibility is that the electron loses energy during the collision with the wall; once again it is no longer possible to calculate the phase, and interference effects are lost because an irreversible energy loss can be viewed as a measurement process; an inelastically scattered wave cannot interfere with an unscattered part. Therefore, we see that two types of phenomena can destroy phase coherence: collisions which intervene at arbitrary times and collisions which do not conserve energy. In our mesoscopic samples such processes are mostly due to phonons, which are propagating vibrating modes of the lattice and which are naturally random, interactions with momentarily trapped electrons, or electron-electron interactions. Since the phonon density increases with temperature, we conclude that to preserve phase coherence it is better to operate at low temperatures. If phase coherence is preserved in the whole device, we can expect to observe interference effects such as those taught during an introductory quantum mechanics course (remember this electron does not go only through one slit, but its wave function propagates through both slits and produces an interference pattern on the screen if we carry out a large number of detection experiments). As in the previous section, to completely lose phase coherence an electron may have to interact with more than one phonon.

Now that we have defined these three characteristic lengths, we are also able to define what a mesoscopic conductor is. Here we shall adopt the point of view developed in the book by Datta [DAT 95], and we define a mesoscopic conductor by a negative property: in a macroscopic conductor (as opposed to mesoscopic), the conductance is ohmic, i.e. we can express it in the form

$$G = \sigma \frac{W}{L} \tag{2.128}$$

where σ is the conductivity, W is the sample width and L is the sample length (this is the 2D version but of course whatever the dimension is the conductance always scales with the reciprocal length). As a matter of fact, for a conductor to be ohmic its dimensions must be greater than all of the three lengths that we defined above: L must be larger than the de Broglie wavelength λ_F, the momentum mean free path λ_M and the phase relaxation length λ_φ. *When the device dimensions get small enough for Ohm's law to break down, the device is called mesoscopic.*

From the above definition, we can expect that mesoscopic transport may largely vary, depending on which characteristic lengths actually exceed the device dimensions. In addition, depending on device materials and physical conditions (temperature, magnetic field, etc.), mesoscopic effects are not necessarily restricted to extremely small devices. For example, we shall see that in the quantum Hall effect regime, transport may be ballistic over millimeters [VON 86], so that mesoscopic transport may require dimensions from a few nanometers up to millimeters! In contrast, in silicon at room temperature, the mean free path is in the range of 10 nm, so that to observe ballistic effects at 300 K we should really produce pretty small devices!

2.14. Mobility: Drude model ●

In semiconductors the collisions experienced by the free electrons or holes are mostly due to phonons (i.e. vibrations of the atomic lattice) and impurities (for impurities and negative charge carriers see the diagram in Figure 2.29). Electron-

electron collisions are much less important. The simplest transport model was developed by Drude a long time ago[14]. Its main assumption is that on average, collisions act as a damping force on the electrons, which prevents them from being indefinitely accelerated by the applied electric field.

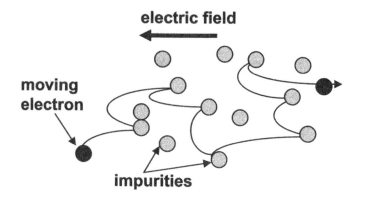

Figure 2.29. *Classical electron trajectories in the Drude model*

We should all remember what a damping force is: something which is proportional to the velocity and acts against movement. Therefore, defining $\langle \vec{v}_D \rangle$ as the drift velocity, we can write Newton's second law of motion as

$$\frac{d}{dt}\langle \vec{v}_D \rangle = -\frac{e\vec{\varepsilon}}{m} - \frac{\langle \vec{v}_D \rangle}{\tau_m} \quad , \tag{2.129}$$

where τ_m is the momentum relaxation time (the "viscous" coefficient), $\vec{\varepsilon}$ is the electric field and e is the absolute value of the electron charge. In the steady state, the average velocity is a constant, and thus the momentum increase rate due to the electric field is equal to the momentum loss rate induced by collisions. Making $d\langle \vec{v}_D \rangle / dt = 0$ gives us

$$\vec{v}_D = -\frac{e\tau_m}{m}\vec{\varepsilon} \tag{2.130}$$

---

14 Drude model is more thoroughly expounded in [ASH 77].

(where we have omitted the averaging brackets to simplify the notation). The electron mobility is defined as the proportionality coefficient between the drift velocity and the electric field:

$$\mu = \frac{e\tau_m}{m} \quad .$$

(2.131)

It is worth noting that at low enough electric fields the relaxation time does not depend on energy, but for higher fields it does, and then the drift velocity is not necessarily proportional to the electric field. The mobility value is of paramount importance to characterize transport. The higher the mobility is, the faster the devices, and the happier the device engineer. Note that the physicist also appreciates a high mobility, because it allows him to observe coherent or interference effects at higher temperatures.

2.15. Conduction in degenerate materials ●

Suppose that we can describe the electron system by running plane waves and BVK boundary conditions. In the absence of electric field, once we have filled all states up to the wavevector values corresponding to the Fermi energy, there are as many electrons which go in one direction as in the opposite one. If $\vec{k}$ is the wavevector of an occupied state, then the state with wavevector $-\vec{k}$ is also occupied. Therefore, the overall current is equal to zero. In addition, if an allowed energy band is fully occupied by electrons (e.g. the valence band in an intrinsic semiconductor at $T=0$), it cannot carry any current, since it is not possible to accelerate any electrons by putting them in a higher energy state (all states are already occupied). However, suppose that the band is only partially filled (e.g. the conduction band). Then it becomes possible to change the electron energies. Thus, applying an electric field will change the momentum of all electrons by an amount equal to the drift velocity times the effective mass. The wavevector is thus modified by an amount $\vec{k}_D = m\vec{v}_D / \hbar = -e\tau_m\vec{\varepsilon} / \hbar$. The current density is the product of the electron charge by the velocity:

$$\vec{j} = -en_S\vec{v}_D$$

(2.132)

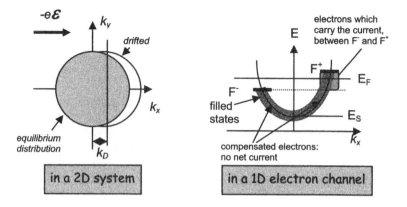

Figure 2.30. *On the left: filled states of a 2D system represented in k-space, with and without electric field. On the right: electron state filling in a 1D system submitted to a longitudinal electric field*

It is important to realize that not all electrons do contribute to the net current. As illustrated by Figure 2.30 in the 2D and 1D cases, if we displace all occupied states in k-space by an amount $\vec{k}_D$, a large fraction of electron pairs still exhibit exactly opposite momentum values, so that it is not necessary to include them in the overall current calculation. Only the momenta of electrons close to the Fermi energy are not counterbalanced and lead to a net current value. This simple analysis reveals an important physical feature: transport properties are determined by those electron states which are located close to the Fermi surface.

If we focus on the 1D case, we see in Figure 2.30 that the electric field makes the electrons going to the right occupy states with positive k values up to a level which we shall denote F^+. Reciprocally, the electrons going towards the left side occupy states up to a level F^-, lower than F^+. Thus, the electrons which carry a net current are going to the right, located between F^- and F^+. F^+ corresponds to a momentum k_F+k_D, F^- to a momentum $-k_F+k_D$. Therefore, we have

$$F^+ \approx \frac{\hbar^2}{2m}\left(k_F + k_D\right)^2 \qquad\qquad F^- \approx \frac{\hbar^2}{2m}\left(k_F - k_D\right)^2 \qquad\qquad (2.133)$$

Calculating the difference between F^+ and F^- from equation (2.133) leads to

$$F^+ - F^- \cong \frac{2\hbar^2}{m} k_F k_D. \qquad\qquad (2.134)$$

Remembering that $v_D = -e\tau_m \varepsilon / m$ and that the momentum mean free path is $\lambda_m = v_F \tau_m$, it will take us a few seconds to find from the formula above that

$$F^+ - F^- \cong 2e\varepsilon\lambda_m. \tag{2.135}$$

Thus, we have obtained that with a uniform electric field, the quasi-Fermi level separation is proportional to the energy gained by an electron in a mean free path [DAT 95] (this energy also has another profound physical meaning, because its ratio with respect to the thermal energy $k_B T$ determines whether current noise is dominated by thermal fluctuations or shot noise contributions, see Chapter 6).

2.16. Einstein relationship ◎

If the electron concentration is constant throughout a device and if the electric field is uniform, electrons are accelerated by the field and move under its action: they "drift" under the action of the electric field, and the resulting electric current is said to be a drift current. There is another situation in which you can have some current, but this time without any externally applied electric field: if, locally, we have a concentration gradient, due to the thermal agitation, carriers will tend to move towards regions where the concentration is lower (their movement tends to make the concentration uniform, just as when we keep a concentrated gas in a small container, and suddenly open it towards a bigger chamber under vacuum; the gas quickly spreads over the two chambers until a uniform concentration is obtained). Of course the larger the concentration gradient is, the larger the current is. The electron current is proportional to the concentration gradient:

$$\vec{j} = eD_n \vec{\nabla} n, \tag{2.136}$$

where D_n is the diffusion coefficient. There is in fact a definite relationship between the two types of transport, which is known as Einstein's relation. This relationship differs depending on whether the electron gas is degenerate or described by a Boltzmann distribution. The relation is in both cases easy to determine, as at thermal equilibrium the current, given by the sum of the drift and diffusion components,

$$\vec{j} = en\mu\vec{\varepsilon} + eD_n \vec{\nabla} n \tag{2.137}$$

must be equal to zero[15]. Consider first the non-degenerate electron gas: the electron concentration being given by the Boltzmann distribution

$$n = N_C \exp\left(-\frac{E_C - eV(x) - E_F}{k_B T}\right) \quad , \tag{2.138}$$

where V is the electrostatic potential, from equation (2.137) at equilibrium we have

$$0 = -en\mu \frac{\partial V}{\partial x} + eD_n \frac{e}{k_B T} \frac{\partial V}{\partial x} n \tag{2.139}$$

which gives the Einstein relation for a non-degenerate electron gas:

$$D_n = \frac{k_B T}{e} \mu \tag{2.140}$$

In the case of a degenerate 2D conductor, we have $n \approx D_{2D}(E_F - E_C + eV(x))$ and canceling the equilibrium current in equation (2.137) leads to

$$0 = -en\mu \frac{\partial V}{\partial x} + eD_n D_{2D} e \frac{\partial V}{\partial x} \tag{2.141}$$

which reduces to

$$D_n = \frac{n\mu}{eD_{2D}} \quad . \tag{2.142}$$

The latter expression also enables us to re-write the 2D conductance as a function of the diffusion coefficient:

$$\sigma = e^2 D_{2D} D_n \tag{2.143}$$

---

15 If the reader is wondering why there should be equilibrium situations in which there are both an electric field and a concentration gradient, this occurs whenever the material is inhomogenous. Think of the *p-n* junction, in which a constant chemical potential throughout the junction implies the existence of a depletion layer and an internal electric field, the latter creating a drift current which is exactly compensated by the diffusion current resulting from the gradient concentration at the junction; see [SZE 81].

Replacing μ by $e\tau_m/m$ and n by $n=D_{2D}(E_F-E_C)=D_{2D}\times mv_F^2/2$ in equation (2.142) gives us an alternative expression for the diffusion coefficient:

$$D_n = \frac{1}{2}v_F^2\tau_m \quad , \tag{2.144}$$

which we will use in some instances.

2.17. Low magnetic field transport ●

If, in addition to the application of a voltage, we put the semiconductor device into a magnetic field B, we have to replace the electric field force by the full Lorentz force

$$\vec{F} = -e\left(\vec{\mathcal{E}} + \vec{v} \wedge \vec{B} \right) \tag{2.145}$$

This situation is mostly important for parameter extraction, because, as we shall see below, this allows us to assess the electron concentration and mobility independently from one another. These two parameters matter a lot for device performance. However, do not believe that this situation is never observed in applications, even if electronic circuits are usually not operated under substantial magnetic field values: in France, for instance, all the home electricity meters of the EDF corporation are now based on the use of a III-V High Electron Mobility Transistor (HEMT), whose transport properties are modified by the magnetic field generated by the main power supply line.

The Lorentz force makes the electrons circle around the axis defined by the magnetic field direction. If the value of B is not high enough to ensure that the electrons can travel over more than one cyclotron orbit before losing their momentum through collisions, then we do not have to take into account new quantum phenomena that we shall describe in another part of this course, but we can content ourselves with applying the same formalism as in section 2.14 (Drude model), just adding the Lorentz force:

$$\frac{d}{dt}\langle\vec{v}_D\rangle = \frac{-e\left(\vec{\mathcal{E}} + \langle\vec{v}_D\rangle \wedge \vec{B}\right)}{m} - \frac{\langle\vec{v}_D\rangle}{\tau_m} \tag{2.146}$$

Operate a 2D device as in Figure 2.31. From equation (2.146) we state as before that in the steady-state regime the drift velocity is locally constant and we obtain without any difficulty (please check this by yourself) that

$$\begin{pmatrix} \varepsilon_x \\ \varepsilon_y \end{pmatrix} = \frac{1}{\sigma} \begin{pmatrix} 1 & +\mu B \\ -\mu B & 1 \end{pmatrix} \begin{pmatrix} J_x \\ J_y \end{pmatrix}.$$

(2.147)

The matrix is known as the resistivity tensor, with resistivity coefficients defined as

$$\begin{pmatrix} \varepsilon_x \\ \varepsilon_y \end{pmatrix} = \begin{pmatrix} \rho_{xx} & \rho_{xy} \\ \rho_{yx} & \rho_{yy} \end{pmatrix} \begin{pmatrix} J_x \\ J_y \end{pmatrix},$$

(2.148)

so that we have

$$\rho_{xx} = \frac{1}{\sigma} \quad and \quad \rho_{yx} = -\rho_{xy} = \frac{-B}{en_S} .$$

(2.149)

Defining a measurable quantity known as the Hall coefficient $R_H = \varepsilon_y / J_x B$, we obtain from equation (2.149) and the condition that the current J_y be zero

$$R_H = -\frac{1}{en_S} ,$$

(2.150)

a remarkable relation which indicates that the Hall coefficient exhibits the same sign as the carrier charge, and only depends on the electron concentration, therefore providing an easy means to determine the latter quantity. From equation (2.149) we also obtain the relationship

$$\mu = \frac{J_x}{en_S \varepsilon_x} ,$$

(2.151)

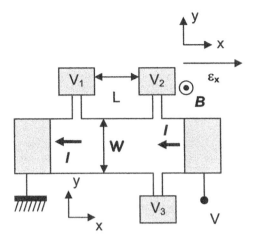

Figure 2.31. *Hall effect in a 2D device with several terminals*

from which we can assess the mobility. With the sign convention of the current I as defined in Figure 2.31 we have $I=-WJ_x$ (positive) and $\varepsilon_y=-V_H/W$, so that we obtain $R_H=V_H/IB$. The electrons, which circulate from the left to the right contact, are repelled by the Lorentz force towards the sample side on the top of the figure; in the permanent current regime there must be an electric field which equilibrates this force so as to cancel out the transverse current flow, and thus the transverse Hall electric field is directed towards the positive y. Thus, V_H is negative and $R_H=V_H/IB$ is negative, as predicted by equation (2.150). The mobility is $\mu=IV/en_SWL$.

Measuring the Hall voltage variation V_H as a function of magnetic field, the slope gives us the electron concentration without any fitting parameter, and then we can obtain the mobility from equation (2.151). This is a convenient way to separate the concentration and the mobility, which in a simple conductivity measurement always come together in a common product.

2.18. High magnetic field transport ●

2.18.1. *Introduction*

This is not the "simplest" part of this book. As a matter of fact, the reader is not required to remember the derivation of Hermite polynomials and Landau wave functions (section 10.3) to understand what is really at the heart of this book, i.e.

ballistic effects and conductance from transmission. Nevertheless, we should try to remember the basic features of high magnetic field transport (summarized at the end of this section) to understand some of the spectacular quantum effects which can be probed in solid-state mesoscopic devices. In addition, such a regime can be used to extract parameters which are essential to device performance. To help us to retain the physics, quite simple heuristic derivations and discussions are given before the more rigorous approach to this problem, in section 2.18.4.2.

2.18.2. Some reminders about the particle Hamiltonian in the presence of an electromagnetic field

Apply an electromagnetic field to a particle. As usual, the Hamiltonian of a particle with charge q is equal to

$$H = T + V = \frac{1}{2} m \left(\frac{d\vec{r}}{dt} \right)^2 + V \qquad (2.152)$$

where the first and second terms are the kinetic and potential energy. However, the particle is not the only entity to exhibit momentum when considering the whole particle+field system. The electromagnetic field associated with the particle also possesses a momentum. Indeed it can be shown from Maxwell's equations and Lagrangian formalism that for such a system the conjugate momentum of $\vec{r}$ is not $m\vec{v}$, but the sum of this quantity and of the momentum due to the field associated with the particle [COH 77]:

$$\vec{p} = m \frac{d\vec{r}}{dt} + q\vec{A}, \qquad (2.153)$$

where $\vec{A}$ is the vector potential, defined by the well known relationship

$$\vec{B} = \vec{\nabla} \times \vec{A}. \qquad (2.154)$$

(for those interested by the demonstration, which takes a few pages of discussion and calculations, they can consult the book by Cohen-Tannoudji *et al.* [COH 77]). Thus, we can write the Hamiltonian as

$$H = \frac{1}{2m} \left(\vec{p} - q\vec{A} \right)^2 + V. \qquad (2.155)$$

In equation (2.155) the conjugate momentum of the position operator is given by the usual quantum-mechanical operator formula

$$\vec{p} = -i\hbar\vec{\nabla} \, . \tag{2.156}$$

2.18.3. *Action of a magnetic field (classical)*

Before going to the quantum, it may be useful to remind ourselves of the classical: if a particle is submitted to a magnetic field B along z and confined to the xy plane, the equation of movement is:

$$m\frac{d\vec{v}}{dt} = q\vec{v} \wedge \vec{B} \quad \Rightarrow \quad \begin{cases} \dfrac{dv_x}{dt} = \dfrac{qB}{m}v_y = \omega_c v_y \\[4mm] \dfrac{dv_y}{dt} = -\dfrac{qB}{m}v_x = -\omega_c v_x \end{cases} \tag{2.157}$$

Differentiate the bottom equation after the bracket and insert the upper one:

$$\frac{d^2 v_y}{dt^2} = -\omega_c\frac{dv_x}{dt} = -\omega_c^2 v_y \, . \tag{2.158}$$

Then if we integrate we obtain

$$\begin{aligned} v_y &= v_0 \sin(\omega_c t + \Phi) \\ v_x &= -v_0 \cos(\omega_c t + \Phi) \end{aligned} \tag{2.159}$$

Integrate once again and we find the equations of movement:

$$\begin{aligned} x &= x_0 + \frac{v_0}{\omega_c}\sin(\omega_c t + \Phi) \\[3mm] y &= y_0 - \frac{v_0}{\omega_c}\cos(\omega_c t + \Phi) \end{aligned} \tag{2.160}$$

From the equations above we also have

$$(x - x_0)^2 + (y - y_0)^2 = \frac{v_0^2}{\omega_c^2}.$$

(2.161)

Thus, as in Figure 2.32, the particle velocity is constant and the trajectory is a closed circular orbit of radius

$$r_C = \frac{v_0}{\omega_C}$$

(2.162)

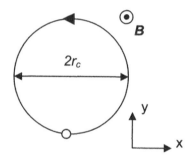

Figure 2.32. *Classical trajectory of a 2D electron submitted to a magnetic field perpendicular to the confinement plane*

2.18.4. *High magnetic field transport*

2.18.4.1. *Simple discussion*

It is not a common situation in electronics, since there is in fact no practical case in which the magnetic field exceeds several Tesla (the background earth magnetic field is in the micro-Tesla range). However, it is an important tool for probing the quality and physical properties of semiconductor structures, and it is also an impressive way to convince yourself of the validity of quantum mechanics: if you see periodic Shubnikov-de Haas conductance oscillations appearing live on the monitor screen when measured, e.g., from a MOSFET operated at low temperatures inside a superconducting magnet (which produces the magnetic field), there is no way to explain it but with the calculations that we are going to study in this section. This theory was first derived by Landau as early as 1930, shortly after the advent of quantum mechanics [LAN 30]. In addition, applying a high magnetic field can put the devices under test in a situation which is known as the quantum Hall regime, which leads to the apparition of ballistic quantum states with can spread over characteristic lengths up to several millimeters [VON 86]. This is thus a striking example of ballistic regime, even if the device dimensions are not reduced down to a nanometer length scale. If you are already accustomed to the quantum-mechanical treatment of the harmonic oscillator, it will not be difficult to retain the basic results of this section because all of them are formally analogous to this celebrated case.

Since the Hamiltonian does not depend on time, the eigenstates are stationary and although what follows is not quite rigorous we can guess much of the underlying physics just by appealing to this feature: remember that in the classical case the electrons follow closed circular orbits. Suppose that they still do so in the quantum case, for the state to remain stationary the electron wave must retain the same phase after having performed a single orbit (imagine not a plane wave, but something which looks like one, except for the fact that it is now localized perpendicularly to the direction of propagation, and follows a circular orbit). Thus, the circumference must be an integer number of de Broglie wavelengths:

$$n\lambda \cong 2\pi r_C \cong \frac{2\pi v}{\omega_C} \quad with \quad \omega_C = \frac{eB}{m} \tag{2.163}$$

This straighforwardly leads to a quantized energy $E=n\hbar\omega_c/2$. This is not exactly the correct answer, and we shall see that the exact values are rather given by

$$E = \left(n+1/2\right)\hbar\omega_C \tag{2.164}$$

However, even with this rudimentary reasoning we have already found the important and correct result that all energy values are separated from one another by a constant energy quantum[16]. Application of a high magnetic field thus transforms the electron states into discrete states, which are known as Landau states. Imagine that we apply the magnetic field very progressively: each initial electron state, with arbitrary energy, will be progressively transformed into a Landau state. Since the number of states is not going to change, this means that each Landau state is degenerate. Assuming that each of these states includes the same number of electron states, in a 2D system we have to share equally the initial states, which exhibit a

---

16 In fact we can deduce from section 8.3 that we must also add a phase shift due to the action of the vector potential, equal to the flux of B through the electron orbit. Thus, it is better to write the overall phase shift over one orbit as $\phi=k\times2\pi r_c-(|e|B/\hbar)\times\pi r_c^2=2n\pi$. In such a case we find the correct energy quantum (check it) and we just miss the ½ constant inside the parenthesis of equation (2.164).

density as given by equation (2.122), between the newly formed Landau states. This immediately gives us the number of states per Landau level:

$$N_L = \hbar\omega_C \times \frac{m}{\pi\hbar^2} = 2\frac{eB}{h} \quad .$$

(2.165)

Of course, such a quantization requires some conditions. If the magnetic field is not high enough, the classical orbits remain very large (see equation (2.162)), and thus an electron supposed to follow them will experience many collisions before achieving a complete circle. Thus, the phase coherence of the wave function cannot be maintained over a full orbit, and we cannot apply the criterion that the phase must be left unchanged after having cycled over one full orbit. Thus, an obvious condition for Landau states to appear is that the mean free path be larger than the cyclotron radius. This gives us the condition

$$\mu B > 1 .$$

(2.166)

If we operate a 2D electron device in a high magnetic field and low temperature, we can also easily extract the electron density n_S. From the results above we know that the electron states are transformed according to the scheme in Figure 2.33. We just replace the discrete Landau states, which ought to be represented by Dirac peaks, by narrow and finite density peaks, because in practice scattering always spreads the initially degenerate states into a larger band a little bit. Then we can use the following trick: suppose that the electron concentration remains constant (fixed by the gate voltage in a MOSFET, or by the modulation doping value in a heterostructure). If we increase the magnetic field, according to equations (2.165) and (2.164) both the density in the Landau bands and the spacing between each band are going to increase. Thus, for filling the same number of states up to the Fermi level the position of the Fermi level cannot always stay in the same Landau band. Various Landau bands will cross the Fermi level until the field becomes high enough for the number of Landau states in the ground band to exceed the number of carriers. This limiting case is called the quantum limit, and then all electrons are contained in the first Landau band, which has become big enough to accommodate all carriers.

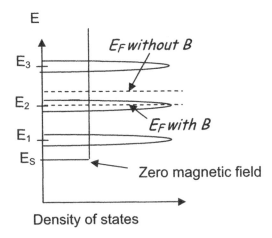

Figure 2.33. *Density of states versus energy in a 2D system
with some disorder-induced broadening; with and without a magnetic field*

Before reaching this limit, suppose that the magnetic field is at a value B_1 such that the Fermi level is just at the top of the Nth Landau band (i.e. the first N Landau bands are totally filled and the following ones are totally empty). Since each Landau band contains $2eB/h$ states, the electron concentration is obviously equal to

$$n_S = N \times \frac{2eB_1}{h} \qquad (2.167)$$

Then, if we increase the magnetic field the number of states in each band increases, so that the Fermi level position decreases in this N^{th} band (since n_s is constant), until this band is completely depopulated and the Fermi level jumps to the $(N-1)^{th}$ Landau band for a value B_2 given by

$$n_S = (N-1) \times \frac{2eB_2}{h}. \qquad (2.168)$$

From equations (2.167) and (2.168) we find that

$$n_S = \frac{2e}{h} \frac{1}{\dfrac{1}{B_1} - \dfrac{1}{B_2}} \qquad (2.169)$$

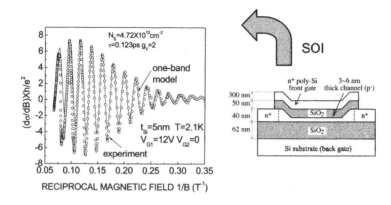

Figure 2.34. *Shubnikov-de Haas conductance oscillations appearing in a very thin SOI MOSFET submitted to a magnetic field (after [OUI 98])*

The spacing $1/B_1-1/B_2$ between two successive reciprocal magnetic field values corresponding to the same filling factor of adjacent Landau bands is thus a constant, proportional to $1/n_S$ (i.e. all those reciprocal magnetic field values exhibit a periodicity inversely proportional to the electron density). It is quite reasonable to suppose that the conductivity exhibits similar properties when the Fermi level is at a given position in a particular Landau band, whatever the Landau band is, and experimentally this is actually the case, as we can check it in the practical example illustrated by Figure 2.34. This means that in 2D the sample conductivity or conductance exhibits periodic oscillations versus the reciprocal magnetic field, and the periodicity gives us the electron concentration without any fitting parameter, according to equation (2.169).

Figure 2.34 corresponds to the situation for which there was only one populated 2D subband before the application of a magnetic field. In addition, in Figure 2.34 the derivative of the conductance is plotted rather than the conductance itself, so as to make the oscillations still more prominent. In good III-V heterostructures this would not be necessary (here the device is made from silicon). Such conductance oscillations are known as Shubnikov-de Haas oscillations. If two subbands are initially populated, the situation is slightly more complicated since both subbands are going to be split into Landau bands which do not coincide. Therefore, in such a case the SdH oscillations will exhibit beating effects just as when we add two periodic functions. Such an example is given in Figure 2.35.

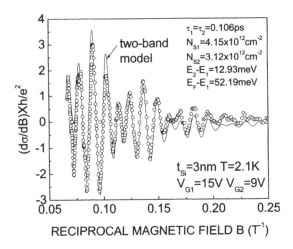

Figure 2.35. *SdH oscillations in the same SOI MOSFET as in the previous figure, but with applied gate voltages such that more than one subband is occupied (after [OUI 98])*

Reciprocally, if we maintain the magnetic field constant but we are able to change the electron concentration through the application of a gate voltage, we shall also obtain conductance oscillations but now as a function of gate voltage. In a MOSFET n_s is proportional to the gate voltage $n_S = C_{ox}(V_G - V_T)/e$ [SZE 81]. Therefore, we can also deduce in a precise way the electron concentration from the periodicity of the conductance oscillations with gate voltage, without any fitting parameter such as the device capacitance or the mobility.

2.18.4.2. *A more involved treatment*

2.18.4.2.1. The problem to be solved

Up to now we have deduced all results from relatively simple considerations. Now we are going to apply a more rigorous approach to our 2D gas. In a general case the 2D system is not infinite, and our device can instead be schematized as in Figure 2.36.

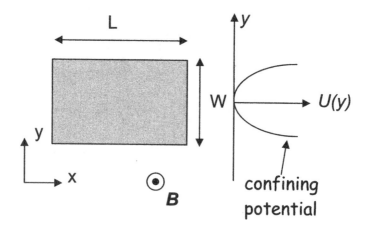

Figure 2.36. *A 2D electron system submitted to a magnetic field and with a confining potential along the y direction*

Here we apply a magnetic field B perpendicularly to the 2D gas plane. Although electrons are free to propagate from one terminal to the next (from left to right), the width W of the sample is finite and thus there is some confining potential $U(y)$ which prohibits the electrons to escape to the edges. Thus, from equations (2.47) and (2.155) the Schrödinger equation can be written

$$\left[E_S + \frac{(i\hbar\nabla - e\vec{A})^2}{2m} + U(y) \right] \Psi(x,y) = E\Psi(x,y)$$

(2.170)

Now we are going to tackle a somewhat subtle characteristic of the vector potential: since the rotational of a divergence is always equal to zero (check it!), from $\vec{B} = \vec{\nabla} \times \vec{A}$ it is clear that in contrast with the magnetic field $\vec{B}$ itself $\vec{A}$ is not uniquely defined. Just add to $\vec{A}$ any function $\vec{\nabla}F(\vec{r})$ which is a divergence and we still find that $\vec{B} = \vec{\nabla} \times \vec{A}$! Choosing a particular function for this divergence so as to carry out the calculations is called to work in a particular gauge. This may seem complicated, or somewhat paradoxical, because when solving equation (2.170) for a particular gauge we expect to find a solution which depends on the gauge, whose divergence term is itself non-physical; and this is in fact the case! However, if the eigenstates of equation (2.170) depend on the gauge, what is physical and must not change with the gauge are the predictions concerning the results of measurements that we can make (e.g. the energy values which can be probed, or a position measurement probability, etc.). Once we admit that such predictions are independent of the gauge (and this can be demonstrated; see Cohen-Tannoudji *et al.* [COH 77]. for a more complete treatment of this fundamental problem), the best thing to do is to carry on with the gauge which gives the simplest results for the wave functions, i.e. the gauge which is best suited to our device geometry[17]. In our case, the gauge which gives us eigenstates which are the simplest for our purposes is the gauge for which

$$A_x = -By \quad and \quad A_y = A_z = 0 .$$
(2.171)

It is known as the Landau gauge. It is trivial to check from $\vec{B} = \vec{\nabla} \times \vec{A}$ that the relationship above does give us a magnetic field $\vec{B}$ directed towards the positive z values. As stated above, *we chose this gauge for convenience*. It shall in fact give us states which are the simplest ones to discuss the variation of electron concentration inside the device. With a little bit of algebra, we then obtain from equation (2.170) that

$$\left[E_S + \frac{(p_x - eBy)^2}{2m} + \frac{p_y^2}{2m} + U(y) \right] \Psi(x,y) = E\Psi(x,y) .$$
(2.172)

From equation (2.172) we can see that the Hamiltonian is invariant under a translation along x, so that we can seek solutions which are plane waves along x and which otherwise depend on y. These solutions are of the form

$$\Psi(x,y) = \frac{1}{\sqrt{L}} e^{ikx} \chi(y) .$$
(2.173)

Introducing such a form in equation (2.172) will then give you

$$\left[E_S + \frac{(\hbar k - eBy)^2}{2m} + \frac{p_y^2}{2m} + U(y) \right] \chi(y) = E\chi(y)$$
(2.174)

---

17 With a bit of effort we can find that if the vector potential is $\vec{A}_1$ in gauge 1 and $\vec{A}_2$ in gauge 2, and if both potentials are connected through the divergence of a function $\lambda(x,y)$, i.e. $\vec{A}_2 = \vec{A}_1 + \vec{\nabla}\lambda$, then for any solution ψ_1 to the Schrödinger equation in gauge *1*, $\psi_2 = exp(-ie\lambda/\hbar c)\psi_1$ is a solution in gauge *2* which differs only by a phase factor. However, ψ_2 is in general not a single eigenstate in gauge *2* and does not respect the symmetry of the corresponding Hamiltonian. It is instead formed by a linear combination of degenerate eigenstates in gauge *2* (see [SWE 89]).

after a few calculation lines. We are going to solve equation (2.174) in various particular cases.

2.18.4.2.2. Confined electrons in the absence of magnetic field

Here $B=0$. To make things simpler we can suppose that the confining potential is *harmonic*, i.e. $U(y)=m\omega_0^2 y^2/2$. Equation (2.174) turns into

$$\left[E_S + \frac{\hbar^2 k^2}{2m} + \frac{p_y^2}{2m} + \frac{m\omega_0^2}{2}y^2 \right]\chi(y) = E\chi(y) \quad . \tag{2.175}$$

This represents nothing but a 1D harmonic oscillator along y, which we solve in section 10.3, and to which we have added an additional degree of freedom along x (k can take any real value). For the sake of clarity and since we are going to make extensive use of the solutions we re-write them below:

$$\chi_{n,k}(y) = H_n(q)\, e^{-\frac{q^2}{2}} \tag{2.176}$$

$$E = E_S + \frac{\hbar^2 k^2}{2m} + \left(n + \frac{1}{2} \right)\hbar\omega_0 \tag{2.177}$$

$$v_{n,k} = \frac{1}{\hbar}\frac{\partial E(n,k)}{\partial k} = \frac{\hbar k}{m} \tag{2.178}$$

where q is given by

$$q = \sqrt{\frac{m\omega_0}{\hbar}}\, y \tag{2.179}$$

and $H_n(q)$ is a Hermite polynomial of degree n (see section 10.3). The energy is quantized along the y axis, and thus leads to the creation of 1D subbands in which the movement along x is free (Figure 2.37). Since the confining potential is harmonic, the bottom values of the 1D subbands are equally spaced. In practice of course the confining potential is not necessarily harmonic, and the energy spacing can differ from this simplistic picture.

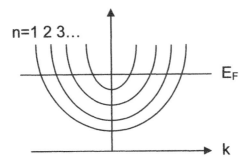

Figure 2.37. *Energy levels as a function of the wavevector in a 2D system confined by a harmonic potential along the y direction*

2.18.4.2.3. Unconfined electrons in a magnetic field

Applying $B{\neq}0$ and $U(y)=0$ to equation (2.174) leads to

$$\left[E_S + \frac{(-eBy + \hbar k)^2}{2m} + \frac{p_y^2}{2m}\right]\chi(y) = E\chi(y) \; .$$

(2.180)

Defining

$$y_k = \frac{\hbar k}{eB}$$

(2.181)

along with

$$\omega_c = \frac{eB}{m} \, ,$$

(2.182)

equation (2.180) becomes

$$\left[E_S + \frac{p_y^2}{2m} + \frac{m\omega_C^2}{2}(y - y_k)^2\right]\chi(y) = E\chi(y) \; .$$

(2.183)

Compare equation (2.183) to equation (2.175). Obviously and as before, we still have an equation which is formally equivalent to that of a harmonic oscillator. The difference is that the parabolic potential is now centered along y_k, which depends on k, and we no longer include the energy term $\hbar^2 k^2 / 2m$. Thus, we just have to shift the known analytical solutions of a harmonic oscillator by an amount y_k! The solutions are analytically described by the following relationships:

$$\chi_{n,k}(y) = H_n(q - q_k) e^{-\frac{(q - q_k)^2}{2}}$$

(2.184)

$$E_{n,k} = E_S + \left(n + \frac{1}{2}\right)\hbar\omega_C$$

(2.185)

$$v_{n,k} = \frac{1}{\hbar}\frac{\partial E(n,k)}{\partial k} = 0$$

(2.186)

where we now have

$$q = \sqrt{\frac{m\omega_C}{\hbar}}y$$

(2.187)

and

$$q_k = \sqrt{\frac{m\omega_C}{\hbar}}y_k.$$

(2.188)

This corresponds to a band structure as schematized in Figure 2.38.

It is worth remembering than in the classical case, since the Lorentz force is always directed perpendicularly to the direction of movement, it exerts no work and thus the absolute velocity cannot change. If the velocity is averaged with time along one orbit it is equal to zero. There is a local current, but the overall, average current is equal to zero. From equation (2.186) we find something which is exactly in line with this result, since *the group velocity is in fact equal to zero*, as expected in the case of a Lorentz force. In any Landau band, the energy remains constant, whatever the value of the quantum number k is. This is somewhat puzzling, because we do have a term like a plane wave in the wave function expression, and the electron state is extended along x. However, equation (2.186) demonstrates that this plane wave term, which ought to be considered with the k-dependent function $\chi_{n,k}(y)$, does not carry any electron current! To pictorially illustrate the shape of such wave functions, we have plotted two of them in Figure 2.39.

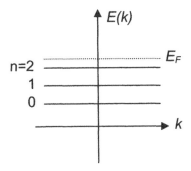

Figure 2.38. *Energy levels of an unconfined 2D system submitted to a perpendicular magnetic field*

Here it may be good to summarize our findings: the magnetic field quantizes the electron energy $E_{n,k} = E_S + (n+1/2)\hbar\omega_c$ with an integer index n and creates corresponding subbands, in which there is a free parameter k, with the dimension of a wavevector. For an infinite sample k can take any real value, and if the device length L is finite it can be fixed by BVK boundary conditions along x. The wave function is localized along y and has a shape along y which depends only on the Landau band number n. It looks like a plane wave along x. The action of k can be described as follows: in a given subband, the higher index k is, the larger the shift y_k of the wave function from $y=0$ is. All these extended states carry no current.

From the results above we can also deduce the number of states in a Landau band in a more rigorous way than in section 2.18.4.1. Suppose that we restrict the device dimensions to a length L along x and a width W along y. Applying BVK boundary conditions to the x axis, restrict the possible k values, which are now separated from one another by $2\pi/L$. This means that in a Landau band with a given n index, the wave functions, which are centered around y_k along y, are separated by an amount

$$\Delta y_k = \frac{2\pi\hbar}{eBL} \quad , \tag{2.189}$$

as illustrated by Figure 2.40. For a width W the total number of states in a Landau band is thus equal to *2(for spin)* $\times W/\Delta y_k$. To obtain the density of states per unit area we must divide by W and L, and then we find that the density of Landau states in one Landau band is equal to *2eB/h*, as predicted in section 2.18.4.1[18].

---

18 Note that with a high magnetic field the spin degeneracy can be lifted by the Zeeman effect, so that the spin down and spin up electrons have well separated energies and can form two separate bands, with a density of states equal to eB/h in each one.

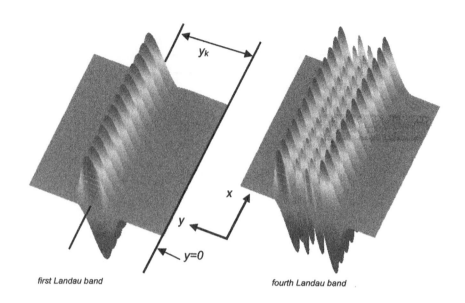

first Landau band   fourth Landau band

Figure 2.39. *Landau wave functions resulting from the application of a magnetic field to an infinite 2D system. Left: real part of the wave function for n=0; right: n=3*

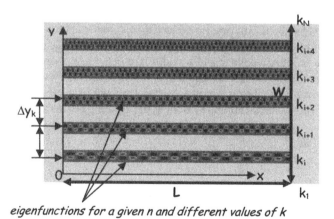

eigenfunctions for a given n and different values of k
(here n=1 and Re(Ψ) is represented)

Figure 2.40. *Landau wave functions for index n=1 and different wavevector numbers k, "viewed from the top" (a bright spot is a node; each wave function is centered around y_k; the higher k is, the more deflected the wave function is towards a sample side)*

2.18.4.2.4. Confined electrons in a magnetic field

Now we have both $B{\neq}0$ and $U(y){\neq}0$. Thus, we still have to solve equation (2.174):

$$\left[E_S + \frac{(p_x - eBy)^2}{2m} + \frac{p_y^2}{2m} + \frac{m\omega_0^2}{2} y^2 \right] \Psi(x,y) = E\Psi(x,y) \,. \tag{2.190}$$

After some cumbersome but not difficult calculation steps, we can turn this equation into

$$\left(E_S + \frac{p_y^2}{2m} + \frac{m}{2} \frac{\omega_0^2 \omega_C^2}{\omega_c^2 + \omega_0^2} y_k^2 + \frac{m}{2} (\omega_c^2 + \omega_0^2) \left(y - \frac{\omega_C^2}{\omega_c^2 + \omega_0^2} y_k \right)^2 \right) \chi(y) = E\chi(y) \,. \tag{2.191}$$

As in the previous sections, from equation (2.191) we can make out a harmonic oscillator-like equation. Here we have to add an energy term independent of y but dependent on k, in a way similar to section 2.18.4.2.2, and a harmonic potential shifted along y as in section 2.18.4.2.3, but with a slightly more complicated shift as a function of k. Thus, although seemingly more complicated, the solutions are given without any difficult step by

$$E_{n,k} = E_S + \left(n + \frac{1}{2} \right) \hbar \sqrt{\omega_C^2 + \omega_0^2} + \frac{\hbar^2 k^2}{2m} \frac{\omega_0^2}{\omega_0^2 + \omega_C^2} \,, \tag{2.192}$$

$$\chi_{n,k}(y) = H_n \left(q - \frac{\omega_C^2}{\omega_c^2 + \omega_0^2} q_k \right) \exp \left(-\frac{1}{2} \left(q - \frac{\omega_C^2}{\omega_c^2 + \omega_0^2} q_k \right)^2 \right) \tag{2.193}$$

and

$$v_{n,k} = \frac{\hbar k}{m} \frac{\omega_0^2}{\omega_0^2 + \omega_C^2} = \frac{\hbar k}{m_{eff}} \tag{2.194}$$

with the same definition for q and q_k as in the previous sections. If we compare these formulae with those obtained in the case of a confining potential without magnetic field and those obtained with a magnetic field and without any confining potential, we can see that equations (2.192), (2.193), (2.193) and (2.194) have borrowed characteristics from both of them. On the one hand, the energy now includes a term proportional to k^2, as in the former case, so that k can actually be considered as a momentum, and the electron states do carry a current. On the other hand, the wave functions are shifted towards one side of the sample as the wavenumber k increases, as in the latter case. With $k>0$ (electrons propagating from the left to the right in Figure 2.33) they are shifted towards the upper side of the sample (see the y shift in the wave function (equation (2.193)), and with electrons propagating from the right to the left, they are shifted to the lower side of the sample. There is one more difference: if we look at equations (2.192) and (2.194) we can see that in addition to those two features, it is just as if the effective mass m had been increased by the magnetic field and replaced by the quantity

$$m \rightarrow m_{eff} = m\left(1 + \frac{\omega_C^2}{\omega_0^2}\right) \tag{2.195}$$

in all of its occurrences, or just as if the magnetic field had increased the electron inertia. From equation (2.192) we can represent the dispersion relation in each band as in Figure 2.41.

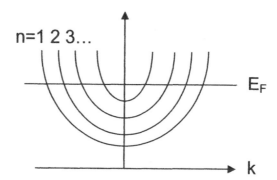

Figure 2.41. *Energy levels of a 2D system confined along y and submitted to a magnetic field, as a function of wavevector k*

From equations (2.193) and (2.194) we straightforwardly find that

$$y_k = \frac{\omega_C^2 + \omega_0^2}{\omega_C \omega_0^2} v_{n,k} \tag{2.196}$$

so that the shift to the sample sides is simply proportional to the electron velocity.

We have thus demonstrated an extremely important feature, which we shall use later on for explaining the ballistic properties of transport in the high magnetic field regime, and this is what we should retain first from this section: *the magnetic field repels the electron states carrying current along +x towards one side of the sample, and reciprocally, the states carrying current along –x are shifted to the opposite side of the sample. The higher the kinetic energy, the larger the shift.* Thus, in this regime and for the states close to the Fermi energy, we have a huge reduction of the spatial overlap between the forward and backward propagating states. This is extremely important for scattering, because this means that forward propagating states around the Fermi energy, which are the states which carry the net current, are spatially separated from the backward propagating states. Thus, *backscattering is totally suppressed* (an electron in such a state has no chance of being backscattered or losing momentum, because to do so it should be kicked to the opposite side of the sample!).

Another observation: we saw that in the classical case the electrons driven by a magnetic field follow circular orbits, so that if we average according to time they do not carry any net current. This is not what we find here with a confining potential. However, suppose that in the classical case we also add boundaries, and to make the things simpler we shall suppose that these boundaries are hard wall potentials on both sides. Clearly, the electrons with a higher kinetic energy circle around the largest orbits, which *in fine* cross the boundary. They will bounce on the hard wall and propagate towards the left or towards the right, depending upon whether they were initially located closest to the "upper" or to the "lower" side of the sample! Qualitatively it actually looks like what we found quantum-mechanically, although in the latter case we are dealing with wave propagation, whereas in the classical case we have well-defined particle trajectories[19].

We shall make a final observation, illustrated by Figure 2.42. We found that even at equilibrium (i.e. no voltage drop applied between the two terminals in Figure 2.42), B induces local current flows ($+k$ states on one side, $-k$ states on the opposite side). This means that if we want to define a local conductivity, it is in no way uniform across the sample (see Figure 2.42).

---

19 If the potential presents a landscape with equipotential lines instead of equipotential planes the analogy between quantum waves and classical trajectories may become even stronger (see exercise 2.19.4).

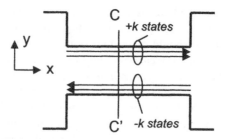

Figure 2.42. *Spatial separation between the forward and backward propagating states in a 2D device submitted to a magnetic field*

If we apply a small electric field, all electrons clearly contribute to local changes in the circulating current patterns induced by an electric field. Electrons with a low kinetic energy create current patterns in the middle part, and electrons with a high kinetic energy create local currents at the sides of the sample. Even without electric field, we have local currents which are different from zero, although the overall conductance is clearly equal to zero. Thus the microscopic conductivity is *not* a Fermi surface property. However, for this book, when we want to calculate the *overall conductance*, we can still assume that it is a Fermi surface property, so that we can just take into account the electrons close to the Fermi energy and sum their local contributions to calculate the transport properties.

We can now summarize: *in a 2D sample with boundaries, application of a high magnetic field leads to a splitting of the electron states into 1D energy subbands, the bottom of each subband being given by Landau energy levels. In each 1D subband, the higher the momentum is, the more important the shift of the corresponding wave function towards one side of the sample is. This shift depends on the sign of the momentum, so that states carrying current in one direction are spatially separated from the states carrying current in the opposite direction.*

Maybe we have slightly suffered to get up to this point, but it was worth doing it, and we can now breathe a little bit, because we are ready to really tackle what is at the heart of this course: ballistic transport and conductance from transmission!

2.19. Exercises

2.19.1. *Exercise*

The purpose of this exercise is the study of a particle in an infinite, 1D potential well (for more details, see [COH 77]).

1) Calculate the uncertainty (i.e. the standard deviation) of the particle position inside the well as a function of the level index n.

2) The momentum p can be defined as the Fourier transform of the position, so that in reciprocal space and 1D the wave function is defined as

$$\varphi(p) = \frac{1}{\sqrt{2\pi\hbar}} \int_{-\infty}^{+\infty} \varphi(x) \exp(-ipx/\hbar)dx \quad .$$

Calculate this wave function in the infinite square well case, with an arbitrary level index n. Show that it can be put in the form

$$\varphi_n(p) = -\frac{i}{2}\sqrt{\frac{L}{\pi\hbar}} \exp\left(i\left(n\frac{\pi}{2} - \frac{L}{2\hbar}p\right)\right)\left(\operatorname{sinc}\left(\frac{L}{2\hbar}\left(p - \frac{n\pi\hbar}{L}\right)\right) + (-1)^{n+1}\operatorname{sinc}\left(\frac{L}{2\hbar}\left(p + \frac{n\pi\hbar}{L}\right)\right)\right)$$

where the function sinc is defined as $\operatorname{sinc}(x) = \sin(x)/x$. Note the dependence of the probability density on p in momentum space. In particular, when the energy value is large how can we compare the quantum and the classical situations?

3) Calculate the average and the variance of the momentum. Then give the momentum uncertainty and comment on the result. Compare the product $\Delta x \Delta p$ with the value given by the Heisenberg uncertainty relation.

2.19.2. *Exercise*

A 1D quantum well is formed by a thin GaAs layer in between two AlGaAs layers, so as to obtain a conduction band profile as in the figure below. The conduction band offset is equal to 0.36 eV, and the isotropic effective mass is equal to $m=0.067\ m_0$. We assume that the effective mass is the same in both GaAs and AlGaAs.

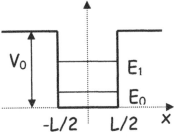

Figure 2.43. *A 1D quantum well*

1) Which width L of the quantum well must be chosen to obtain a ground state energy equal to $E_0=50$ meV? *This is the length to be considered in all subsequent questions* (check your result by considering the adequate figure in the book).

2) Give an approximate value for the second energy level by using the adequate figure in the book.

3) Calculate the first two levels in the hard wall approximation and discuss.

4) Give the approximate numerical value of the temperature for which quantization is wiped out by thermal fluctuations.

5) Give the analytical form taken by the wave function outside and inside the well. Which conditions must be fulfilled by these expressions? We can take the same notations as in the book.

6) Using the fact that the wave functions are normalized and the relations determined in question 5) calculate the analytical expression of the coefficients entering into the wave function expression inside and outside the well. In particular, show that the coefficient B in front of the cosine function which represents the wave function inside the well is given by

$$B = \sqrt{\frac{2}{L + 2/\beta}}$$

where $\beta = \dfrac{\sqrt{2m(V_0 - E)}}{\hbar}$. Comment on the dependence on L and β.

Then calculate the coefficient in front of the exponentially decreasing wave functions outside the well. Show that it is equal to

$$D = \exp\left(\frac{\beta L}{2}\right) \sqrt{\frac{2}{\left(L + \dfrac{2}{\beta}\right)\left(1 + \dfrac{\beta^2}{k^2}\right)}}$$

with $k = (2mE)^{1/2}/\hbar$.

7) Give the numerical values of the two coefficients B and D for the ground state and the second energy level (assume without demonstrating it that factor A in front of the sine function is the same as factor B determined in question 6) for the cosine function, and that factor D is also the same a+s factor D in question 6), but introducing for A and D the correct energy value).

8) Consider all space dimensions. The well is now a cube of size L^3 (equal to the value determined in 1). What is the ground state energy? What is the value of the next level and what is its degeneracy?

2.19.3. *Exercise*

The band structure of silicon presents six minima in the conduction band, and the surfaces of constant energy are ellipsoids as depicted in Figure 2.44. Two effective masses $m_L=0.98m_0$ and $m_T=0.19m_0$ are needed to describe the dispersion relation. Consider the *Si-SiO$_2$* system with an interface directed along the (100) direction (i.e. perpendicular to k_x or x in the figures). The conduction band offset is high, equal to 3.2 eV, and here is assumed to be infinite. The oxide and silicon permittivities are $\varepsilon_{ox}=3.9\varepsilon_0$ and $\varepsilon_S=11.9\varepsilon_0$, respectively.

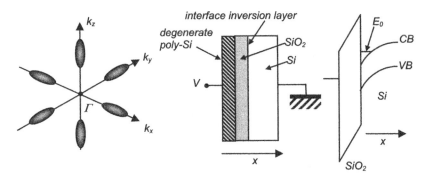

Figure 2.44. *From left to right: (i) surfaces of constant energy in the Si conduction band, represented in reciprocal space and in the first Brillouin zone, (ii) MOS structure and (iii) energy bands as a function of position*

1) Which valleys are populated in the ground state? What is the mass to be used in the Schrödinger equation in order to calculate the corresponding quantized energy levels? What is the valley degeneracy?

2) Focus on the valleys determined in the previous question. Write the Schrodinger equation, assuming a constant transverse electric field value $\mathcal{E}$. What is neglected in the latter assumption?

3) Show that, by making an appropriate variable change, the Schrödinger equation can be turned into the form

$$\frac{\partial^2 \varphi}{\partial u^2} + u\varphi = 0 \quad .$$

4) By solving the equation above in reciprocal space, show that the solution of the equation above is of the form

$$\varphi(u) = CAi(-u)$$

where $Ai(x)$ is defined as

$$Ai(x) = \frac{1}{2\pi} \int_{-\infty}^{+\infty} \exp(v^3/3 + xv)dv ,$$

which is a well tabulated integral better known as the *Airy function*, also plotted below (Figure 2.45).

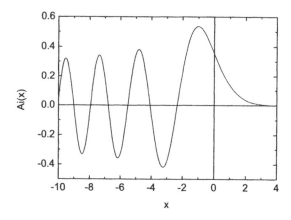

Figure 2.45. *Numerical plot of the Airy function Ai(x)*

5) Assume that the zeroes a_p of the Airy function can be approximated as

$$a_p \cong -\left(\frac{3\pi}{8}(4p-1)\right)^{2/3}$$

where $p=1$, 2, 3, etc. Calculate the quantized energy levels as a function of the electric field. Draw the shape of the wave functions corresponding to the first levels.

6) Neglect the depletion charge due to the doping impurities. Express the average electric field in the potential well as a function of electron concentration. The centroid of the first lobe from the right of $Ai^2(u)$, which can be calculated numerically, is $\alpha \cong -0.779$. Calculate the quantum capacitance associated with the surface layer if just one subband is occupied.

7) Numerical application: take $n_s = 10^{12} cm^{-2}$. How many subbands are populated? Is the quantum capacitance negligible against the oxide capacitance if the insulator thickness is equal to *2nm*? Conclude.

2.19.4. *Exercise*

A 2D AlGaAs-GaAs-AlGaAs heterostructure is patterned in the form of a Hall bar as in Figure 2.46. All measurements are carried out at low temperature. In a first step we apply a perpendicular magnetic field $B=0.1T$ and inject a current $I=1$ μA flowing from the right to the left contact. The two voltage drops $V_2 - V_1 = +0.312$ mV and $V_2 - V_3 = -0.208$ mV are measured. The device width is $W=20$ μm and the length between contacts 1 and 2 is equal to $L=150$ μm. The quantum well width is $t_W = 20$ *nm*.

1) The carriers are of what type?

2) Calculate the carrier concentration in the 2D gas and the surface mobility.

3) What is the Fermi level value and how many subbands are populated?

4) The magnetic field is progressively increased. Calculate the value B_C above which all carriers are contained in the first Landau subband.

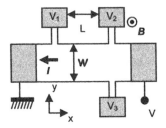

Figure 2.46. *A Hall bar patterned in a 2D gas*

5) Since the Landau band is 1D and the states are described by a longitudinal wavevector with BVK conditions, is it possible to use the 1D density of states given by equation (2.123)?

6) Assume that the potential along y can be represented by a very smooth parabolic potential until the sample edges. Show that to a line, along which is centered a wave function with a given group velocity, corresponds a classical trajectory with the same velocity. What does this line become when the parabolic potential is infinitely smooth? How does this result justify the choice of the Landau gauge?

7) What do you expect if, due to some defect, a potential peak is induced somewhere in the sample?

8) We apply $B=2B_C$. Define the area actually covered by the quantum states.

2.20. Bibliography

[AND 82] ANDO T., FOWLER A.B., STERN F., "Electronic properties of two-dimensional systems", *Reviews of Modern Physics*, vol. 54, no. 2, 1982, p. 437-672.

[ASH 76] ASHCROFT N.W., MERMIN N.D., *Solid State Physics*, New York, Saunders College, 1976.

[BOH 54] BOHM D., *Quantum Theory*, London, Constable, 1954.

[BELL 87] BELL J.S., *Speakable and Unspeakable in Quantum Mechanics*, Cambridge, Cambridge University Press, 1987.

[BRE 81] BREWS J.R., *MOS Physics and Technology*, New York, Wiley, 1981.

[COH 77] COHEN-TANNOUDJI C., DIU B., LALOE F., *Quantum Mechanics*, Hermann and Wiley & Sons, 1977.

[DAT 95] DATTA S., *Electronic Transport in Mesoscopic Systems*, Cambridge, Cambridge University Press, 1995.

[DIR 58] DIRAC P.A.M., *The Principles of Quantum Mechanics* (4th Ed.), Oxford, Oxford University Press, 1958.

[FEY 65] FEYNMAN R.P., LEIGHTON R.B., SANS M., *The Feynman Lectures of Physics, vol. 3: Quantum Mechanics*, Reading, Addison-Wesley, 1965.

[FOU 94] FOUILLANT C., ALIBERT C., "A new approach to the symmetric rectangular quantum well: Analytic determination of well width from energy levels", *American Journal of Physics*, vol. 62, no. 6, 1994, p. 564-565.

[HOL 93] HOLLAND P.R., *The Quantum Theory of Motion*, Cambridge, Cambridge University Press, 1993.

[JAC 98] JACKSON J.D., *Classical Electrodynamics*, New York, Wiley (3rd edition), 1998.

[JON 74] JONSSON C., "Electron diffraction at multiple slits", *American Journal of Physics*, vol. 42, 1974, p. 4-11.

[JOR 75] JORDAN T.F., "Why $-i\nabla$ is the momentum", *American Journal of Physics*, vol. 43, no. 12, 1975, p.1089-1093.

[KEL 95] KELLY M.J., *Low-Dimensional Semiconductors*, Oxford, Clarendon Press, 1995.

[LAN 30] LANDAU L., "Paramagnetism of metals", *Zeitschrift für physic*, vol. 64, no. 9-10, 1930, p. 629-637.

[LAN 77] LANDAU L.D., LIFSHITZ E.M., *Quantum Mechanics*, Oxford, Pergamon, 1977.

[MER 70] MERZBACHER E., *Quantum Mechanics*, New York, Wiley, 1970.

[MES 66] MESSIAH A., *Quantum Mechanics*, New York, Wiley, 1966.

[OUI 98] OUISSE T., MAUDE D.K., HORIGUCHI S., ONO Y., TAKAHASHI Y., MURASE K., CRISTOLOVEANU S., "Subband structure and anomalous valley splitting in ultra-thin silicon-on-insulator MOSFETs", *Physica B*, vol. 249-251, p. 731-734, 1998.

[OUI 00] OUISSE T., NASSIOPOULOU A., "Dependence of the radiative recombination lifetime upon electric field in silicon quantum dots embedded into SiO_2", *Europhys. Lett.*, vol. 51, 2000, p. 168-173.

[PAU 40] PAULI W., "The connection between spin and statistics", *Physical Review*, vol. 58, 1940, p. 716-722.

[RID 82] RIDLEY B.K., "The electron-phonon interaction in quasi-two-dimensional semiconductor quantum-well structures", *Journal of Physics C: Solid-state Physics*, vol. 15, 1982, p. 5899-5917.

[SCH 68] SCHIFF L.I., *Quantum Mechanics*, New York, McGraw-Hill, 1968.

[SMI 69] SMITH R.A., *Wave Mechanics of Crystalline Solids*, 2nd ed., London, Chapman and Hall, 1969.

[STY 02] STYER D.F., BALKIN M.S., BECKER K.M., BURNS M.R., DUDLEY .E., FORTH S.T., GAUMER J.S., KRAMER M.A., OERTEL D.C., PARK L.H., RINKOSKI M.T., SMITH C.T., WOTHERSPOON T.D., "Nine formulations of quantum mechanics", *American Journal of Physics*, vol. 70, no. 3, 2002, p. 288-297.

[SWE 89] SWENSON R.J., "The correct relation between wave functions in two gauges", *American Journal of Physics*, vol. 57, no. 4, 1989, p. 381-382.

[SZE 81] SZE S.M., *Physics of Semiconductor Devices*, New York, Wiley, 1981.

[TON 89] TONOMURA A., ENDO J., MATSUDA T., KAWASAKI T., EZAWA H., "Demonstration of single-electron buildup of an interference pattern", *American Journal of Physics*, vol. 57, no. 8, 1989, p. 117-120.

[VAL 04] VALLEE O., SOARES M., *Airy Functions and Applications to Physics*, World Scientific, 2004.

[VON 86] VON KLITZING K., "The quantized Hall effect", *Reviews of Modern Physics*, vol. 58, no. 3, 1986, p. 519-531.

[ZIM 65] ZIMAN J.M., *Principles of the Theory of Solids*, Cambridge, Cambridge University Press, 1964.

Chapter 3

Ballistic Transport
and Transmission Conductance

3.1. Conductance of a ballistic conductor ●

Electrical resistance characterizes the energy dissipation imposed upon the carriers when they pass through a device under the action of an electric field. It reflects the power lost in the device. Usually, when we want to study the transport properties of a macroscopic conductor, or a device which is not mesoscopic, we expect the resistive contribution of the device active layer to represent a non-negligible part of the overall measured resistance, and most often to be the prevailing part. Even if the contacts between the metal pads and the sample exhibit a susbtantial resistance value, we can get rid of it by using four-probe measurements, in which we pass the current through some terminals and we measure the voltage from the others. Then we obtain something which is not zero, which depends on the resistivity value of your sample and which scales with the device length according to Ohm's law. Now suppose that we want to measure the transport properties of a really small device, in which the electrons are truly ballistic. If nothing impedes the electron passing through the device we might expect the device resistance to go down to zero, and not to depend on the device length any longer. In addition, even if the conductance was scaling inversely with the device length following Ohm's law, we might also expect the measured conductance to continuously increase as we reduce the device length. However, this is not the case. In fact, if we continuously reduce the resistance of something, we can expect that sooner or later the active layer resistance must become smaller than the contact resistance, which we need to link the small device to the macroscopic wires. Thus, we can expect that after a substantial enough reduction of device length, all the losses are due to the contact rather than to the "device" itself. We shall investigate this point in the case of quantum, ballistic devices, and we shall see that this picture actually fits the reality: we cannot observe a vanishing resistance, and dissipation occurs at the contact between the ballistic sample and the rest of the circuit; however, the conductance itself is still fixed by the ballistic device properties! When the conductance does not scale with length, it is no longer appropriate to define a conductivity, and the conductance must be calculated through the probability for an electron impinging at the device boundary to be transmitted through the device to the opposite terminal. This is not so new a feature if we have already calculated the tunneling current through a square barrier. We already know that for such a device the resistance is all but proportional to the barrier length. However, it took a certain number of years for physicists to be accustomed to the fact that it was the correct approach in the case of mesoscopic and ballistic devices. This field of research has greatly benefited from the theoretical ideas and concepts put forth by, among many others, P.W. Anderson, R. Landauer and M. Büttiker. However, as for any other solid-state physics topics, advances could not have been obtained without a huge progress in technological processes and experimental work. The formulae expressing the conductance as a function of the transmission probabilities are called the Landauer-Büttiker formalism.

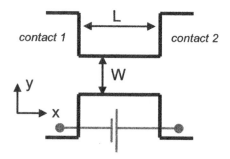

Figure 3.1. *Ballistic device in which electrons cannot be scattered, connected to two terminals*

We shall first consider a ballistic device as in the scheme of Figure 3.1. We make the assumption that the device in between the contacts is a perfect, 1D conductor composed of a given number of 1D channels (i.e. a quantum wire as described in section 2.6.1, with a number of populated 1D channels).

In addition, to make the things simpler, we suppose that the temperature T is equal to zero, and that there is no electron-electron interactions leading to energy transfer between the various electrons. This means that there is no thermal spreading of the carriers in energy. Thus, if the left terminal 1 is maintained at a given electrochemical potential μ_1, and the right terminal 2 is maintained at an electrochemical potential μ_2, we expect the current to be carried by states between μ_2 and μ_1. In addition, we shall also suppose that the difference between μ_1 and μ_2 is small, so that the transmission probability of an electron impinging on the device does not depend on energy (this is a rather drastic assumption, and later on we shall study devices in which this is not realized).

The central assumption is that *once an electron has entered into the ballistic conductor, it cannot be reflected back.* As a matter of fact, this is not a completely new assumption, but a consequence which arises from the assumed ballistic character of the device: electrons do not experience any collision which can backscatter them; if we put one electron into the device going left, it cannot do anything but go to the left side since its momentum cannot change. What is new but quite reasonable is that we also suppose that the electron cannot be scattered when moving out of the ballistic device into the opposite terminal. If we look at Figure 3.1, this does not seem so foolish an assumption: it is just as if we are driving on the highway and there is a constriction from three lanes to one lane (someone has crashed his car into the one in front as he was driving too fast and his vehicle is blocking two lanes); we do not have any trouble finding the way out once we are in the single lane and the highway recovers three lanes. The problem is getting in. Thus, if we consider an electron with a momentum k directed from the left to the right, and from now we shall call such a state a $k+$ state, it cannot originate in the right contact: in order to come from the right contact its momentum should have been initially directed from the right to the left, and backscattered at a certain time, to make it go from the left to the right. However, this is impossible with our set of assumptions (electrons cannot be reflected back). This has a very important consequence: since no $k+$ state can originate in the right contact, this means that all $k+$ states necessarily originate from the left contact, and thus if the "big" terminal reservoir with chemical potential μ_1 is the only one which is in contact with the $k+$ states of the ballistic device, *the right propagating states must be in equilibrium with the left contact.* Of course the reasoning is exactly the same for the $k-$ states: *the left propagating states must be in equilibrium with the right contact.* Thus the $k+$ states are filled up to the chemical potential μ_1, and the $k-$ states are filled up to the chemical potential μ_2. This corresponds to the energy band diagram of Figure 3.2.

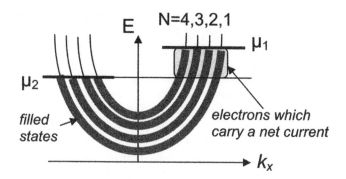

Figure 3.2. *Energy band diagram of a ballistic device with a finite number of populated 1D subbands, biased as in the previous figure*

We are now ready to calculate the overall conductance. The $k+$ states are filled up to $\mu^+ = \mu_1$, whereas the $k-$ states are filled up to $\mu^- = \mu_2$, so that the net contribution to the overall current is clearly given by the electrons lying in the $k+$ states between μ_2 and μ_1. Let us assume that the number of open channels is equal to M (i.e. M channels are such that their subband bottom energy E_n is lower than μ_2). To simplify the calculations we shall make the hypothesis that no 1D channel opens between μ_2 and μ_1. In one channel, the density of electrons per unit length and per unit energy is $D_{1D}(E)$, as given by equation (2.123), and only half of these available states correspond to $k+$ states. Thus, between μ_2 and μ_1 the current density per unit energy is obviously the product of the electron charge by the velocity $v^+(E)$ (dependent on energy) and by $D_{1D}(E)/2$. Summing over energy gives us the current in one channel:

$$I^+ = e \int_{\mu_2}^{\mu_1} \frac{D_{1D}(E) \times v^+(E)}{2} dE \qquad (3.1)$$

If we replace $D_{1D}(E)$ and $v^+(E)$ by their energy-dependent expressions equations (2.123) and (2.27), we easily find out that the product is independent of energy. The electron velocity increases as the square root of the energy, but the density of available states decreases as the reciprocal square root, so that the current density is a constant. This is specific to 1D systems. We immediately find that for one channel the overall current is

$$I^+ = \frac{2e}{h}(\mu_1 - \mu_2) \qquad (3.2)$$

(note that we have adopted the convention that current I is positive when the electrons go from 1 to 2, and negative when they go from 2 to 1; this corresponds to the usual current sign, which is positive when the current flows from the higher to the lower electrostatic potential, and the conductance is $G=V/I$). The voltage difference in Figure 3.1 is equal to $(\mu_1-\mu_2)/e$, and summing over the number M of available channels immediately gives us the conductance

$$G_C = M\frac{2e^2}{h}.$$

(3.3)

Index c is aimed at indicating the contacts, since the losses do not occur inside the ballistic conductor (we shall discuss that point below). The corresponding contact resistance is given by

$$R_C = \frac{h}{2e^2 M} = \frac{12.9k\Omega}{M}.$$

(3.4)

As we increase the number of open channels, the resistance decreases, but we should have already noticed that the contact resistance of one single channel is far from being negligible. $h/2e^2$ is the resistance quantum and we should retain its value.

An interesting point is that since the chemical potential is constant for the $k+$ states inside the channel, and since it is also the case for the $k-$ states, we do not have to consider any voltage drop inside the device (indeed things are a little bit more complicated in reality because the electrons do create an electrostatic potential which modifies the potential landscape inside the device, but for the sake of simplicity here we shall neglect it). If we assimilate a voltage drop to a chemical potential variation, for the $k+$ states it should occur after the right contact, and for the $k-$ states it should occur after the left contact, since these two groups of states are in equilibrium either with the left or with the right terminal reservoirs. As discussed in further detail in the book by Datta [DAT 95], no matter how we define the potential (i.e. just consider the $k+$ states, or the $k-$ states, or take the average between the potential corresponding to both), we always find that the potential drop occurs at the contacts, and not inside the conductor. Thus, the dissipation, which occurs when we have both a current and a voltage drop, takes place at the contacts. This is a somewhat striking result, because we were able to calculate the current and found something which only depends on universal constants, and yet we did not make any attempt to explain in a microscopic way how the carriers lose their energy in the resistive part. The explanation for the existence of a resistance can be formulated in other words: as electrons in the $k+$ states travel through the conductor, the situation is such that only a few modes are populated and they are all occupied up to μ_1. Backscattering is prohibited because the conductor is perfect (and it can be made perfect because it is small), and since all states are filled below μ_l no electron can ever lose energy inside the conductor. However, as they find their way out, they are in a situation in which there are infinitely more forward (and backward) propagating states up to μ_l than the number of electrons which come out. Thus, they can re-distribute themselves in the wide contact so as to minimize their overall energy, falling into the available states with a lower energy. To do this they have a lot of available space, much larger than the inelastic scattering length. The space region over which the electrons are re-distributed in energy corresponds to a varying chemical potential for the $k+$ states (Figure 3.3). Reciprocally, consider the $k-$ states at the entrance of the constriction: if there was just a wall instead of a constriction, they would be obtained by the total reflection of the $k+$ states impinging onto the wall. However, with a constriction, many of the $k+$ states are now transmitted through the 1D channels, so that there is a depletion of $k-$ states close to the entrance of the constriction. This is the reason why the chemical potential of the $k-$ states increases up to μ_l at the constriction entrance (Figure 3.3).

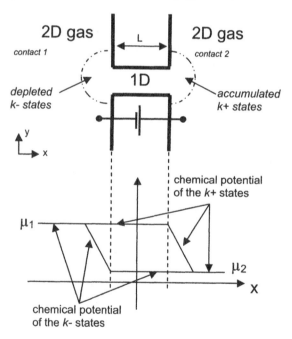

Figure 3.3. *A perfect ballistic device (top) and the corresponding variation of the chemical potentials (bottom)*

Now of course we wonder whether this quantized conductance has already been experimentally observed. The answer is yes. In 2D devices such as Si MOSFETs or heterostructures, if we fabricate a local constriction such as schematized in the right part of Figure 3.4, we obtain something which is 1D. In addition, if we can operate the gates above or by the sides of the constriction, we can change its effective width by depleting its sides. As we deplete the edges the free space becomes more constrained and the 1D subband bottom energies E_n correspondingly increase, so that we can really control the number of open channels below the Fermi energy. Thus, whenever we change the gate voltage so as to make one energy E_n go below the Fermi level, we add one channel and expect a quantized conductance increase by a factor $2e^2/h$. This was first achieved in 1988[1], and is now routinely observed in many laboratories dedicated to research on semiconductor nanostructures. A typical example is given in Figure 3.4. Provided that the constriction is truly ballistic, we obtain the same conductance steps, whatever the material used for fabricating the device, silicon, gallium arsenide, or even simple metal pads with only a few atoms remaining in the constriction.

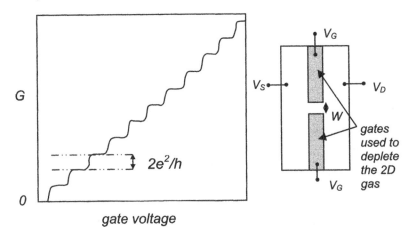

Figure 3.4. *At low temperature the conductance of a quantum point contact exhibits steps close to $2e^2/h$*

3.2. Connection between 2D and 1D systems ●

In a narrow constriction there are only a few numbers of occupied 1D channels. As the device width W is enlarged we decrease the values of the bottom energies in the 1D subbands. If the Fermi level is kept constant, this means that more and more 1D subbands are populated. As a matter of fact, when W is large we can still use the

---

1 Simultaneously reported in [VAN 88] and [WHA 88].

1D formalism with many populated subbands. To calculate the number of such 1D subbands in a wide sample we can just apply BVK boundary conditions to the y axis (the reader may object as the wave functions are stationary along y, which is true, but remember that the density of states is the same if we calculate them with stationary states in a box or with propagating states with BVK boundary conditions, so that we do not make a mistake by calculating them in the BVK approach; and it is indeed simpler to do so). The maximum kinetic energy is $E_F=\hbar^2 k_F^2/2m$, and the momentum k_y along y can thus take values from $-k_F$ to $+k_F$, $-k_F<k_y<k_F$. Along with the BVK condition which states that the k_ys are separated by intervals of $2\pi/W$, we find that the number of possible k_y values, and thus the number of 1D subbands, is equal to $M=2k_F/(2\pi/W)$. The number M of 1D channels in a wide sample of width W, Fermi energy E_F and Fermi wavevector k_F is thus

$$M(wide\ sample) = \frac{k_F W}{\pi}.$$

(3.5)

Note that we did not introduce the spin degeneracy in the calculation because we have already taken it into account when calculating the current and conductance of a 1D channel. We shall use the relation equation (3.5) in some instances of this book.

3.3. A classical analogy O

In order to calculate the resistance of a macroscopic conductor we must use a microscopic conductivity, in which is implemented the momentum loss rate induced by the collisions. Thus, to calculate the conductivity from the first principles we need to know how the electrons dissipate the energy gained by the electric field. We required nothing like this in the previous section. However, it is worth noting that this is not specifically due to the quantum nature of the problem, but it is instead a general property of systems in which the transport is controlled by a lossless "constriction". What is quantum is the available 1D density of states, which ultimately determines the quantized conductance value. To illustrate the idea that we can calculate a flow limited by constrictions without losses even if we do not have to take into account the detailed process of energy loss, we can also consider a system for water flow as schematized in Figure 3.5. Although the comparison should not be pushed too far, we can see some striking similarities with our ballistic device: the energy dissipation does not occur in the constrained part between the left and right water tanks, but the overall water flow is clearly a function of the pipe characteristics such as diameter and the respective water levels in the left and right tanks.

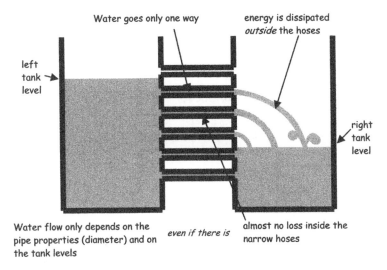

Figure 3.5. *Two water reservoirs connected by narrow pipes*

In this case energy dissipation occurs either in the right reservoir through water molecules which pass through the pipes located above the right tank level and which fall into the right tank, or, in a more complicated way, by the balance between the potential energy lost by all the molecules which are all translated towards the bottom of the left water tank, and the energy gained by the molecules which are pushed to the top by the molecules which come out of the hoses located below the right tank level. Of course, in this case we do not obtain universal values because the hose diameter and thus the water volume in the hoses can be continuously varied, but the situation is not so different from the one we studied in the previous section.

3.4. Transmission conductance: Landauer's formula ●

In section 3.1 we made the assumption that there was no scattering inside the conductor, which was assumed to be perfect. In practice this is not necessarily the case, and we have to take into account the fact that depending on the electron wave function, an impurity or a defect located inside the mesoscopic device can backscatter some of them. In particular, two incident electrons with exactly the same characteristics are not necessarily detected in the same place after having interacted with the mesoscopic device! This is the same as when we study the transmission of a wavepacket through a tunneling barrier: part of the wave is transmitted, and part of the wave is reflected back. The electron which was initially associated with the left side propagating wave can be detected either on the left or on the right side of the barrier after the wave has been split by the energy barrier. To introduce such a possibility we shall try to calculate the conductance from the probability for an electron to be transmitted through the mesoscopic device. A systematic use of this method was first advocated by Rolf Landauer.

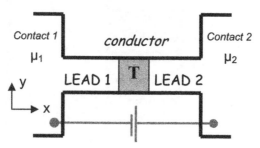

Figure 3.6. *A device with a scattering center in its middle part*

To make things as simple as possible we suppose that the transmission probability T that an incoming electron injected into one lead is transmitted to the other side does not depend on energy. In practice, this is correct only if the difference between μ_1 and μ_2 is kept small. If we remember the transmission coefficient calculated in the square barrier case, we know that T does vary a lot with energy. However, for a small applied voltage it is a reasonable approximation. We model the device as illustrated in Figure 3.6. The mesoscopic sample includes an obstacle which can reflect impinging carriers, and is perfect otherwise, so that we can describe it as two perfect, ballistic leads separated by the scattering part, and connected to wide conductors at their ends.

If we introduce the possibility of backscattering, the $k+$ states and the $k-$ states in one of these two leads are not necessarily all occupied up to the chemical potential of only one reservoir, since they can now exchange carriers with the two terminals. However, in a steady-state regime the occupation function does not vary with time, so that an obvious generalization of equation (3.1) is to weight the current density per unit energy by the occupation function. The current corresponding, e.g., to the $k+$ states in lead 1 and for one 1D channel is given by the general formula

$$I^+ = e \int_{E_N}^{+\infty} \frac{D(E) \times v^+(E)}{2} f^+(E) dE \qquad (3.6)$$

which immediately reduces to

$$I^+ = \frac{2e}{h} \int_{E_N}^{+\infty} f^+(E) dE . \qquad (3.7)$$

Summing over all channels gives us the general expression

$$I^+ = \frac{2e}{h} \int_{-\infty}^{+\infty} f^+(E) M(E) dE .$$ (3.8)

These formulae also allow us to take into account a temperature dependence of the distribution function.

In the case of Figure 3.6, at $T=0$, we know that in lead 1 the electrons in the $k+$ states can only originate in terminal 1 (suppose that one of these $k+$ states originates in terminal 2; initially, to come from the right this electron ought to be in a $k-$ state; then it passed the scattering center and was obviously in a $k-$ state when going into lead 1; but since lead 1 is perfect, this electron cannot be reflected back in a $k+$ state; hence an electron originating from terminal 2 cannot be found in a $k+$ state in lead 1). Thus, the $k+$ states in lead 1 are in equilibrium with terminal 1 and filled up to μ_1. Now, when impinging on the scattering center, some of the electrons in lead 1 in the $k+$ states are transmitted (a part T of them), and some of them are reflected back in $k-$ states of lead 1. Let us define I_1^+ as the current corresponding to the impinging electrons in lead 1, I_1^- as the current corresponding to the electrons reflected back into lead 1, and I_2^+ as the current corresponding to the electrons transmitted in $k+$ states to lead 2. These currents are schematized in Figure 3.7.

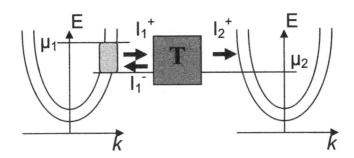

Figure 3.7. *Energy band diagram and schematized currents corresponding to the device from the previous figure*

Obviously, we have

$$I_1^+ = \frac{2e}{h} M(\mu_1 - \mu_2),$$ (3.9)

$$I_2^+ = \frac{2e}{h} MT(\mu_1 - \mu_2),$$

(3.10)

and

$$I_1^- = \frac{2e}{h} M(1 - T)(\mu_1 - \mu_2).$$

(3.11)

The net current is the difference $I = I_1^+ - I_2^+ = I_1^-$:

$$I = I_1^+ - I_1^- = I_2^+ = \frac{2e}{h} MT(\mu_1 - \mu_2),$$

(3.12)

so that the conductance is given by

$$G = \frac{eI}{\mu_1 - \mu_2} = \frac{2e^2}{h} MT .$$

(3.13)

This is the celebrated Landauer formula (and if we retain only a couple of things from this book this expression should be one of them!):

$$G = \frac{2e^2}{h} MT .$$

(3.14)

The conductance does not depend on device length, and if the transmission probability is equal to unity (i.e. the conductor is perfectly ballistic), we recover the quantized conductance value that we previously derived.

3.5. What if the device length really does go down to zero? ○

The quantized conductance given by equation (3.3) is derived assuming that the ballistic device is 1D, so that the product of the velocity and the density-of-states appearing in the current integral does not depend on energy. However, if device length L is reduced down to dimensions smaller than the electron Fermi wavelength we can no longer use the 1D density of states. In fact, equation (3.3) is *not* correct if $L=0$.

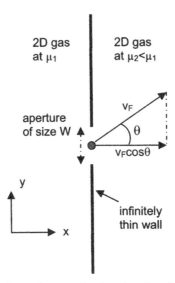

Figure 3.8. *A ballistic device with a length reduced down to zero*

Consider two 2D gases connected together by an aperture of width W made in an infinitely thin wall, as in Figure 3.8. It is not possible to calculate the current analytically if W becomes smaller than the Fermi wavelength, but if W is much larger than λ_F (semi-classical approximation), something can be done: at each boundary point in the aperture there are carriers going in all directions with a uniform angular distribution, and their density is given by the 2D density-of-states formula. If there is a small chemical potential difference at the boundary the current corresponds to all uncompensated carriers going towards the right. However, of course individual carriers do not contribute to the current with the same amount, because a wave with a propagating angle θ obviously gives a contribution to the current along x equal to $ev_F \sin\theta$ (see Figure 3.8). Thus, the current is given by

$$I = \int_{E=\mu_2}^{E=\mu_1} \int_{\theta=-\pi/2}^{\theta=+\pi/2} \int_{y=-W/2}^{y=W/2} \underbrace{\frac{1}{2\pi}}_{\substack{\text{angular}\\\text{distribution}\\\text{of the states}}} \underbrace{\frac{m}{\pi\hbar^2}}_{\substack{\text{2D}\\\text{density}}} e \underbrace{v_F \cos\theta}_{\substack{\text{electron}\\\text{velocity}\\\text{along x}}} \, d\theta dE dy$$

(3.15)

so that the conductance reduces to

$$G = \frac{2e^2}{h} \frac{k_F W}{\pi}.$$

(3.16)

This expression is known as the Sharvin conductance. It is worth noting that if you consider Landauer formula with a number of channels as given by equation (3.5) (i.e. with a wide constriction of finite length), you recover exactly the same expression. However, in the infinitely thin wall case k_F may be continuously varied.

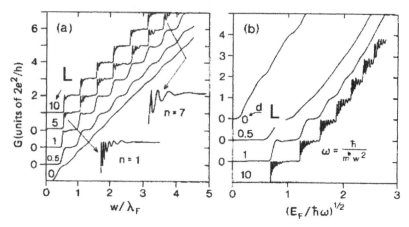

Figure 3.9. *Numerically calculated conductance of a quantum point contact separating two 2D electron gases for various device lengths L in units of the Fermi wavelength λ_F: (a) is for a hard-wall potential and (b) is for a parabolic confining potential. Reproduced with permission from E. Tekman and S. Ciraci, Phys. Rev. B 43, 7145 (1991), copyright (1991) by the American Physical Society*

As W approaches the Fermi wavelength, accurate numerical calculations do not give a conductance obeying Landauer formula, even if some interference effects are still present [TEK 91]. As illustrated by Figure 3.9, for small W and $L=0$ some oscillations can be seen, which tend to vanish as W gets larger than a few λ_F. Then the conductance approaches Sharvin formula. There is also a minimum W value below which the conductance vanishes (see the bottom curve of the left Figure 3.9): due to the uncertainty principle, the momentum p_y cannot take arbitrarily small values and we have $\Delta W \Delta p_y \sim \hbar$. Since p_y cannot exceed the Fermi momentum $p_F = \hbar k_F$, transmission is suppressed for an aperture smaller than $1/k_F$. A better semi-classical approximation is thus to consider an effective channel aperture $W_{eff} = W - 1/k_F$:

$$G = \frac{2e^2}{h} \frac{k_F}{\pi} \left(W - \frac{1}{k_F} \right)$$

(3.17)

It is also worth noting that as soon as the proportionality factor between device length and Fermi wavelength exceeds a few units, the conductance exhibits the same quantized steps as in the pure 1D case (see Figure 3.9), so that even with quite short devices we can still make use of the 1D formalism.

3.6. A smart experiment which shows you everything ○

There is something rather disturbing in all the results discussed in the previous sections: if we measure a ballistic conductor in general the only quantity with which we can play is the overall current. Although this quantity is clearly a function of the interference pattern of a quantum device, it is a little bit difficult to demonstrate experimentally that this unique quantity is actually the result of all the mechanisms that we described, and most often to observe such effects people also have to vary another macroscopic quantity, such as the magnetic field. To illustrate this in a somewhat simple way, remember the famous double-slit experiment: the large plate on which we can detect the impinging electrons is equivalent to the situation of the left figure below, in which we would be free to put many, many small detectors filling the detection plane.

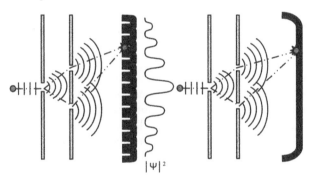

Figure 3.10. *A double-slit experiment with electrons; a) with many small detectors and b) with a very big and single detector*

However, unfortunately our situation instead looks like the one depicted by the right part of Figure 3.10. We can only measure the total current, and it is a little bit as if we wanted to recover the interference pattern by using only a very large detector, as in the right scheme! The big detector is our current meter. So is there any way to go somewhat further? Here I will just quote one of two possible solutions because it corresponds to the experiment that I want to introduce. Suppose that we cannot change the big detector, but that we are allowed to use a small, opaque screen. Just proceed as in Figure 3.11: move the small piece in front of the detector, so that we locally suppress the possibility of the impinging electron wave reaching the detector. On average we still have a large detection flow, but with a small decrease which exactly corresponds to the local intensity of the interference pattern. This is in fact the principle followed by a research team from Harvard University a few years ago, but with real ballistic conductors instead of double-slit set-ups [TOP 00].

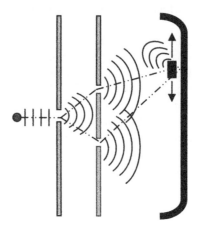

Figure 3.11. *A possible solution to the problem illustrated by the previous figure*

This team carried out a nice (and quite difficult) experiment using an atomic force microscope to investigate the conductance of a point contact. Here I would like to describe and summarize it to show you that thanks to the progress in the experimental dexterity that can be exerted by some researchers, we can not only measure the quantized conductance, but indirectly visualize the electron wave functions corresponding to the 1D channels of the quantum point contact. The principle is to drive a polarized metallized AFM probe just above a 2D gas and a constriction, so as to assess the variation in the constriction resistance as a function of the cantilever probe location, as illustrated by Figure 3.12 (left).

What we obtain experimentally is a map of the measured conductance as a function of the metallized probe position. There are many experimental difficulties for producing this, and although we shall not describe them into much detail, let us mention two of them. First we must have an unpassivated device with a 2D gas very close to the surface, so as to be able to electrostatically influence it with the probe. In addition, we must be able to put the probe at a given and very short (a few 10 nm) distance from the device. The reader will have probably also followed a course on AFM techniques; and should know that in general this distance is assessed by measuring the deflection of a laser beam on a four-quadrant photodiode, the optical beam being itself reflected by the cantilever. Here this method is forbidden, because the stray part of the laser beam would excite carriers inside the 2D gas and would perturb the measurements. Thus the cantilever is made from a special piezo-electric material, whose deformation as we approach the sample creates a voltage that we measure in turn on electrodes located by the sides of the cantilever, so as to calculate the distance from the 2D gas.

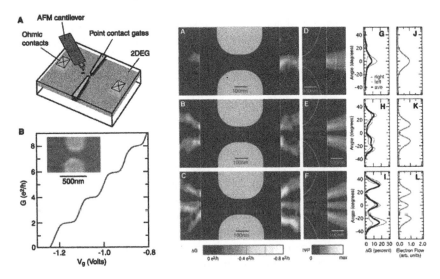

Figure 3.12. *Surface probe experiment reported in [TOP 00] and conducted on a point contact between two 2D electron gases (see the explanations in the text); from Topinka et al., Science, vol. 289, p. 2323 (2000), reprinted with permission from AAAS [TOP 00]*

Once the researchers succeeded in tackling all those difficulties, by applying a voltage onto the AFM probe so as to locally deplete the semiconductor layer, they obtained images like those in Figure 3.12 (middle). Their results were first published in 2000 [TOP 00]. What is the meaning of the resistance variations outside the point contact? In fact, the AFM probe acts as a localized scatterer, and if it is directed along the trajectory followed by some electron waves it can backscatter them into the point contact. Thus, if the probe is located at a position for which the square of the electron wave function is initially close to zero, it does not affect the overall resistance, and the image pixel remains black. However, if the probe is located at a point where the square of the wave function is not initially negligible, then it can scatter a susbtantial part of the outcoming electrons back into the point contact, and the resistance is going to rise. The team was thus able to visualize the electron densities just out of a point contact! By adjusting the gate voltage they were also able to monitor situations in which only the first (A), the first two (B) or the first three (C) 1D channels were populated. In the first image we can check that the electrons emerge from the middle of the point contact, as expected for a ground wave function with only one maximum in the y direction. By substracting the first image from the second we obtain the insert (B) in which only the second occupied subband is selected, and we can observe two lobes! Selecting the third 1D subband we have three lobes, etc. The left side shows the experiment conducted on both terminals, and the right side the predictions (D, E and F). This is nothing but the experimental visualization of all the concepts that we have developed up to the current page: quantum well quantization, quantized ballistic conductance and reflection by a scatterer!

3.7. Relationship between the Landauer formula and Ohm's law ⦿

What is the link between Ohm's formalism, in which the resistance is proportional to the device length, and Landauer's formalism, in which the resistance no longer scales with length? Here we shall reproduce simple arguments that can be found in Datta's book to show that it is possible to derive Ohm's law from Landauer's formula [DAT 95]. We shall start from very small devices, and as we increase the length we shall recover Ohm's law. As a matter of fact, as we increase the device length the sample is going to be less and less perfect, and the length will exceed the phase coherence length. This means that we can no longer describe an electron passing through the device by an unperturbed, propagating wave function. It is just as if the electron was experiencing some measurement process when crossing the device: it looses its phase, and sometimes its energy. A way to reproduce this in a simple analytical treatment is to put incoherent conductors in series. By incoherent conductors we mean that we neglect any interference effect between one to the next. Consider first two incoherent conductors with transmission probabilities T_1 and T_2, as in Figure 3.13.

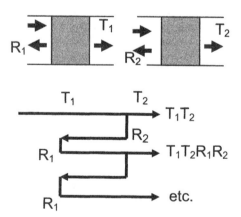

Figure 3.13. *Incoherent multiple transmission and reflections of an electron wave impinging on two mesoscopic conductors put in series*

We define R_1 and R_2 as the reflection probabilities, and of course we have $R_1+T_1=R_2+T_2=1$. If we examine the case of an electron impinging onto the first scattering center (device 1), if it is transmitted through 1 (with probability T_1) it can either be transmitted through 2 or reflected from 2. If it is reflected from 2 then it can be transmitted back through 1, or reflected again from 1. Thus, to calculate the overall transmission probability of an electron using the two devices put in series we have to take into account the possibility of multiple reflections using the system, as schematized in Figure 3.13. The probability of being transmitted without reflection is obviously equal to T_1T_2. The probability of being transmitted after bouncing once on obstacle 2 and once on obstacle 1 is equal to $T_1 \times R_2 \times R_1 \times T_2$, etc. To calculate the overall probability we just have to sum the probabilities corresponding to all these incompatible events (no reflection, two reflections, four, six, etc.). The transmission probability between the two terminals T_{12} is thus given by

$$T_{12} = T_1T_2 + T_1T_2R_1R_2 + T_1T_2R_1^2R_2^2 + \ldots \qquad (3.18)$$

in which it is easy to factorize a geometric series, so that it reduces to

$$T_{12} = \frac{T_1T_2}{1-R_1R_2} \qquad (3.19)$$

We just need a one line calculation to obtain from the expression above that

$$\frac{1-T_{12}}{T_{12}} = \frac{1-T_1}{T_1} + \frac{1-T_2}{T_2}. \qquad (3.20)$$

This is quite a good starting point, because we only find one quantity, $(1-T)/T$, which is additive. This is just what we are looking for if we want to reproduce Ohm's law, which is additive with respect to the device length: put in series two macroscopic conductors of lengths L_1 and L_2, Ohm's law states that the overall resistance is the sum of the two resistances, and thus is proportional to L_1+L_2. To simplify the discussion suppose that we add N identical scatterers. Then from equation (3.20) we immediately find that

$$\frac{1-T(N)}{T(N)} = N\frac{1-T}{T} \qquad (3.21)$$

and thus that

$$T(N) = \frac{T}{T + N(1-T)}.$$

(3.22)

We just have to pass to the continuous limit: take a linear density of scatterers equal to n (number of scattering centers per unit length), and define the length $L_0 = T/n(1-T)$. From equation (3.22) we have

$$T(L) = \frac{L_0}{L + L_0}$$

(3.23)

Can we assign a physical meaning to the length L_0? We recall that the mean free path λ_m is the average distance traveled by an electron before being scattered. If the transmission is high, i.e. with T close to 1, the electron has a unit probability of being scattered if $n(1-T) \lambda_m \approx 1$. This can be explained by the scheme of Figure 3.14.

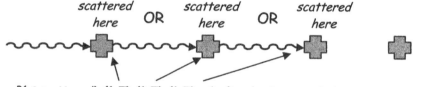

P(e⁻ scattered)=(1-T)+(1-T)+(1-T)...=1 after having traveled at max $\lambda_m = n\lambda_m(1-T)$.

Figure 3.14. *Possible scattering events of a traveling electron*

Thus, by comparison we find that L_0 is of the order of a mean free path $L_0 \approx \lambda_m$. Now we can use some already obtained formulae, along with equation (3.23), the combination of which will give us the desired result. Replace in the Landauer formula equation (3.14) the transmission probability by expression equation (3.23), and the number of channels M by its wide sample expression equation (3.5). We obtain

$$R = \frac{1}{G} = \frac{h}{2e^2} \frac{\pi}{k_F W \lambda_m}(L + \lambda_m).$$

(3.24)

Now transform the usual conductivity expression $\sigma = n_s q \mu$ slightly; using $\lambda_m = v_F \tau_m$, $\mu = q\tau_m/m$, $E_F = \hbar^2 k_F^2/2m$ and $n_S = D_{2D}E_F$, we will easily find that the conductivity can be expressed as

$$\sigma = \frac{e^2}{h} \lambda_m k_F, \tag{3.25}$$

so that the product $k_F \lambda_m$ is also given by $\sigma(h/e^2)$. Eventually, replace this product by the corresponding expression in equation (3.24) and we arrive at

$$R = G^{-1} = G_B^{-1} + G_C^{-1} = \frac{\pi}{2} \left(\frac{L}{\sigma W} + \frac{L_0}{\sigma W} \right). \tag{3.26}$$

This is not exactly Ohm's law, since we have an additional factor $\pi/2$, but remember that we have identified L_0 to the mean free path λ_m, and there are many other simplifying assumptions in our toy model. Owing to its simplicity and to the rough character of our assumptions, our heuristic demonstration is not that bad. Within this additional factor, Ohm's law is obtained as the sum of a "conventional" reciprocal conductivity, which scales with the conductor length, and of a contact resistance which does not scale with length.

3.8. Dissipation with a scatterer ⓜ

Here again we shall develop arguments already discussed and more thoroughly expounded in the book by Datta [DAT 95]. We investigate a structure as schematized in Figure 3.15. Introducing the possibility of the electrons being reflected by a scatterer has an important consequence: this means that in lead 1, above μ_1 we not only populate the $k+$ propagating states but also some $k-$ states with reflected electrons. Thus, in lead 1 the chemical potential of the $k-$ states is no longer equal to μ_2, but is higher. Reciprocally, the electrons transmitted from lead 1 into lead 2 do not completely fill the $k+$ states in lead 2 up to μ_1, since only a fraction T of them are transmitted. Thus, just after the scatterer we have many empty $k+$ states up to μ_1. As a consequence, coherence is not maintained as easily as when the Fermi exclusion principle prevents the electrons being scattered in already occupied states. If quantum coherence is *not* maintained, on average the electrons are going to lose some kinetic energy, fall into the lower-lying $k+$ states and re-distribute themselves after energy relaxation. Thus, there is some possibility of the scatterer inducing dissipation, which must be studied case by case, and what we shall examine in this section is how these processes shape the chemical potential variations around the scatterer, and where dissipation takes place.

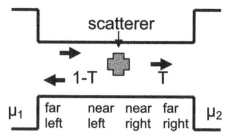

Figure 3.15. *A mesoscopic device including a scattering center which may partially reflect the electron waves*

On the one hand, since not all the right propagating electrons are transmitted there are empty states on the right part of the scatterer (energy relaxation). Thus, transmitted electrons re-distribute themselves between μ_1 and μ_2. On the other hand, some of the electrons are reflected and since initially there was no filled state in lead 1 above μ_1 they will also re-distribute themselves. Let us define μ'' as the chemical potential of the $+k$ states to the far right in lead 2 and μ' the chemical potential of the $k-$ states to the far left in lead 1, after the energy re-distribution is complete. The distribution functions are now of the form schematized in Figure 3.16.

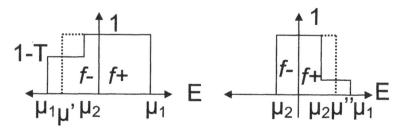

Figure 3.16. *Occupancy function of the forward and backward propagating states, before (left scheme) and after (right scheme) the scatterer*

Since the number of electrons is conserved we obviously have

$$N_R^{-k} = (1-T)N_I^{+k} \qquad N_T^{+k} = TN_I^{+k}$$

(3.27)

k- reflected e-

K+ incident e- above μ_2

k+ transmitted e-

where N_R^{-k} is the density of $k-$ electrons reflected into lead 1 and N_T^{+k} is the density of $k+$ electrons transmitted into lead 2. Taking into account the distributions of Figure 3.16, and summing over energy gives us the following relationships in one 1D channel:

$$N_I^{+k} = \frac{1}{2} \int_{\mu_2}^{\mu_1} D_{1D}(E) dE \quad N_T^{+k} = \frac{1}{2} \int_{\mu_2}^{\mu''} D_{1D}(E) dE$$

(3.28)

$$N_R^{-k} = \frac{1}{2} \int_{\mu_2}^{\mu'} D_{1D}(E) dE$$

where D_{1D} is the 1D density of states, given by the expression $D_{1D}(E) = (1/\pi\hbar)(2m/E)^{1/2}$. Integrating leads to

$$\mu' = \left(\sqrt{\mu_2} + (1-T)\left(\sqrt{\mu_1} - \sqrt{\mu_2}\right)\right)^2 \quad \mu'' = \left(\sqrt{\mu_2} + T\left(\sqrt{\mu_1} - \sqrt{\mu_2}\right)\right)^2 \quad (3.29)$$

If we apply a small voltage between contacts 1 and 2, i.e. if $\mu_1 - \mu_2$ is small, we can make a first-order expansion of both expressions and then we obtain after a few lines of calculations that

$$\mu'' \cong \mu_2 + T(\mu_1 - \mu_2)$$

(3.30)

$$\mu' \cong \mu_2 + (1-T)(\mu_1 - \mu_2)$$

(3.31)

This is the so-called linear approximation. To make things even simpler, we can normalize the potentials and choose an energy scale such that $\mu_1 = 1$, $\mu_2 = 0$. Then from equations (3.30) and (3.31) we have $\mu_1^+{}_{left} = 1$, $\mu^+{}_{right} = T$, $\mu^-{}_{left} = 1-T$ and $\mu^-{}_{right} = 0$, so that the variation of chemical potential for the $k+$ and $k-$ states can be schematized as in Figure 3.17.

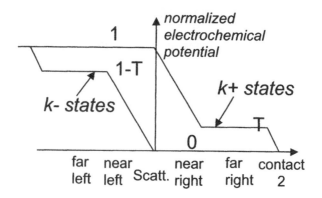

Figure 3.17. *Variation of the chemical potentials as a function of space around the scatterer*

The potential drop ΔV_S at the scatterer is the variation of chemical potential, either for the $k+$ or the $k-$ states, and in both cases we find that

$$e\Delta V_S = (1-T)(\mu_1 - \mu_2),$$ (3.32)

so that we have three parts in which dissipation takes place: the two contacts and the scatterer. Since there is dissipation around the scatterer, we can see the overall resistance as the sum of the contact and the scatterer resistance, and if we re-arrange the Landauer formula as

$$G^{-1} = \frac{h}{2e^2}\frac{1}{MT} = G_C^{-1} + G_S^{-1} = \frac{h}{2e^2 M} + \frac{h}{2e^2 M}\frac{1-T}{T}$$ (3.33)

we can easily see that in fact, the overall resistance can be thought of as the resistance association of a ballistic point contact and a scatterer. The voltage dissipation around the scatterer takes place over the length required for the transmitted electrons to re-arrange themselves in the available states, i.e. the length over which they are going to give up energy to the lattice. In contrast, if coherence is maintained, a scatterer does not induce dissipation, and it may seem better to formalize the problem in terms of what is known as scattering states, which are briefly discussed in section 3.12.3.

3.9. Voltage probe measurements ●

In a macroscopic device if we carry out a four-probe resistance measurement we expect the nature of the probes not to affect the results, and even to allow us to get rid of the "usual" contact resistance value. Thus, we do measure the sample conductivity. This is no longer the case in a mesoscopic sample, which works more as a waveguide for the electron wave functions, and can thus couple in a different way the $k+$ and the $k-$ states of one branch, or one lead, to another.

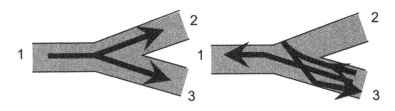

Figure 3.18. *Possible paths that can be followed by an electron in a branching node*

Imagine that we have some branching as in Figure 3.18. It is reasonable to predict that ballistic electrons coming from the left ($k+$ states) will be equally shared between the two right channels. However, consider now electrons coming from the right, in the lower branch ($k-$ states). It is quite reasonable to predict that in such a case, the ballistic electrons will not be equally shared between 1 and 3, but rather that most of them will be transferred to lead 1 or reflected back into lead 3. In fact, we will see that the transmissions from 3 to 1 and from 1 to 3 must be equal in the absence of a magnetic field; otherwise we would observe a current even without any chemical potential difference. This implies that a substantial fraction of the electrons originating in contact 3 is reflected back and even that terminals 2 and 3 are not uncoupled as dictated by intuition. Thus, it is clear that leads 1 and 3 do not couple the $k+$ and $k-$ states in the same way. More generally, if we forbid any net current to flow into one lead, the chemical potential of this lead will tend to adjust itself closer to the chemical potential imposed to the lead which is more strongly coupled to it, and this may correspond to $+k$ rather than $-k$ states. Thus, although located at the same distance from two other leads, a terminal can be imposed on a chemical potential closer to one of those two leads, rather than the average between both. In the same spirit, probes can affect the carriers by making them lose quantum coherence, or can change a resistance measurement result by imposing a chemical potential at a part of the device which in a macroscopic view could not be affected by the probe presence, but which in a mesoscopic sample does change the way the electrons are transmitted through the whole device. In a mesoscopic device, probes are invasive, and they can impede electrons which otherwise would have been ballistic to cross the device without being scattered, because they can force those electrons to transit through the probe reservoir and to lose phase coherence. Let us now apply these arguments to a four-probe measurement configuration.

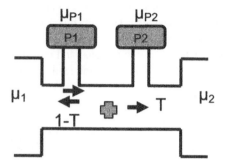

Figure 3.19. *A four-probe device with a scattering center in its middle part*

Inject a current I between contacts 1 and 2, and measure the voltage drop V between probes 1 and 2. The four-probe resistance R_{4p} is obtained by measuring the ratio V/I:

$$R_{4p} = \frac{V_{P2} - V_{P1}}{I} = \frac{\mu_{P1} - \mu_{P2}}{eI} = \frac{h}{2e^2 MT} \frac{\mu_{P1} - \mu_{P2}}{\mu_1 - \mu_2} \qquad (3.34)$$

and in this case (probes perpendicular to the main device between 1 and 2), it is quite reasonable to assume that probes 1 and 2 are identically coupled to the $k+$ and $k-$ states. Thus, we find that in this case

$$R_{4p} = \frac{h}{2e^2 M} \frac{1-T}{T}. \qquad (3.35)$$

However, things are different if we consider a new structure such as in Figure 3.20. In this case, we can instead assume that probe 1 is only coupled to the $k-$ states, and that probe 2 is only coupled to the $k+$ states. Since all the $k+$ states on the left part are in equilibrium with contact 1 we expect that probe 2 is at the chemical potential $\mu''=\mu_2+T(\mu_1-\mu_2)$ (equation (3.30)), and for similar reasons that probe 2 is at the chemical potential $\mu'=\mu_2+(1-T)(\mu_1-\mu_2)$ (equation (3.31)). Thus, in this case we expect that

$$R_{4p} = \frac{h}{2e^2 M} \frac{1-2T}{T}. \qquad (3.36)$$

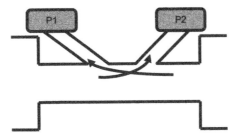

Figure 3.20. *A four-probe device with asymmetrically branched voltage probes*

This is quite puzzling, because if you suppose that T is better than *0.5*, then the four-probe resistance is negative! In a macroscopic sample this would be quite impossible, but in a mesoscopic sample this has indeed already been observed. This example should prevent you from straighforwardly applying well-established macroscopic results to mesoscopic devices.

3.10. Comment about the assumption that T is constant ●

In section 3.4 and equation (3.14) we assumed that the transmission coefficient was identical for all 1D channels. In fact this is true for "ideal" simple point contacts but this is not necessarily applicable to any case. In practice it turns out that the transmission may not only vary with energy, but also with the considered 1D channel: an incident electron in one channel does not have the same transverse energy as in another channel, even if those two electrons are both at the Fermi energy. For instance, an electron lying in a "lower" 1D channel has more longitudinal kinetic energy, but less "transverse" energy. Thus, if submitted to a bend it may be more easily reflected because there is less transverse energy available to make it easier to turn inside the bend. Thus, in practice the transmission T has to be adjusted, depending on each considered channel. This is easily understandable if we consider an example as in Figure 3.21.

If we introduce an obstacle such as a square antidot in the middle of the guide, the electrons in the first mode, which are preferentially found in the middle of the wire, are more susceptible to being reflected back, whereas incident electrons lying in the second 1D subband, whose probability density exhibits two maxima by the sides of the wire, have a higher probability of being transmitted through the obstacle. Figure 3.21 represents a numerical solving of Schrödinger's equation for each occupied mode, and obviously confirms these expectations, as seen either in the wave function picture or in the calculated transmission coefficient. The first mode is substantially reflected, giving rise to an interference pattern on the left side and to a smaller transmitted part. In contrast, the second mode has a transmission much closer to 1. In addition, a given incident mode is not transmitted only in the form of an outgoing wave pertaining to the same channel. It is instead mixed by the obstacle and this results in transmitted parts in the other 1D channels as well. This possibility is taken into account by generalizing the Landauer formula as described in the next section. This extended formalism will also allow us to treat the case of devices with multiple contacts.

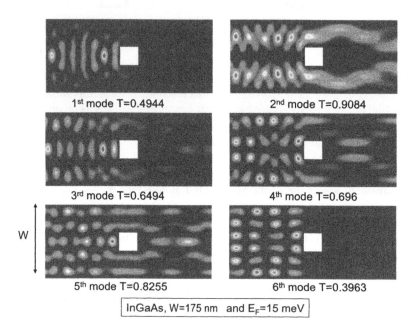

Figure 3.21. *An electron waveguide with a square antidot; each figure represents the square of the wave function modulus for an incident electron at the Fermi energy and in a given 1D mode; each wave is the combination of the reflected and transmitted parts. Wave functions are determined by the scattering matrix formalism and wave matching technique; seven input modes are occupied*

3.11. Generalization of Landauer's formula: Büttiker's formula ●

3.11.1. *Büttiker's formula*

A simple and mechanical way to treat an example such as in the previous section is to generalize the Landauer formula to an arbitrary number of terminals. The way of doing this was first developed by M. Büttiker [BUT 86]. As before, we shall first assume that the various 1D channels of one lead are independent from one another.

Then the generalization is quite simple. Just consider the formula previously established for two contacts

$$I = \frac{2e}{h} \bar{T} \left(\mu_1 - \mu_2 \right)$$
(3.37)

where $\bar{T} = MT$ and extend it to an arbitrary number of terminals,

$$I_P = \frac{2e}{h} \sum_q \left(\bar{T}_{qp} \mu_p - \bar{T}_{pq} \mu_q \right),$$
(3.38)

where I_P is the net current through lead p and a term such as $\bar{T}_{pq}$ has a straightforward meaning: it is the probability that electrons outgoing from terminal q and located at energies between μ_q and μ_p are transmitted into terminal p. Defining

$$G_{pq} = \frac{2e^2}{h} \bar{T}_{pq}$$
(3.39)

we obtain the Büttiker formula

$$I_p = \sum_q \left(G_{qp} V_p - G_{pq} V_q \right)$$
(3.40)

Provided that we know the transmission coefficients, we can calculate the voltages and currents in any configuration. From equation (3.40) we can check or deduce a number of properties. First, if we apply the same voltage to all terminals the current in any terminal is obviously equal to zero. Thus we at once find that the G_{pq}s obeys the equality

$$\sum_q G_{pq} = \sum_q G_{qp}$$
(3.41)

and the Büttiker formula can be simplified to

$$I_p = \sum_q G_{pq} \left(V_p - V_q \right).$$
(3.42)

If we use terminal p as a voltage probe, i.e. we have $I_p=0$, we obtain from the Büttiker formula that

$$
V_p\Big|_{I_p=0} = \frac{\sum_{q \neq p} G_{pq} V_q}{\sum_{q \neq p} G_{pq}} .
\tag{3.43}
$$

The conductance coefficients exhibit a final important property, which we shall not demonstrate, and which is known as the reciprocity or Onsager's relation [ON 31]. If we apply a magnetic field B perpendicular to the sample, and if we revert this magnetic field, the G_{pq}s obey the relation

$$
G_{pq}\Big|_{B} = G_{qp}\Big|_{-B},
\tag{3.44}
$$

which implies in particular that *in the absence of magnetic field B=0, we have* $G_{pq}=G_{qp}$. Although we shall admit the reciprocity relations without derivation, some simple considerations will help us to see them as quite a reasonable result. Consider a symmetric device with a branching node between three terminals:

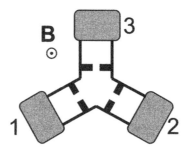

Figure 3.22. *A ballistic device with three symmetric contacts and a perpendicularly applied magnetic field*

If we apply a magnetic field $B>0$, it is clear that ballistic electrons outgoing from lead 1 into the branching node (i.e. in a "$k+$ state" for lead 1) tend to be preferentially deflected towards lead 3 under the action of the magnetic field (remember the Lorentz force). Now suppose that we revert field B. It is obvious that for such a symmetric structure electrons outgoing from lead 3 into the branching node (i.e. in a "$k+$ state" for lead 3) are deflected towards lead 1 in exactly the same way as the electrons from lead 1 to lead 3 with $B>0$. Thus, such a symmetric structure clearly obeys the relationship equation (3.44), and it would be impossible to obtain something like, $G_{pq}(B)=G_{pq}(-B)$ for example, which is obviously wrong in this case (and in any!). The point is that if this type of symmetry is a sufficient condition for obtaining the reciprocity relation, it is by no means a necessary one, and equation (3.44) is in fact always verified.

3.11.2. Three-terminal device

As can be seen in equation (3.40), the various conductance coefficients which are found in the Büttiker relation form a matrix, which links the currents in all the terminals to the voltages. In the case of a three-terminal device, we have from equation (3.40)

$$
\begin{pmatrix} I_1 \\ I_2 \\ I_3 \end{pmatrix} = \begin{pmatrix} G_{12}+G_{13} & -G_{12} & -G_{13} \\ -G_{21} & G_{21}+G_{23} & -G_{23} \\ -G_{31} & -G_{32} & G_{31}+G_{32} \end{pmatrix} \begin{pmatrix} V_1 \\ V_2 \\ V_3 \end{pmatrix} \tag{3.45}
$$

and if we set $V_3=0$ for convenience (we can always arbitrarily fix the origin of voltage), the matrix relation reduces to

$$
\begin{pmatrix} I_1 \\ I_2 \end{pmatrix} = \begin{pmatrix} G_{12}+G_{13} & -G_{12} \\ -G_{21} & G_{21}+G_{23} \end{pmatrix} \begin{pmatrix} V_1 \\ V_2 \end{pmatrix}. \tag{3.46}
$$

If we invert the matrix, we now obtain the relationships which give each voltage as a function of the currents passing through each terminal, and the corresponding matrix is called the resistance matrix:

$$
\begin{pmatrix} V_1 \\ V_2 \end{pmatrix} = \begin{pmatrix} R_{11} & R_{12} \\ R_{21} & R_{22} \end{pmatrix} \begin{pmatrix} I_1 \\ I_2 \end{pmatrix}. \tag{3.47}
$$

Suppose that we pass a current between 1 and 3 and measure the voltage between 2 and 3, as in the left part of Figure 3.23. Then the three-probe resistance R_{3p} is given by

$$
R_{3p} = \left[\frac{V_2}{I_1} \right]_{I_2=0} = R_{21} \tag{3.48}
$$

where R_{21} is obtained by inverting the conductance matrix:

$$R_{21} = \frac{G_{21}}{G_{12}G_{23} + G_{13}G_{21} + G_{13}G_{23}}. \tag{3.49}$$

Suppose that we pass a current between 3 and 2 and measure the voltage between 1 and 3. The three-probe resistance is now given by

$$R_{3p} = \left[\frac{V_1}{I_2}\right]_{I_1=0} = R_{12}. \tag{3.50}$$

A knowledge of the various transmission probabilities allows us to calculate any quantity related to a particular voltage or current configuration.

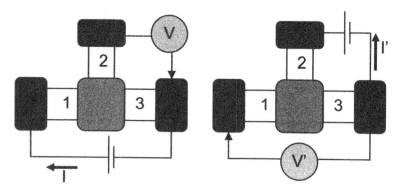

Figure 3.23. *Two different measurement configurations of a three-terminal device*

3.11.3. *Four-terminal device*

Generalization of the results obtained in the previous section to the four terminal case is trivial. We now have, applying Büttiker's formula and with the convention that $V_4=0$:

$$\begin{pmatrix} I_1 \\ I_2 \\ I_3 \end{pmatrix} = \begin{pmatrix} G_{12}+G_{13}+G_{14} & -G_{12} & -G_{13} \\ -G_{21} & G_{21}+G_{23}+G_{24} & -G_{23} \\ -G_{31} & -G_{32} & G_{31}+G_{32}+G_{34} \end{pmatrix} \begin{pmatrix} V_1 \\ V_2 \\ V_3 \end{pmatrix}. \tag{3.51}$$

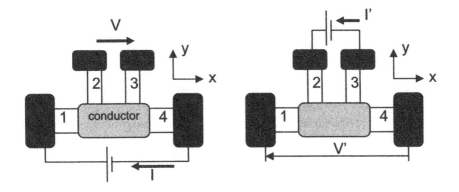

Figure 3.24. *A four-terminal device in two different measurement configurations*

With a measurement configuration as in the left part of Figure 3.24, we should easily find that the four probe resistance is given by

$$R_{4p} = \frac{V}{I} = \left[\frac{V_2 - V_3}{I_1} \right]_{I_2 = I_3 = 0} = R_{21} - R_{31}, \tag{3.52}$$

whereas in a measurement configuration as in the right scheme of Figure 3.24 the four-probe resistance is now equal to

$$R_{4p} = \left[\frac{V_1}{I_2} \right]_{I_1 = 0, I_2 = -I_3} = R_{12} - R_{13}. \tag{3.53}$$

3.12. Non-zero temperature ●

3.12.1. *Large applied bias $\mu_1 - \mu_2 \gg 0$*

Up to now we conveniently assumed that $T=0$ and that the transmission coefficient T is a constant, which is reasonable with a small applied bias. However, if the carriers in each contact spread in energy due to their interaction with a thermal bath ($T \neq 0$), this has a number of important consequences: on one hand, carriers in contact 1 now spread up to energies higher than μ_1, and some energy levels below μ_1 remain empty (roughly on an energy interval equal to a few $k_B T$ around μ_1), so that the carriers available for transmission from contact 2 to 1 spread over a larger energy interval. In particular this means that the transmission probability exhibits an energy dependence. On the other hand, since on the left there are now some empty states below μ_2 and since the electrons in the right contact also spread on a large energy interval (with average μ_2), we have also to take into account that some electrons from the right contact can be transmitted through the device even for energies above μ_2. As a consequence, we have to take into account electron flow both from contacts 1 to 2 and from 2 to 1 in the overall energy range. If we assume that electrons with substantially different energies cannot exchange it with one another, we can divide the energy interval relevant to electron transport into small slices, and sum over them, taking into account that (i) the current flows in the energy range extending from μ_1+k_BT to μ_2-k_BT, and not only from 1 to 2 but also from 2 to 1 and (ii) the transmission probability T depends on energy (if $\mu_1-\mu_2>>0$ the latter must also be taken into account even with $T=0$). However, before doing so we have to tackle a subtlety that we have left hidden up to now, but that we have to face if we want to handle in a correct way the fact that current-carrying electrons now originate in both contacts.

3.12.2. *Incoherent states*

First, suppose that before and after the scatterer we have incoherent states, i.e. that an incoming electron in a $k+$ state in lead 1 can be "transferred" or can "jump" into a $k+$ state in lead 2, which is different from the $k+$ state in lead 1, so that there is no definite phase relationship between these two states. The probability current for incoming electrons with energy E to be transmitted is proportional to the number of available electrons in the corresponding $k+$ states from lead 1, for example $D_{1D}(E)f_1(E)$, where $f_1(E)$ is the occupation function of the electron states in contact 1. However, initially the transmission coefficient T_{21} from 1 to 2 corresponds to the probability for an electron to be transmitted in any of the available $k+$ states with energy E in lead 2. Since a fraction $f_2(E)$ of these states are now already occupied, the habit was taken to assert that the transmission probability must be reduced to $T_{21}(1-f_2(E))$. Thus, for incoherent states people usually stated that the current i^+ of the electrons transmitted from 1 to 2 at an energy E and in an infinitesimal interval dE had to be given by

$$i^+ = MeD_{1D}(E)v_G(E)T_{21}(E)f_1(E)(1-f_2(E))dE = \frac{2e}{h}MT_{21}(E)f_1(E)(1-f_2(E)). \quad (3.54)$$

Reciprocally, the current corresponding to electrons from 2 to 1 was taken as

$$i^- = -MeD_{1D}(E)v_G(E)T_{12}(E)f_2(E)(1-f_1(E))dE = \frac{2e}{h}MT_{12}(E)f_2(E)(1-f_1(E)). \quad (3.55)$$

Summing in the relevant energy interval then leads to

$$I = \frac{2e}{h} \int_{\mu_2 - k_B T}^{\mu_1 + k_B T} \left(\overline{T}_{21} f_1(E) - \overline{T}_{12} f_2(E) + \left(\overline{T}_{12} - \overline{T}_{21} \right) f_1(E) f_2(2) \right) dE . \qquad (3.56)$$

With $B=0$, two terminals and no exchange between the various energy intervals, i.e. without inelastic scattering, we have $T_{12}=T_{21}=T$, so that the overall current is given by

$$I = \frac{2e}{h} \int_{\mu_2 - k_B T}^{\mu_1 + k_B T} \overline{T}(E) \left(f_2(E) - f_1(E) \right) dE \quad \textit{(elastic scattering)}. \qquad (3.57)$$

This formula turns out to be right in the coherent case, but is in fact obtained from a false reasoning, and in general it is in fact *not* correct in the incoherent case. The discussion above is simply not rigorous enough, and the correctness of the result is rather fortuitous. Let us point out the underlying contradiction, which lies in the fact that we manipulate current-carrying states which cannot be treated as stationary, as implicitly assumed in our reasoning[2]. Suppose that all k^+ levels are populated. We have $1-f_2=0$ and from our first formula the current i^+ should be zero. However, since the k^+ states *carry a current*, we might also say that if $f_2=1$ the current i^+ should be equal to $2Me/h$. Depending on how we turn things, the current is either zero or maximum! Our approach is certainly not rigorous enough, and to treat the problem in this way we should instead deal with a situation in which electrons hop between sites, but pass more time in a site than during the jump. Therefore we should introduce something such as a hopping rate, and not only for passing from either side of the barrier, but also from the leads into the contacts. Or we should define time-ordered, non-stationary states. In fact, we shall not treat the problem this way, because we are mostly interested in deriving a formula in the quantum-coherent case, and in such a case such a complex reasoning is not required.

What does occur in our mesoscopic device if the whole sample is coherent? *We cannot say any longer that an electron jumps from one lead to the other, because the wave functions are coherent over the whole device, and extend from lead 1 to lead 2.* Thus, how can we calculate the current with coherent states? We shall investigate this point in the next section. We shall see that for elastic scattering the final formula is the same as equation (3.57), but that it may differ if there is some inelastic scattering.

---

2 We shall encounter the same problem in Chapter 6 dealing with quantum shot noise.

3.12.3. *Coherent states*

First we shall treat a simple 1D case as schematized in Figure 3.25. If we have an energy barrier as in Figure 3.25, we know that an incoming plane wave will be partially transmitted and partially reflected, with a complex transmission coefficient *t* and a complex reflection coefficient *r*, so that we can seek for running wave solutions of the form

$$\begin{cases} \varphi(x) = e^{ikx} + re^{-ikx} & \text{before the barrier} \\ \quad \varphi(x) = te^{ikx} & \text{after the barrier} \end{cases} \tag{3.58}$$

In most textbooks, it is stated that the e^{ikx} part represents an incoming electron, which has a probability $T = |t|^2$ to be transmitted, or a probability $R = |r|^2$ to be reflected. Although this view is more manageable for coffee break discussions and physical reasoning, it does not entirely do justice to the full subtlety of this seemingly simple case. Provided that we do not make a position measurement, we cannot say that the electron is on either side of the barrier, at any time. *More specifically, with a wave function such as equation (3.58) it is not possible to say that, e.g., the electron as a particle traveled towards the positive x, then was reflected by the obstacle so that now it propagates towards the negative x.* We cannot even say that a momentum measurement will give either k or $-k$ (if we calculate the Fourier transform of a wave such as equation (3.58) in addition to the Dirac peaks at $+k$ and $-k$ we will find other terms, far from $+k$ and $-k$ and very small (but not equal to zero), which cancel only if $t=1$, i.e. if the wave function becomes a plane wave).

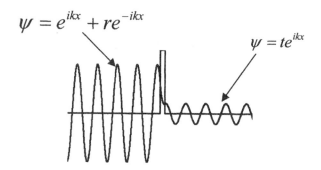

$$\psi = e^{ikx} + re^{-ikx}$$

$$\psi = te^{ikx}$$

Figure 3.25. *A coherent wave impinging on a tunneling barrier*

The only thing that we can say is that if we measure the electron position we can find it on either side of the barrier, and that we can calculate the average velocity which would be obtained after many momentum measurements on identically prepared systems. The most useful information we can get about the dynamics of the system is given by the value of the probability current. In this precise case, *it is a constant on both sides of the barrier.* We can check this point by directly calculating

$$J = -\frac{i\hbar}{2m}(\psi * \nabla \psi - \psi \nabla \psi *)$$

(3.59)

using the wave expression equation (3.58). It will cost us just a little more time to calculate it than for us to type the result here:

$$J = \frac{\hbar k}{m}|t|^2.$$

(3.60)

(note that even with a wavepacket, the picture cannot be totally particle-like, though we can say that before the wavepacket has reached the barrier, the electron will be found with certainty on the left side after a position measurement). Thus, if we want to describe a system with a potential such as in Figure 3.25 with extended stationary states, as we did when there was no scatterer, we have to use wave functions such as equation (3.58), and assign to them a probability current as given by equation (3.60). We are now ready to calculate the overall current. In contrast to the incoherent state case, we no longer have to wonder whether the states in lead 2 are full or empty. The states with a $k+$ "incident" part are still in equilibrium only with contact 1 and the states with a $k-$ "incident" part are in equilibrium only with contact 2, because the probability current of a coherent $k+$ state in lead 2 is positive, so that an electron can be injected into such a state only from lead 1 (in the Bohm's picture that we have already mentioned in the introduction, it is still clearer as we would find that *it is impossible to obtain an electron trajectory which bounces onto the barrier, even though you have a "reflected wave" part; all electrons described by this state go from the left to the right*). Although we shall not demonstrate it, such "scattering" states are all orthogonal and form a complete basis, so that we just have to sum the current carried by all those states in the relevant energy interval [KRI 87] (note that the density of coherent states does not change with respect to the one we calculated with plane waves, since an incident plane wave e^{ikx} obviously gives rise to only one coherent state). The coherent states which carry current towards the right contain electrons up to μ_1+k_BT with the equilibrium density of contact 1, and the coherent states which carry current towards the left are filled up to μ_2 with the equilibrium density of contact 2, so that the net current, which is the difference between these two quantities, is given by

$$I = \frac{2e}{h} \int_{\mu_2-k_BT}^{\mu_1+k_BT} \left(\overline{T}_{21}f_1(E) - \overline{T}_{12}f_2(E) \right) dE$$

(3.61)

and reduces to

$$I = \frac{2e}{h} \int_{\mu_2 - k_B T}^{\mu_1 + k_B T} \overline{T}(E)\big(f_1(E) - f_2(E)\big) dE \quad \text{(elastic scattering)} \tag{3.62}$$

if there is no inelastic scattering and $B=0$. Note that if $T_{12} \neq T_{21}$, the coherent state formula equation (3.62) does not reduce to the incoherent state formula equation (3.56). However, if $T_{12} = T_{21}$ is verified, a straightforward comparison between equations (3.62) and (3.57) shows us that both formalisms give the same result.

The generalization to multi-terminal devices is straightforward. The system is described by states with an incident wave in lead p which lead to transmitted waves in all other leads and to backscattered waves from all other leads. Such states are called scattering states in the literature and form an orthogonal eigenstate basis. A scattering state is in equilibrium only with one reservoir (the one with the incident and all reflected waves), so that the current I_p in one lead is now given by the sum over energy of two different contributions. On the one hand, we have to take into account the current due to the scattering states in equilibrium with lead p, which for one such state is the sum of an incident part minus all backscattered waves in lead p. On the other hand, we must add to this contribution that from all the scattering states in equilibrium with the other leads, each of them having a transmitted part with factor T_{pq} to the lead p. Thus, the overall current in lead p is equal to

$$I_p = \frac{2e}{h} \int \left(\underbrace{\sum_q \overline{T}_{qp} f_p(E)}_{\substack{\text{current carried by the scattering} \\ \text{states in equilibrium with lead p}}} - \underbrace{\sum_q \overline{T}_{pq} f_q(E)}_{\substack{\text{current carried by the transmitted} \\ \text{parts of the scattering states in} \\ \text{equilibrium with all other leads.}}} \right) dE \tag{3.63}$$

and with equation (3.41) this formula simplifies to

$$I_p = \frac{2e}{h} \int \left(\sum_q \overline{T}_{pq} \big(f_p(E) - f_q(E)\big) \right) dE . \tag{3.64}$$

We should apply equation (3.64) whenever transport is coherent. Note that if there is no inelastic scattering, this formula can also be applied to approximate incoherent transport. It is worth noting that this type of expression is not restricted to research lab devices, and has in fact already been applied to assess the characteristics of nano-scale silicon MOSFETs, for which calculations are complicated by the geometry and the determination of the transmission coefficient as a function of energy, because it is not easy to estimate in a precise way how the channel barrier depends on the voltages applied to drain, source and gate.

3.12.4. *Physical parameters included in the transmission probability*

In the previous section, we stated that the transmission probability T was only the squared modulus of the complex coefficient t appearing in factor of the transmitted wave part (equation (3.58)). In fact this is only true if the ingoing channel under consideration has exactly the same width and the same transverse quantized energy as the outgoing channel. The correct formula is slightly different in the general case. For example suppose that the input and output leads have a different width. Since the overall energy is conserved, the longitudinal electron momentum cannot be the same in the two leads. Suppose in addition that only one 1D subband is involved in both leads. The momentum is k_2 in the output lead and k_1 in the input lead. The probability current J_2 in the outgoing lead is given as before by

$$J_2 = \frac{\hbar k_2}{m}|t|^2,$$

(3.65)

and applying equation (3.59) to the upper formula in equation (3.58) gives the probability current in the input lead:

$$J_1 = \frac{\hbar k_1}{m}\left(1-|r|^2\right).$$

(3.66)

Conservation of the probability current implies that $J_1=J_2$. However, since $k_1{\neq}k_2$, it is not possible to write as in the previous section that $|r|^2+|t|^2=1$. If we just consider the current in lead 1, since the absolute momentum is conserved after reflection, we can still define $R=|r|^2$ as the reflection probability. However, using equations (3.65) and (3.66) to write $J_1=J_2$ leads to

$$T = 1 - R = \frac{k_2}{k_1}|t|^2.$$

(3.67)

Thus, to obtain the transmission probability the complex part of the transmitted wave must be weighted by the ratio of the square roots of the momenta in the input and output leads. This is easily generalizable to the Büttiker formula equation (3.38) with an arbitrary number of input and output channels.

3.12.5. *Linear response (μ_1-μ_2<k_BT or T(E)=Cst)*

If we make the assumption that a two-terminal mesoscopic device exhibits only a small departure from equilibrium, i.e. that $\mu_1 \approx \mu_2$, we can write the distribution functions of both contacts as $f_1 = f_{10} + \delta f_1$ and $f_2 = f_{10} + \delta f_2$. In such a case we can rewrite the current formula equation (3.64) as

$$I = \frac{2e}{h} \int \overline{T}(E)_{eq} \, \delta(f_1 - f_2) dE \tag{3.68}$$

and the quantity δf_1-δf_2 can be replaced by a first order development with respect to the difference of chemical potential in the two contacts such as

$$\delta(f_1 - f_2) = (\mu_1 - \mu_2) \left(\frac{\partial f}{\partial \mu} \right)_{eq} = (\mu_1 - \mu_2) \left(-\frac{\partial f}{\partial E} \right)_{eq}, \tag{3.69}$$

so that the current formula eventually reduces to

$$G = \frac{e \delta I}{\mu_1 - \mu_2} = \frac{2e^2}{h} \int \overline{T}(E) \left(-\frac{\partial f}{\partial E} \right)_{eq} dE \tag{3.70}$$

Fortunately, if we make *T=0* the derivative of the distribution function is a Dirac peak at the Fermi energy E_F, and we again find the usual expression $G = (2e^2 / h)\overline{T}(E_F)$.

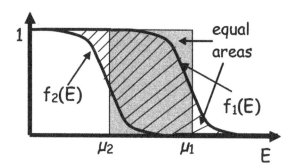

Figure 3.26. *Integration interval over which the distribution function difference between forward and backward propagating states must be integrated to give the overall current*

In addition, expression equation (3.64) can also be greatly simplified in another case: even if the chemical potentials are susbtantially different, if the transmission coefficient does not vary much in the relevant energy interval then it can be factorized out of the integral in equation (3.64). The remaining problem is to calculate the integral of the distribution function difference:

$$I = \frac{2e}{h}\overline{T}\int(f_1(E) - f_2(E))dE \quad (if\ T\ does\ not\ depend\ on\ E).\qquad(3.71)$$

However, this is quite easy, as illustrated by Figure 3.26. We can fold the area which must be calculated above μ_1 and the area below μ_2 inside the interval $[\mu_1,\mu_2]$ as shown in Figure 3.26. Because $f(E_F+\Delta E)=1-f(E_F-\Delta E)$, it becomes evident that the desired integral is given by the blue rectangle area, so that the overall current is equal to

$$I = \frac{2e}{h}\overline{T}(\mu_1 - \mu_2).\qquad(3.72)$$

It is worth noting that in order to obtain this result, we not only neglect the energy dependence of the transmission coefficient, but its electric field dependence as well. This is generally not the case for devices such as resonant tunneling diodes biased at a "high" voltage.

3.13. The integer quantum Hall effect ●

3.13.1. *The experiment*

To conclude this chapter we shall describe a system which is quite a good example, as it exhibits the particular and extraordinary property that ballistic transport can be observed on dimensions which are in the millimeter range (i.e. the electrons relevant for electrical transport do not suffer any collision and propagate "freely" over distances in the mm range). This is the regime of integer Quantum Hall Effect (QHE), which can be observed in 2D gases formed in a MOSFET or in a *III-V* heterostructure. It is not a regime relevant to usual device operation, since it is observed at low T and a high magnetic field, so that we are not going to develop revolutionary device applications from such a phenomenon. However, it gives rise to so accurate a resistance value that it has now replaced the ancient international resistance standard. Although somewhat difficult to explain to non-physicists in mundane situations, it fully deserves to be part of the scientific background to be acquired by any solid-state physicist/engineer. This effect was discovered by K. von Klitzing in Grenoble in 1979, from the MOSFET Hall bar shown in Figure 3.27

[VON 80]. A current I was injected between the lower and upper contacts, and the voltages were monitored from the transverse contacts. A typical Hall bar is schematized in the right part of Figure 3.27.

What von Klitzing observed was curves as in Figure 3.28. The longitudinal resistivity $R_{xx}=(V_1-V_2)/I$ exhibited peaks and in between those peaks it reduced to exactly zero, meaning that there was no dissipation at all. Whenever there was no dissipation, the transverse resistivity $R_{xy}=V_H/I=(V_2-V_3)/I$ was quantized, equal to an integer number times the quantum resistance h/e^2:

$$R_H = \frac{h}{ne^2} \left(\cong \frac{12,906\ \Omega}{n} \right) \tag{3.73}$$

The more surprising fact was that quantization was realized with an accuracy better than 10^{-8}. Later on, checking the effect on different materials showed that it was possible to measure the fine structure constant e^2/h with such good accuracy and whatever the device that this quantized resistance was adopted as the international resistance standard. There was no fluctuation due to impurities or defects, which do vary from one sample to the next.

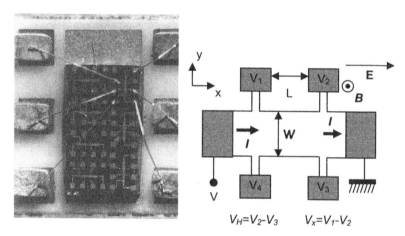

$$V_H=V_2-V_3 \qquad V_x=V_1-V_2$$

Figure 3.27. *Original MOSFET Hall bar used by von Klitzing and scheme of a typical Hall bar used to observe the integer quantum Hall effect; by courtesy from Prof. von Klitzing*

Still later on another effect was discovered in *very* pure samples, in which the resistance is now quantized in fractions of the resistance quantum (fractional quantum Hall effect), but this is a more complicated phenomenon which is explained in terms of many-body effects, which are beyond the scope of this introductory course. Von Klitzing was awarded the Nobel prize in physics for the discovery of the integer quantum Hall effect.

Remember the shape of the Landau bands that we defined in the case of an infinite 2D gas as in section 2.18.4. We discovered that the magnetic field quantizes the energy levels into Dirac peaks, which are somewhat broadened by the topological disorder and the thermal agitation present in any real sample. Intuitively, we would expect that the resistance is minimum when the Fermi energy E_F is located in a Landau band, because to obtain a transport we must have a partially filled band, neither full nor empty. However, a Hall bar has a finite width, and in the case of a confined potential we also learnt that the Fermi level can be positioned in edge states, which are physically located by the sides of the sample, the forward propagating states and the backward propagating states being separated by the sample width. This spatial separation totally suppresses backscattering, and it is this effect that is at the origin of the QHE. Whenever the Fermi level is located in edge states, a resistance drop down to zero is observed and the Hall resistance is quantized. This is what we are going to show below in further detail.

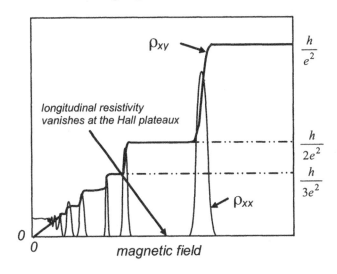

Figure 3.28. *Shape of the transverse and longitudinal resistivities versus magnetic field in the Integer Quantum Hall Effect regime*

3.13.2. The explanation

In a wide Hall bar such as in Figure 3.27 the potential shape is as in Figure 3.29: it is flat in the middle of the device, but close to the edge there is a rising potential which prevents the electrons from escaping from the sample. With a magnetic field perpendicular to the 2D gas plane we saw that the Schrödinger equation takes the form

$$\left[E_S + \frac{(i\hbar\nabla - e\vec{A})^2}{2m} + U(y) \right]\Psi(x,y) = E\Psi(x,y) \tag{3.74}$$

and the wave functions and Landau levels are given by

$$\Psi_{n,k}(x) = \frac{e^{ikx}}{\sqrt{L}} e^{-\frac{(q-q_k)^2}{2}} H_n(q - q_k) \tag{3.75}$$

and

$$E_{n,k} = E_S + \left(n + \frac{1}{2}\right)\hbar\omega_C , \tag{3.76}$$

respectively, with the same definitions as in section 2.18.4:

$$\omega_c = \frac{eB}{m} \qquad q_k = \sqrt{\frac{m\omega_0}{\hbar}}y_k \qquad q = \sqrt{\frac{m\omega_0}{\hbar}}y \qquad y_k = \frac{\hbar k}{eB}. \tag{3.77}$$

Here $U(y)$ is flat over a large area (see Figure 3.29), and to solve equation (3.74) we can consider that the edge potential simply acts as a perturbation onto the Landau states of an infinite 2D gas. Thus, we know from perturbation theory (section 10.4.1) that to the first order, the Landau levels will be modified as

$$E_{n,k} \cong E_S + \left(n + \frac{1}{2}\right)\hbar\omega_C + \left\langle \psi_{n,k} \left| U(y) \right| \psi_{n,k} \right\rangle. \tag{3.78}$$

We also know that the Landau functions (equation (3.75)) are sharply peaked around y_k, so that we can assume that the potential $U(y)$ does not vary much in the part where the wave function is not negligible. Taking $U(y) \approx U(y_k) = const$ in equation (3.78) immediately yields

$$E_{n,k} \cong E_S + \left(n + \frac{1}{2}\right)\hbar\omega_C + U(y_k). \tag{3.79}$$

In the middle of the device, since $U(y)$ does not vary and is almost constant, the corresponding energy levels are not modified and do not depend on k. However, as k increases the wave functions are moved towards the sample edges, and close to those edges we see from equation (3.79) that now the energy increases with k. Thus, we can draw a schematic as in Figure 3.29 (bottom) for the energy levels: in each 1D Landau subband the dispersion curve is flat for low k values, and rises for high k values which correspond to electron states spatially located close to the edges of the samples. We can easily calculate the velocity:

$$v_{n,k} = \frac{1}{\hbar}\frac{\partial E_{n,k}}{\partial k} = \frac{1}{\hbar}\frac{\partial U(y_k)}{\partial k} = \frac{1}{eB}\frac{\partial U(y)}{\partial y} \ . \tag{3.80}$$

Thus, the states in the middle of the sample, where the potential $U(y)$ is flat, do not carry current, as for an infinite 2D gas, but the electron in the edge states exhibit a velocity which is proportional to the transverse potential gradient. Equation (3.80) also shows that the "left" and "right" edge states carry current in opposite directions. These edge states are responsible for the QHE. As we can see from Figure 3.29, if the chemical potentials μ_1 and μ_2 of the left and right contacts are located in these edge states, the net current will be carried by electrons which are located at the edges of the sample. Since the forward propagating states are at the top edge and the backward propagating states are at the bottom edge in the figure, backscattering and dissipation are totally suppressed (Figure 3.29). *To lose momentum an electron should be scattered from one side to the opposite!* Transport becomes ballistic over lengths which can be in the millimeter range. This is the reason why the resistance falls to zero.

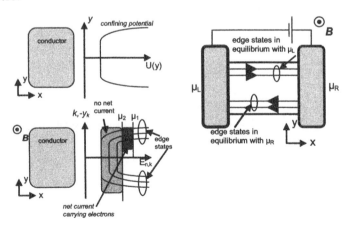

Figure 3.29. *(a) Potential profile of a 2D gas device with boundaries submitted to a magnetic field and energy band diagram of the corresponding 1D Landau subbands; (b) schematic representation of the edge states carrying the current in the same device*

Provided that the chemical potentials lie in the edge states, the resistance falls down to zero so that if we measure the longitudinal voltage in a Hall bar, we should find $V_L=0$. Since the "left" states at the top edge are in equilibrium with the left contact and the "right" states at the bottom edge are in equilibrium with the right contact (Figure 3.30), if we measure the voltage V_H from a Hall bar such as in Figure 3.30, V_2 and V_3 are at the same potential as the left contact 1 and V_5 and V_6 are at the same potential as the right contact 2. Thus, the Hall voltage is equal to $V_H=(\mu_L-\mu_R)/e$.

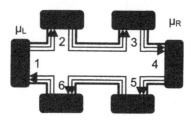

Figure 3.30. *A 2D Hall bar in the integer quantum Hall regime and the edge states*

It is worth noting that all of this is true only if E_F does not lie in a Landau level of the bulk. Otherwise the bulk states contribute to transport, backscattering is allowed and the resistance rises! When the current is carried by edge states, since transport is ballistic and the current density depends on position as indicated by equation (3.80), for one channel we have to sum the net current contribution which occurs between the positions y_L and y_R, which correspond to the crossing between the considered Landau state level and the chemical potentials μ_L and μ_R, respectively (see Figure 3.29). Since we know that the density of such states per unit area is given by $2eB/h$ (see section 2.18.4.2.3), we obtain by summing over all 1D Landau channels

$$I = e \sum_{all\,channels} \int_{y_R}^{y_L} D(x,y) \times v(y) dy = e \sum_{all\,channels} \int_{y_R}^{y_L} \frac{eB}{h} \times \frac{1}{eB} \times \frac{\partial U}{\partial y} dy = \frac{e}{h} M(\mu_L - \mu_R) \qquad (3.81)$$

where M is the number of Landau bands giving edge states between μ_R and μ_L. Since $V_2=V_3$ the longitudinal resistance is equal to zero, and from equation (3.38) the Hall resistance is equal to a quantized value:

$$R_L = \frac{V_L}{I} = 0 \text{ and } R_H = \frac{V_H}{I} = \frac{h}{e^2 M}. \qquad (3.82)$$

If we examine the two previous equations we should see that we did not use the usual spin degeneracy factor of 2. The reason is that depending on the materials used, this degeneracy can be significantly lifted by the magnetic field through the

Zeeman effect. Thus, at high magnetic field a 1D band is either with spin up or spin down, and the two bands can be clearly separated.

We can put all of this on a more formal footing and use the Büttiker formalism. In the QHE regime incoming $k+$ states in contact 2 originate only from contact 1, incoming $k+$ states in contact 3 come from contact 2, incoming $k+$ states in contact 4 originate only in contact 3, etc., so that it is straightforward to establish the various transmission coefficients in the conductance matrix, which is equal to

$$\begin{pmatrix} I_1 \\ I_2 \\ I_3 \\ I_5 \\ I_6 \end{pmatrix} = \begin{pmatrix} G_C & 0 & 0 & 0 & -G_C \\ -G_C & G_C & 0 & 0 & 0 \\ 0 & -G_C & G_C & 0 & 0 \\ 0 & 0 & 0 & G_C & 0 \\ 0 & 0 & 0 & -G_C & G_C \end{pmatrix} \begin{pmatrix} V_1 \\ V_2 \\ V_3 \\ V_5 \\ V_6 \end{pmatrix} \tag{3.83}$$

if we take $V_4=0$, and where G_C is the contact resistance $G_C=e^2M/h$ (to draw a figure such as Figure 3.30 remember that with a perpendicular magnetic field directed towards the top, the electrons turn left). Then the results already derived above follow mechanically from the conductance matrix relations. Since we have $I_2=I_3=I_5=I_6=0$, we immediately find from equation (3.83) that $V_2=V_3=V_1$ and $V_5=V_6=0$. Thus, we also have $I_1=G_CV_1$ and we find the expected relationships

$$R_L = (V_2 - V_3)/I_1 = (V_6 - V_5)/I_1 = 0 \tag{3.84}$$
$$R_H = (V_2 - V_6)/I_1 = (V_3 - V_5)/I_1 = G_C^{-1} .$$

So far, so good, except for one point: if we examine Figure 3.28 we see that the quantized plateaux are quite extended, and that the resistive intervals are quite narrow. However, if we consider the schematics of Figure 3.29 the edge states are in fact very few, and the "bulk" Landau states are very numerous. With such a configuration, we should expect the Fermi level to stay a much longer time in the bulk regime, and not in the QHE regime. If we vary E_F through the application of a gate voltage, it should jump from bulk states from a given Landau band to the next, because their filling requires many electrons, whereas the edge states contain almost none and should be filled at once. The same type of phenomenon should occur if we vary the magnetic field. However, in the experiment it is clear that more time is spent with a chemical potential seemingly lying in the edge states. Initially this was a bit puzzling but the reason for this paradoxical behavior was put forward soon after the discovery of the effect: in such big samples we always have many defects, mostly impurities. Thus, the sample exhibits local disturbances of the uniform potential, as depicted by Figure 3.31.

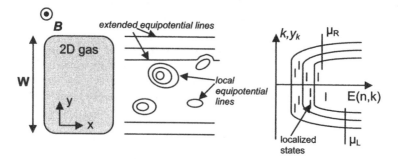

Figure 3.31. *2D gas with lateral boundaries in a magnetic field; the middle scheme represents the local equipotential lines due to inhomogenities and defects, and the right figure is the energy band diagram of the structure; localized states appear in between the bulk Landau states*

Around a given local potential, there is the possibility for some electrons to become trapped in an equipotential orbit around the potential "peak" or "hole" (see exercise 2.19.4). These closed cyclotron orbits are of course quantized and they do not lead to current, but they can lie in the intervals outside the bulk Landau bands. If we modify the Fermi level we also sweep across all these localized states and we have to fill them. Therefore, if the sample contains many such states, they can control the way with which we populate all the levels, and prohibit the Fermi level pinning in the bulk Landau states (E_F is pinned in the regions where the density of states is high, i.e. the energy domains which require more electrons to be filled). As a matter of fact, we can see the quantized Hall resistance because potential fluctuations and defects in the bulk create many such localized states, which help to maintain E_F between the bulk Landau levels on a large range of electron concentration or magnetic field, and allow us to observe such an extraordinarily precise resistance quantum.

3.14. Exercises

3.14.1. *Exercise*

A thin GaAs layer is grown in between two AlGaAs layers. The upper AlGaAs layer is delta-doped, so as to create a 2D electron gas inside the GaAs film. By etching the upper layers we define a nanostructure as illustrated by Figure 3.32. All arms are defined with the same width. Temperature is low and electron transport is perfectly ballistic. Doping is such that M 1D channels are populated in one constriction.

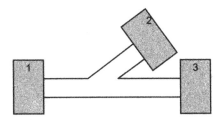

Figure 3.32. *A three-terminal ballistic device*

1) There is no magnetic field. With the notation $\bar{T}_{21}=T_{SC}$, $\bar{T}_{23}=T_{WC}$ and $\bar{T}_{31}=T_D$, write down the conductance matrix and simplify the number of independent coefficients by using the reciprocity relations. "*SC*" is for "strongly coupled" and "*WC*" stands for "weakly coupled".

2) Discuss the magnitudes expected for the various transmission coefficients with respect to one another, as a result of the device geometry.

3) A voltage V_1 is applied between 1 and 3 and contact 2 is left unconnected. Calculate the voltage V_2. Comment. Compare the result with what would be obtained with a macroscopic conductor.

4) Calculate the resistance $R_{21}=V_2/I_1$. Give a physical comment.

5) Calculate the difference between the reflection coefficients corresponding to electrons originating either in contact 1 or in contact 3. Simplify the result and give a physical comment. In particular give simple arguments involving wave propagation and the reciprocity relations to explain the result.

3.14.2. *Exercise*

1) Consider a semiconductor for which the dispersion relation between energy and wavevector is given by a power law $E=ak^n$ where n is an integer. What is the expression of the velocity as a function of wavevector?

2) Consider a 1D constriction fabricated from the same semiconductor. What is the expression of the density of states as a function of energy?

3) What is the value of the current as a function of the difference of chemical potential between the left and right contacts? Comment on the result and compare it with the usual case with a parabolic dispersion relationship.

4) Calculate the current for a 1D constriction with a dispersion relationship of the form $E=ak^2+bk^4$. Comment.

5) Calculate the current for a 1D constriction with an arbitrary dispersion relationship $E=f(k)$. Comment on the universality of the result.

3.14.3. Exercise

A thin GaAs layer is grown in between two AlGaAs layers. The upper AlGaAs layer is delta-doped, so as to create a 2D electron gas inside the GaAs film. By etching the upper layers we define a nanostructure as illustrated below. A magnetic field is applied perpendicularly to the GaAs layer plane. Temperature is low and electron transport is perfectly ballistic. Doping is such that only one 1D channel is populated in one constriction.

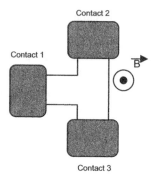

Figure 3.33. *A three-terminal ballistic device with magnetic field*

1) Explain qualitatively how a 2D gas can be formed in the GaAs layer.

2) Using the reciprocity relationships and from the device symmetry simplify the number of independent transmission coefficients. Take the notation $\bar{T}_{21}=T_L$, $T_{31}=T_R$ and $\bar{T}_{32}=T_C$. Carefully justify the simplifications made.

3) Assume that contact 3 is connected to the ground so that its potential is $V_3=0$. Calculate the conductance matrix (use the same indices as for the transmission coefficients).

4) A voltage V_1 is applied between *1* and *3*, and contact *2* is left unconnected. Calculate the conductance between contacts *1* and *3*.

5) Suppose that the magnetic field is equal to zero. Simplify the formula found above.

6) $B=0$. By considering the shape of the device, discuss qualitatively the magnitude of G_L with respect to G_C.

7) *B=0*. Factorize G_L in the formula above and give the first and zero[th] order development of the conductance with respect to G_L/G_C.

8) Why and how does contact *2*, though not connected, modify the overall conductance? Explain the zero[th] order development result by a simple reasoning and considering the various possible electron transmission processes which can describe the propagation of an electron coming from contact *1*.

9) *B=0*. A voltage V_2 is applied between contact *2* and contact *3* and contact *1* is left unconnected. Calculate the conductance between contact *2* and contact *3*.

10) Explain the result above by a simple reasoning in the same spirit as in question 7.

11) Devise an experimental method to extract the conductance matrix coefficients. Explain which quantities we should measure and how we can express the conductance coefficients from those quantities.

12) A high magnetic field *B* is applied as in the figure above. Assuming that one channel is populated at the Fermi level, what are the values of G_L, G_R and G_C? Give the conductance matrix. Justify your answer.

13) With the same magnetic field condition as in the previous question, a small voltage V_1 is applied between contacts *1* and *3*, and contact *2* is left unconnected. What is the value of V_2? What is the conductance between contacts *1* and *3*?

3.14.4. *Exercise*

A 15 nm thick GaAs layer is grown in between two AlGaAs layers. The upper AlGaAs layer is delta-doped, so as to create a 2D electron gas inside the GaAs film. By etching the upper layers we define a nanostructure as in Figure 3.34. The constrictions are all identical. A magnetic field is applied perpendicularly to the GaAs layer plane. We assume that temperature is low and electron transport is perfectly ballistic. M is defined as the number of populated 1D channels in one constriction, whether B is high or equal to zero.

1) What is a 1D conductor? What are the conditions to be fulfilled to obtain such a device?

2) Explain the physical meaning and consequences of ballistic transport. Why does the conductance always remain of finite value?

3) From the device symmetries simplify the number of independent transmission coefficients. Take the notation $\bar{T}_{21}=T_R$, $\bar{T}_{31}=T_F$ and $\bar{T}_{41}=T_L$.

4) Calculate the conductance matrix (define the indexes with the same notations as those used for the transmission coefficients). To simplify the notations we can define $G_0=G_R+G_L+G_F$ (or $T_0=T_R+T_L+T_F=M$).

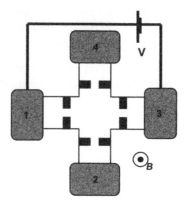

Figure 3.34. *A four-terminal ballistic device with magnetic field*

5) A small voltage is applied between *3* and *1* as in the figure above, and contacts *2* and *4* are left unconnected. For convenience take $V_3=0$. Demonstrate that the conductance between contacts *1* and *3* is given by

$$G = \frac{e^2}{h}\left(T_0 + T_F + \frac{(T_R-T_L)^2}{2T_F + T_R + T_L}\right).$$

6) If there is no magnetic field, how can you simplify the formula above?

7) There is no applied magnetic field. Why and how do contact 2 and 4, though not connected, modify the overall conductance? Would the situation be identical if the conductors had a macroscopic size?

8) Make the assumption that there is no magnetic field, only one occupied channel in the entrance constriction and that no carrier is straightforwardly transmitted from 3 to 1. From question 6) what is the conductance? Comment on the result physically by comparing it to the conductance of a simple two-terminal ballistic constriction.

9) Make the assumption that there is no magnetic field, only one occupied channel in the entrance constriction and that all carriers are straightforwardly transmitted from 3 to 1. From question 6) what is the conductance? Give a physical

comment about putting in series two identical constrictions, if the transport remains ballistic between both of them.

10) Explain qualitatively what happens if a high magnetic field is applied to the structure as in the figure, and how the transmission coefficients are modified (which increase, which decrease?).

11) Suppose that the magnetic field is high enough to put the device into the integer quantum Hall regime, so that the current is carried by edge states, not reflected by the constrictions. Give the values of the transmission coefficients. What is the overall conductance? What are the values of V_4 and V_2? Any physical comment is welcome.

3.14.5. *Exercise*

A 15 nm thick GaAs layer is grown in between two AlGaAs layers. The upper AlGaAs layer is delta-doped, so as to create a 2D electron gas inside the GaAs film. By etching the upper layers we define a nanostructure as in Figure 3.35. The constrictions are all identical. A magnetic field is applied perpendicularly to the GaAs layer plane. We assume that temperature is low and electron transport is perfectly ballistic. M is defined as the number of populated 1D channels in one constriction, whether B is higher than or equal to zero.

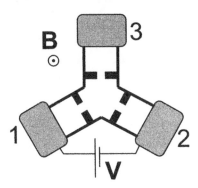

Figure 3.35. *A three-terminal ballistic device with magnetic field*

1) If the device was fabricated from silicon would the ballistic effects be more prominent or not? Why?

2) From the device symmetries simplify the number of independent transmission coefficients. Take the notation $\bar{T}_{21}=T_R$ and $\bar{T}_{31}=T_L$.

3) Calculate the conductance matrix (define the indexes with the same notations as those used for the transmission coefficients).

4) A voltage is applied between 2 and 1 as in Figure 1, and contact 3 is left unconnected. Calculate the conductance between contacts 1 and 2. For convenience take $V_2=0$.

5) Suppose that the magnetic field is high enough to put the device into the integer quantum Hall regime, so that the current is carried by edge states, not reflected by the constrictions. Give the values of the transmission coefficients and justify them. What is the overall conductance? What is the value of V_3? Any physical comment is welcome.

6) Suppose that $B=0$. What is the conductance? Compare it with the conductance of a simple two-terminal constriction. Explain the difference in a qualitative way. Why and how does contact 3, though not connected, modify the overall conductance? Would the situation be identical if the conductors had a macroscopic size?

We now consider the nanostructure defined by Figure 3.36.

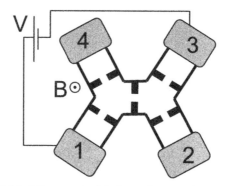

Figure 3.36. *A four-terminal ballistic device with magnetic field*

7) Put the device symmetries to good use so as to simplify the number of independent transmission coefficients. Use the notations $\bar{T}_{21}=T_R$, $\bar{T}_{12}=T_L$, $\bar{T}_{41}=T_b$ $T_{14}=T_r$, $T_{31}=T_C$ and $T_{42}=T_C'$. Give the conductance matrix.

8) A small voltage V is applied as in Figure 3.36 and the magnetic field is high enough for electronic transport to be ensured by M edge conducting channels in one constriction. What are the values of V_4 and V_2? What is the conductance?

3.15. Bibliography

[BUT 86] BUTTIKER M., "Four-terminal phase-coherent conductance", *Physical Review Letters*, vol. 57, no. 14, 1986, p. 1761-1764.

[DAT 95] DATTA S., *Electronic Transport in Mesoscopic Systems*, Cambridge, Cambridge University Press, 1995.

[KRI 87] KRIMAN A.M., KLUKSDAHL N..C., FERRY D.K., "Scattering states and distribution functions for microstructures", *Physical Review B*, vol. 36, no. 11, 1987, p. 5953-5959.

[ON 31] ONSAGER L., "Reciprocal relations in irreversible processes. I.", *Physical Review B*, vol. 37, 1931, p. 405-426.

[TEK 91] TEKMAN E., CIRACI S., "Theoretical study of transport through a quantum point contact", *Physical Review B*, vol. 43, no. 9, 1991, p. 7145-7169.

[TOP 00] TOPINKA M.A., LEROY B.J., SHAW S.E.J., HELLER E.J., WESTERVELT R.M., MARANOVSKY K.D., GOSSARD A.C., "Imaging coherent electron flow from a quantum point contact", *Science*, vol. 289, no. 5488, 2000, p. 2323-2326.

[VAN 88] VAN WEES B.J., VAN HOUTEN H., BEENAKER C.W.J., WILLIAMSON J.G., KOUWENHOVEN L.P., VAN DER MAREL D., FOXON C.T., "quantized conductance of point contacts in a two-dimensional electron gas", *Physical Review Letters*, vol. 60, no. 9, 1988, p. 848-850.

[VON 80] VON KLITZING K., DORDA G., PEPPER M., "New method for high accuracy determination of the fine-structure constant based on quantized Hall resistance", *Physical Review Letters*, vol. 45, no. 6, 1980, p. 494-497.

[WHA 88] WHARAM D.A., THORNTON T.J., NEWBURY R., PEPPER M., AHMED H., FROST J.E.F., HASKO D.G., PEACOCK D.C., RITCHIE D.A., JONES G.A.C., "One-dimensional transport and the quantisation of the ballistic resistance", *Journal of Physics C*, vol. 21, no. 8, 1988, p. L209-L214.

Chapter 4

S-matrix Formalism

4.1. Scattering matrix or S-matrix ●

A convenient way to formulate transport through transmission is known as the S-matrix formalism. The S-matrix of a coherent device relates all the incoming wave amplitudes to the outgoing wave amplitudes. Its use is of interest for two main reasons: on the one hand, if instead of relating the incoming wave amplitudes to the outgoing wave amplitudes we express the outgoing wave amplitudes times the square root of the corresponding wavevector as a function of the ingoing ones, the current conservation requirement imposes well-defined relations between the S-matrix coefficients. On the other hand, the knowledge of the S-matrix of different samples allows us to combine them and calculate the overall S-matrix by applying appropriate combination rules.

Typically, with a coherent device as schematically represented in Figure 4.1, we can write the relation between outgoing amplitudes, the S-matrix and incoming amplitudes under the form

$$\begin{pmatrix} b_1 \\ a_2 \end{pmatrix} = \begin{bmatrix} r & t' \\ t & r' \end{bmatrix} \begin{pmatrix} a_1 \\ b_2 \end{pmatrix}. \tag{4.1}$$

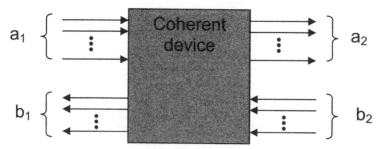

Figure 4.1. *Quantum device represented by a scattering matrix with incoming and outgoing amplitudes of each 1D channel*

where a_1 and b_2 correspond to vectors including all incoming mode amplitudes at the device input and output, respectively, b_1 and a_2 correspond to all outgoing modes at the device input and output, the submatrices r and r' correspond to wave reflection, and t and t' to wave transmission.

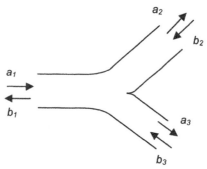

Figure 4.2. *A Y-junction with one channel input and two symmetric output arms*

To introduce important S-matrix properties, we shall first consider a simple example. Consider a three-terminal junction (or a "Y-junction") as in Figure 4.2, with only one populated mode in all three input leads. Terminals 2 and 3 are symmetric. We wish to determine if it is possible to find such a junction with a complete decoupling between terminals 2 and 3, i.e. no part of a wave at input 2 can be transmitted at output 3, and reciprocally, no part of an incoming wave at input 3 can be transmitted to output 2. Initially we can write the S-matrix as

$$\begin{pmatrix} b_1 \\ a_2 \\ a_3 \end{pmatrix} = \begin{bmatrix} s_{11} & s_{12} & s_{13} \\ s_{21} & s_{22} & s_{23} \\ s_{31} & s_{32} & s_{33} \end{bmatrix} \begin{pmatrix} a_1 \\ b_2 \\ b_3 \end{pmatrix} \tag{4.2}$$

(here the a_is and b_is are coefficients and not vectors). To simplify the analysis we assume that all coefficients are real. In the absence of magnetic field the reciprocity relations and the symmetry between terminals 2 and 3 imply $T_{31}=T_{13}=T_{12}=T_{21}=\alpha$. In addition, current conservation obviously requires $T_{21}+T_{31}+R_1=1$, where R_1 is the reflection probability of an incoming wave at contact 1. Thus, $R_1=1-2\alpha$. We also want $T_{23}=T_{32}=0$ as we want contacts 2 and 3 to remain uncoupled. Current conservation also implies that $T_{12}+T_{32}+R_2=1$, i.e. $R_2=1-\alpha$. The *S*-matrix coefficients are square roots of the transmission or reflection coefficients, so that the *S*-matrix must be of the form

$$
\begin{pmatrix} b_1 \\ a_2 \\ a_3 \end{pmatrix} = \begin{bmatrix} \sqrt{1-2\alpha} & \sqrt{\alpha} & \sqrt{\alpha} \\ \sqrt{\alpha} & \sqrt{1-\alpha} & 0 \\ \sqrt{\alpha} & 0 & \sqrt{1-\alpha} \end{bmatrix} \begin{pmatrix} a_1 \\ b_2 \\ b_3 \end{pmatrix}.
\tag{4.3}
$$

We can note that both the sum of the squared coefficients over any line or over any column is equal to *1*, due to the current conservation requirement. If we inject a wave of amplitude $a_1=1$ in contact *1*, the sum of the squared amplitudes of all resulting outgoing waves is equal to *1*, and the same is obtained with any input channel number. So it would seem that our *S*-matrix is physically meaningful. However, now try to inject one wave of amplitude $a_1=1$ at input 1 and one wave of amplitude $b_2=1$ at input 2, *simultaneously*. The sum of the squared output amplitudes is clearly larger than two, except for the trivial case $\alpha=0$, when the three arms are in fact not connected. Thus we have an output which exceeds the input, so that current conservation is not satisfied by this matrix! From this simple example we first realize that it is not possible to completely uncouple two identical arms if they are also coupled to a third one in a quantum-coherent *Y*-junction[1] (i.e. it is not possible to have $s_{23}=s_{32}=0$). However, most of all we also demonstrated that the condition

$$
\sum_{i \; or \; j} s_{ij} s_{ij}{}^* = 1 \; ,
\tag{4.4}
$$

albeit necessary, is *not* sufficient to obtain a physically acceptable *S*-matrix, i.e. which ensures current conservation. Let us now determine the full requirement. The sum of the square modulus carried over all output channels must be equal to the sum of the square modulus over all input channels, so that if the input is a vector *c* and the output a vector *d* they must fulfill the condition

---

1 A similar discussion with complex *S*-matrix coefficients would lead to the same conclusion.

$$\sum_i |d_i|^2 = \sum_i |c_i|^2 . \tag{4.5}$$

This is equivalent to

$$\begin{pmatrix} d_1^* & d_2^* & \cdots & d_n^* \end{pmatrix} \begin{pmatrix} d_1 \\ d_2 \\ \cdots \\ d_n \end{pmatrix} = \begin{pmatrix} c_1^* & c_2^* & \cdots & c_n^* \end{pmatrix} \begin{pmatrix} c_1 \\ c_2 \\ \cdots \\ c_n \end{pmatrix} \tag{4.6}$$

where the first term of the left-hand side is called the adjoint of d and is in general denoted as $d^\dagger$, so that equation (4.6) can be written in the more compact form

$$d^\dagger d = c^\dagger c . \tag{4.7}$$

Since we have $d=Sc$, equation (4.7) can be turned into

$$(Sc)^\dagger (Sc) = c^\dagger c . \tag{4.8}$$

Defining the adjoint matrix $S^\dagger$ as the conjugate transpose of S, so that we can write

$$\left(S^\dagger\right)_{ij} = S_{ji}^* , \tag{4.9}$$

we can easily check the equality

$$(Sc)^\dagger = \left(\sum_j s_{1j}^* c_j^* \quad \cdots \quad \sum_j s_{nj}^* c_j^* \right) = c^\dagger S^\dagger . \tag{4.10}$$

Thus combining equations (4.8) and (4.10) we obtain

$$(Sc)^\dagger (Sc) = c^\dagger S^\dagger Sc = c^\dagger c . \tag{4.11}$$

The equation above implies that the *S*-matrix must be unitary, i.e. it verifies the relation

$$S^\dagger S = I .$$

(4.12)

where *I* represents the identity matrix.

4.2. *S*-matrix combination rules ●

Consider two coherent samples put in series, each characterized by its own *S*-matrix, as illustrated by Figure 4.3. We would like to obtain the *S*-matrix corresponding to the two devices put in series, $S=S_1\otimes S_2$. We start from the set of relations corresponding to the two supposedly known *S*-matrices:

$$\begin{aligned}
b_1 &= r_1 a_1 + t_1' b \\
a &= t_1 a_1 + r_1' b \\
b &= r_2 a + t_2' b_2 \\
a_2 &= t_2 a + r_2' b_2
\end{aligned}$$

(4.13)

From the second and third equation above we can express *b* as a function of a_1 and b_2 and write $b(I\text{-}r_2 r_1')=r_2 t_1 a_1 + t_2 'b_2$. Then, substituting *b* by this expression in the first equation above leads to

$$b_1 = \left(r_1 + t_1'(I - r_2 r_1')^{-1} r_2 t_1\right) a_1 + t_1'(I - r_2 r_1')^{-1} t_2' b_2 .$$

(4.14)

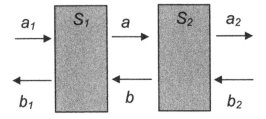

Figure 4.3. *Two devices put in series and their corresponding S-matrices*

Quite similar algebraic manipulations allow us to express a_3 as a function of a_1 and b_2, so that the combination rules which give the overall S-matrix coefficients as a function of that of S_1 and S_2 can be written as below:

$$r = r_1 + t_1'\left(I - r_2 r_1'\right)^{-1} r_2 t_1$$
$$t' = t_1'\left(I - r_2 r_1'\right)^{-1} t_2'$$
$$t = t_2\left(I - r_1' r_2\right)^{-1} t_1$$
$$r' = r_2' + t_2\left(I - r_1' r_2\right)^{-1} r_1' t_2' \quad . \tag{4.15}$$

Now looking at these rather cumbersome relations maybe the reader is wondering why not use matrices which are easier to combine, for instance L-matrices relating the left coefficients to the right ones, since in the latter case we just have to multiply the two matrices to combine them. The point is that with such a matrix we cannot calculate what we want, because what we want is precisely to inject something at one input and see what we get at all outputs. With an L-matrix we should already know what is reflected at the input to get all other input and output values.

4.3. A simple example: the S-matrix of a Y-junction ◍

Consider a Y-junction with symmetric outputs and just one occupied channel as in Figure 4.2. We already demonstrated in section 4.1 that it is not possible to uncouple terminals 2 and 3. This illustrates how counter-intuitive the result imposed by requiring S-matrix unitarity may be, since for such a junction we might naively expect terminals 2 and 3 to remain uncoupled. As an example let us determine the general form of such a junction with the help of equation (4.12). We already know that s_{23} and s_{32} are different from zero. Thus the symmetry of the outputs and equation (4.4) dictate the following form for the S-matrix coefficients:

$$S = \begin{bmatrix} \sqrt{1-2t^2}\,e^{i\theta_2} & t e^{i\theta_1} & t e^{i\theta_1} \\ t & \sqrt{1-t^2-s^2}\,e^{i\theta_4} & s e^{i\theta_3} \\ t & s e^{i\theta_3} & \sqrt{1-t^2-s^2}\,e^{i\theta_4} \end{bmatrix} \tag{4.16}$$

The matrix is defined within an overall phase factor not written above. t and s are positive numbers, the θ_is are arbitrary phase factors. We have $0 < t < 1/\sqrt{2}$ and $s^2 < 1-t^2$. Calculating the product $S^\dagger S$ and imposing $S^\dagger S = I$ leads to two equations:

$$t^2 + 2s\sqrt{1-t^2-s^2}\,\cos\!\left(\theta_3 - \theta_4\right) = 0$$

$$\left(1-t^2\right)e^{i\left(\theta_2-\theta_1\right)} + \sqrt{1-2t^2}\left(e^{-i\theta_4}\sqrt{1-t^2-s^2} + se^{-i\theta_3}\right) = 0 \qquad (4.17)$$

(the calculation of $S^{\dagger}S$ will show us that all diagonal elements are already equal to 1 and the two equations above correspond to the cancellation of the off-diagonal coefficients). The first equation above imposes $cos(\theta_3-\theta_4)<0$. Restricting our choice to real coefficients leads to $cos(\theta_3-\theta_4)=-1$, so that $exp(i\theta_3)=-exp(i\theta_4)=\pm1$, and the other exponential terms must be equal to ±1. The S-matrix is thus given by

$$S = \begin{bmatrix} \varepsilon_1\sqrt{1-2t^2} & \pm t & \pm t \\ t & -\varepsilon_2\sqrt{1-t^2-s^2} & \varepsilon_2 s \\ t & \varepsilon_2 s & -\varepsilon_2\sqrt{1-t^2-s^2} \end{bmatrix}. \qquad (4.18)$$

where $\varepsilon_1=\pm1$ and $\varepsilon_2=\pm1$. In each possible case, and considering s as the unknown, the upper equation (4.17) has four roots (which you can easily calculate), and only one of those roots is acceptable and satisfies the lower equation (4.17). Calculating all these roots leads to the general result

$$S = \begin{bmatrix} \varepsilon_1\sqrt{1-2t^2} & \pm t & \pm t \\ t & -\dfrac{1}{2}\left(\varepsilon_2 + \varepsilon_1\sqrt{1-2t^2}\right) & \dfrac{1}{2}\left(\varepsilon_2 - \varepsilon_1\sqrt{1-2t^2}\right) \\ t & \dfrac{1}{2}\left(\varepsilon_2 - \varepsilon_1\sqrt{1-2t^2}\right) & -\dfrac{1}{2}\left(\varepsilon_2 + \varepsilon_1\sqrt{1-2t^2}\right) \end{bmatrix} \qquad (4.19)$$

with $0<t<1/\sqrt{2}$, $\varepsilon_1=\pm1$ and $\varepsilon_2=\pm1$. For instance, taking an input reflection coefficient equal to zero corresponds to the four possible solutions

$$S = \begin{bmatrix} 0 & \pm1/\sqrt{2} & \pm1/\sqrt{2} \\ 1/\sqrt{2} & 1/2 & -1/2 \\ 1/\sqrt{2} & -1/2 & 1/2 \end{bmatrix} \text{ or } S = \begin{bmatrix} 0 & \pm1/\sqrt{2} & \pm1/\sqrt{2} \\ 1/\sqrt{2} & -1/2 & 1/2 \\ 1/\sqrt{2} & 1/2 & -1/2 \end{bmatrix}. \qquad (4.20)$$

From equation (4.19), we can also deduce that the ratio $|t/s|$ is a continuously decreasing function of t in the interval $0<t<1/\sqrt{2}$.

4.4. A more involved example: a quantum ring ⓪

Consider a ring as given in Figure 4.4. The quantum ring is formed by two Y-junctions as defined in section 4.3 and the two arms of the ring incorporate a quantum device, each characterized by an S-matrix (S_U and S_D, for "up" and "down"). Here we just consider the case without magnetic field ($B=0$), and assume that the junctions are symmetric and exhibit identical transmission parameters.

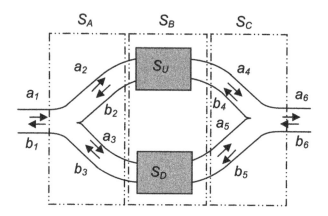

Figure 4.4. *A quantum ring with a quantum device in each arm*

The S-matrix of the junction at the input is given by equation (4.19), where we take $\varepsilon_1=\varepsilon_2=1$, and the S-matrix of the output Y-junction can also be easily derived from equation (4.19) by taking into account the fact that the input is now formed by the two symmetric branches. We have

$$S_A = \begin{bmatrix} c & t & t \\ t & -\frac{1}{2}(1+c) & \frac{1}{2}(1-c) \\ t & \frac{1}{2}(1-c) & -\frac{1}{2}(1+c) \end{bmatrix} \quad and \quad S_C = \begin{bmatrix} -\frac{1}{2}(1+c) & \frac{1}{2}(1-c) & t \\ \frac{1}{2}(1-c)) & -\frac{1}{2}(1+c) & t \\ t & t & c \end{bmatrix} \quad (4.21)$$

where $c=(1-2t^2)^{1/2}$. The S-matrix S_B giving parameters b_2, b_3, a_4 and a_5 as a function of a_2, a_3, b_4 and b_5 is also easily found and given by

$$S_B = \begin{bmatrix} r_u & 0 & t_u' & 0 \\ 0 & r_d & 0 & t_d' \\ t_u & 0 & r_u' & 0 \\ 0 & t_d & 0 & r_d' \end{bmatrix}. \tag{4.22}$$

Combining the three *S*-matrices S_A, S_B and S_C according to the rules prescribed in section 4.2 is rather tedious. However, it is not conceptually difficult. If the reader is fond of sudoku games, just be informed that carrying out the entire calculation is of the same difficulty level, but, owing to the time required, it can perhaps save you the purchase of a full sudoku magazine. Thus, we leave the calculation details as a useful exercise to the student, and just quote the final outcome. The complex transmission coefficient is given by

$$t_T = \frac{4t^2(t_u \alpha_d + t_d \alpha_u)}{\begin{aligned} &2(\alpha_u + \alpha_d) - s_u - s_d + s_{ud} + s_{du} \\ &- (1 - 2t^2)\big(s_d(1 - 2\alpha_u) + s_u(1 - 2\alpha_d) - s_{ud} - s_{du}\big) \\ &- 2\sqrt{1 - 2t^2}\big((\alpha_d - s_d)(1 - \alpha_u) + (\alpha_u - s_u)(1 - \alpha_d) + s_{ud} + s_{du}\big) \end{aligned}} \tag{4.23}$$

with the following notations for the symbols with index *u*

$$\begin{aligned} s_u &= r_u r_u' - t_u t_u' \\ \alpha_u &= 1 + r_u + r_u' + s_u \\ s_{ud} &= r_u r_d' - t_u t_d' \\ s_{du} &= r_d r_u' - t_d t_u' \end{aligned} \tag{4.24}$$

and taking the same conventions for index *d*. Note that the overall transmission is *not* the sum of the transmission through the two arms. If it was the case, we would expect t_T to be given by something like $t_T = t^2(t_u + t_d)$, and our result is clearly more complicated than that. In fact, we already demonstrated in section 4.1 that it is not possible to completely uncouple the two arms. Thus, an electron wave traveling in the upper arm will not be entirely transmitted to the output, but will be shared between a reflection part in the upper arm, a part transmitted to the output and a part transmitted into the lower arm. If the device is perfectly quantum-coherent there is clearly an infinite number of internal reflections and transmissions inside the ring, and our calculation result is just the product of this multiple wave scattering. It is all but obvious that equation (4.23) may actually look like a simple sum of two transmissions, and most often this is in fact not the case. However, by adequately choosing the value of the parameter t which results in a weaker coupling between the two arms, is it possible to find something which looks like a simple addition? By inspecting the S-matrix of the Y-junction we can see that t gives the smallest transmission between the two arms when it gets close to zero, i.e. when the input transmission is weak (see the discussion at the end of section 4.3). In such a case the input reflection c is close to 1, and can be approximated by $c \cong 1 - t^2$. Making a zero[th] order expansion of equation (4.23) gives us

$$t_T \cong t^2 \left(\frac{t_u}{\alpha_u} + \frac{t_d}{\alpha_d} \right)$$

(4.25)

and the outcome of a second order expansion is

$$t_T \cong \frac{4 t^2 \left(t_u \alpha_d + t_d \alpha_u \right)}{4 \alpha_u \alpha_d + 2 t^2 \left(\alpha_u \left(1 - s_d - \alpha_d \right) + \alpha_d \left(1 - s_u - \alpha_u \right) \right)}$$

(4.26)

From equation (4.25) we can see that *we can actually dissociate the two arms and then observe an output which is the quantum interference between the two momentarily separated paths, provided that the input transmission is small and the α factors do not vanish or become comparable to t.* This justifies the fact that in the literature, and in some other parts of this book (e.g. in the case of the Aharonov-Bohm effect), physically meaningful results can be obtained by modeling an interference resulting from two spatially separated paths simply by summing the complex transmissions corresponding to each path. However, equation (4.25) also shows that we do not necessarily obtain a result corresponding to a straightforward sum of the transmissions, but instead this sum is weighted by physical parameters depending on each arm. The situation closest to a simple summation is obtained when the reflection coefficients in each arm are in phase, large and their modulus is close to 1. Then $\alpha_u \cong \alpha_d \cong 4$ (omitting a common phase factor) and the overall transmission reduces to

$$t_T \cong \frac{t^2}{4} \left(t_u + t_d \right).$$

(4.27)

In addition, if the denominator depends on a parameter such as the electron energy (and we shall study such a point in the next chapter), care must be exercised to ensure that the denominators in equation (4.25) do not vanish at a particular value. If the factors α_u and α_d are as small as t, which can be the case if the transmissions in each arm are large, the zeroth order approximation is no longer valid. We shall see that this can be the case if one arm is subject to resonant tunneling, a very important phenomenon in the field of mesoscopic physics. When this is so, we must use the next non-vanishing order result, given by equation (4.26), or even the full equation (4.23). In such a case the contributions from the two arms cannot be dissociated. As a consequence, when we carry out similar calculations with specific scattering matrices corresponding to each arm, we should always keep in mind the possibility of obtaining results which can be *strongly* counter-intuitive. For example, we shall see in section 5.8, devoted to an important interference effect known as Fano resonance, that by identifying one of the devices in the two arms with a resonant tunneling structure, and if the phase coherence length is much larger than the ring dimension, a completely quantum-coherent device such as our quantum ring does not necessarily give rise to such a specific resonance. However, if both dimensions are similar, the electrons are "partially" coherent but cannot experience multiple reflections and transmissions, so that it is better to sum independently the two transmissions, and in this case we can almost always predict a Fano-like resonance. We shall thus conclude this section by insisting upon the fact that the *nature* of the observed quantum interference crucially depends on the ratio between the phase coherence and device lengths.

4.5. A final more complex example: solving the 2D Schrödinger equation ○

4.5.1. *Calculation principle*

The *S*-matrix formalism is involved in one of the simplest numerical methods developed for solving the Schrödinger equation in an open 2D structure. Although it does not enable us to account for electron-electron interactions in a simple way, it constitutes a numerical procedure which is fast, efficient and relatively simple. It allows us to obtain a first estimation of the transmission coefficient magnitude, and to assess the wave function amplitude *versus* position in a 2D quantum-coherent device. If the reader is fond of programming, with a bit of effort and some time this section will allow them to develop their own numerical program, to be implemented in softwares such as Mathlab™, Mathematica™ or Mathcad™.

The calculation principle can be depicted as follows: consider a 2D device as illustrated in Figure 4.5. To calculate the wave function inside the structure, we are going to cut this device into thin slices oriented perpendicularly to the direction of propagation in the input and output leads (*x* axis). Then, requiring the continuity of the wave function and its derivative between two slices will give us a relation between the coefficients of the linear combinations of eigenstates in two successive slices, and give us the S-matrix defining the transmission between two slices. For this reason this technique is often called wave-matching analysis, and is in fact straightforwardly inspired by the numerical methods developed to calculate the electromagnetic field in waveguides[2]. In our case this method is slightly simpler because we only want to determine one scalar field (the wave function), instead of the electromagnetic fields $\vec{E}$ and $\vec{B}$ with three components and polarization.

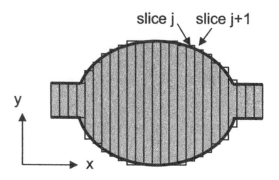

Figure 4.5. *A quantum device subdivided into parallel slices*

Combining all these intermediate S-matrices in series using the calculation rules given by equation (4.15), we finally obtain the overall S-matrix of our samples, and the wave function variations inside it. This section is in fact not essential to understand the physics of electron transport in nano-devices, but if the reader is fond of programming with a bit of effort it will allow them to obtain good images of 2D Schrodinger waves inside quantum-coherent devices. Not all details are discussed as is usual in the other sections of this book, and we just keep the reasoning essential to obtain the equations useful for the numerical solving of the Schrödinger equation; in addition, it is worth noting that a rigorous treatment should incorporate electron-electron interactions, and this is better achieved using Green's function techniques, which are more difficult to understand and are not treated in this introductory book. Thus, this section can be skipped on the first reading and might even receive a double symbol ∞.

If the slices are thin enough, in each particular slice we can neglect the variation of the potential in the x direction. If we know how to solve the Schrödinger equation

---

2 A more formal and complete presentation, enabling us to solve more complex cases, is given in [FER 97].

in the direction of the slice of index j, i.e. along the y axis, we can calculate all energy levels due to quantization of the movement along y, and affect to each of them an index p. Thus we can decompose any eigenfunction with energy $E_{p,j}$ in one slice as the product of a plane wave corresponding to wave propagation along $+x$ or $-x$ by a stationary wave, the solution of the Schrödinger equation along y:

$$\varphi_{j,p}(x,y) = \chi_{j,p}(y)\frac{e^{\pm ik_{j,p}x}}{\sqrt{k_{j,p}}} = \chi_{j,p}(y)\phi_{j,p}(\pm x) .$$

(4.28)

Here we have divided the plane wave by an additional factor $k_p^{1/2}$ to ensure that all eigenstates carry the same unit current, so as to simplify the S-matrix calculation. Let us now justify this procedure: a wave injected in an input channel with index i, with an amplitude α_i (i.e. of the form $\alpha_i exp(ik_i x)$), will be partly transmitted as a wave of amplitude α_j in an output channel of index j. The unitary S-matrix transmission coefficient is $t_{ij} = \alpha_j k_j^{1/2}/\alpha_i k_i^{1/2}$. To calculate t_{ij}, we see that if we define the coefficient $a_i = \alpha_i \times k_i^{1/2}$, so that the wave is of the form $a_i exp(ik_i x)/k_i^{1/2}$, t_{ij} is simply given by the ratio a_j/a_i. Now why should we affect a unit current to all input channels? The quantity which we are interested in is the signal intensity of a detector which could be placed anywhere inside the sample. If the electron waves were stationary this intensity would be proportional to the density of states at the considered energy times the square modulus of the normalized wave function. However, in our mesoscopic device we are dealing with running waves and in such a case the collecting signal would instead be proportional to the probability flow at a given point. For a given input channel and at a given point this is proportional to the product of the density of states by the electron velocity and the square modulus of the wave function. In one dimension the product of the first two terms is independent of the channel. Thus, we must affect the same initial weight to any channel, and this is just what is achieved with a wave as defined by equation (4.28).

If the potential is constant inside the structure $\chi_{j,n}(y)$ can be replaced by the particularly simple sine eigenfunctions of a 1D quantum well. We have $E = E_{j,n} + \hbar^2 k_{j,n}^2/2m$. Any particular wave in one slice is a linear combination of these eigenfunctions,

$$\psi_j(x,y) = \sum_p \left(a_p \phi_p(x)\chi_p(y) + b_p \phi_p(-x)\chi_p(y)\right),$$

(4.29)

and in the next slice (of index $j+1$) a wave function is a linear combination of the form

$$\psi_{j+1}(x,y) = \sum_n \left(a_n \phi_n(x) \chi_n(y) + b_n \phi_n(-x) \chi_n(y) \right), \tag{4.30}$$

where the integer n now indexes the quantized energy levels along y in the $(j+1)^{th}$ slice. n varies between 1 and N and p from 1 to P.

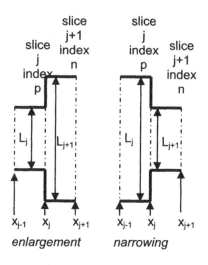

Figure 4.6. *Two possible situations for the boundary between two successive slices. Left: enlargement and right: narrowing*

The $(j+1)^{th}$ slice can be wider or narrower than the jth slice (enlargement or narrowing). First we shall consider the case of enlargement, as illustrated in Figure 4.6. Each function $\chi_{j,p}$ can be expanded as a function of the eigenfunctions of the next slice:

$$\chi_p(x,y) = \sum_{m=1}^{N} \langle \chi_n | \chi_p \rangle \chi_n(y) = \sum_{m=1}^{N} \chi_n(y) \int_0^{L_p} \chi_n^*(y) \chi_p(y) dy, \tag{4.31}$$

where we dropped the indexes j and $j+1$ since we affect index n only to slice $j+1$ and the index p only to slice j. This gives for the function ψ_j

$$\psi_j(x,y) = \sum_{p=1}^{P} \left(\sum_{n=1}^{N} \langle \chi_n | \chi_p \rangle \chi_n(y) \right) \left(a_p \phi_p(x) + b_p \phi_p(-x) \right). \tag{4.32}$$

By inverting the order of the summation this can be turned into

$$\psi_j(x,y)=\sum_{n=1}^{N}\left(\sum_{p=1}^{P}\langle\chi_n|\chi_p\rangle\big(a_p\phi_p(x)+b_p\phi_p(-x)\big)\right)\chi_n(y).$$

(4.33)

Here we must focus on an important point. If only a few number of modes are populated in one slice, we cannot expand the corresponding wave function from one slice as a function of the propagating eigenfunctions of the next slice only, as their number is too small to obtain a correct expansion. Thus, we must also involve wave functions with a higher index, so as to obtain a good numerical precision. For typical device geometries this may involve a few hundred waves. This means that many of these waves are evanescent, since above a given value of the index n the bottom subband energy lies above the Fermi level, and the wavevector $k_n=(2m(E_F-E_n))^{1/2}/\hbar$ is purely imaginary (in other words, the corresponding wave vanishes exponentially with the distance instead of propagating). The good point is that none of the formulae that we present here have to be changed, whether the wave is evanescent or propagating.

The continuity of the wave function at the boundary between slices j and $j+1$ imposes that the coefficients in front of each function $\chi_n(y)$ in equations (4.30) and (4.33) be equal for $x=x_j$, which leads to the linear system

$$\sum_{p=1}^{P}\big(a_p\langle\chi_n|\chi_p\rangle\phi_p(x_j)+b_p\langle\chi_n|\chi_p\rangle\phi_p(-x_j)\big)=a_n\phi_n(x_j)+b_n\phi_n(-x_j)$$

(4.34)

which includes N equations. Why expand the eigenfunctions of the narrower section as a function of that in the larger slice, and not the opposite? There is in fact a pretty good reason for this: at the enlargement, the wave function in the larger slice must vanish at the segments which correspond to the difference between the narrower and the wider section (see Figure 4.7), as they represent a hard wall for the wave function.

Since any wave in the narrower section takes values only in the narrower section, the scalar products in its expansion, calculated on the smaller section, are the same as those which would be obtained by prolongation of this wave function on an interval equal to the wider section, but affecting zero to any point not belonging to the narrower section. Thus this expansion is also that of the wave represented at the bottom of Figure 4.7, which correctly enforces the boundary condition requiring wave function vanishing along the hard wall segments.

The remaining missing equations are obtained by imposing the continuity of the wave function derivative. From equation (4.30), the derivative of the wave function in slice $j+1$ is given by

$$\frac{\partial \psi_{j+1}(x,y)}{\partial x} = i\sum_{n=1}^{N}\left(a_n k_n \phi_n(x) - b_n k_n \phi_n(-x)\right)\chi_n(y).\tag{4.35}$$

Projecting this derivative onto the eigenfunctions of slice j and inverting the summation order leads to

$$\frac{\partial \psi_{j+1}(x,y)}{\partial x} = \sum_{p=1}^{P}\chi_p(y)\sum_{n=1}^{N}\left(a_n \phi_n(x) - b_n \phi_n(-x)\right)k_n\left\langle \chi_n | \chi_p \right\rangle.\tag{4.36}$$

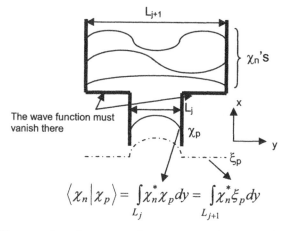

$$\left\langle \chi_n | \chi_p \right\rangle = \int_{L_j}\chi_n^* \chi_p\, dy = \int_{L_{j+1}}\chi_n^* \xi_p\, dy$$

Figure 4.7. *The expansion of a wave χ_p of a narrower section versus the eigenfunctions of the wider section is also that of its prolongation ξ_p in the wider section with zero values at the hard wall segments*

Note that in this case the normal derivative at the hard wall segments can take any value; this has been properly accounted for by expanding the wave function derivative in the wider slice *versus* the eigenfunctions in the narrower slice. The derivative matching equations are thus multiplied by zero within the hardwall segment, and avoid to impose any derivative value along it. Requiring the continuity of the derivative between the two slices leads to a second set of linear equations, which completes that given by equation (4.34):

$$k_p\left(a_p \phi_p(x_j) - b_p \phi_p(x_j)\right) = \sum_{n=1}^{N}k_n\left\langle \chi_n | \chi_p \right\rangle\left(a_n \phi_n(x_j) - b_n \phi_n(x_j)\right).\tag{4.37}$$

We have thus obtained a set of $(N+P)$ equations with $(N+P)$ unkowns, as we want to express the b_ps and the a_ns as a function of the a_ps and b_ns, in order to obtain the *S*-matrix between the two slices. Equations (4.34) and (4.37) can be re-expressed under the matrix form

$$M_1 \begin{pmatrix} b_p \\ a_n \end{pmatrix} + M_2 \begin{pmatrix} a_p \\ b_n \end{pmatrix} = 0 \quad , \tag{4.38}$$

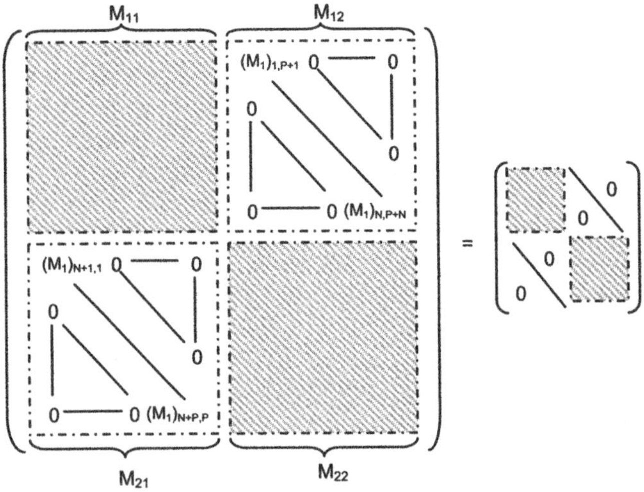

Figure 4.8. *Schematic representation of matrix M_1 introduced by equation (4.38)*

where M_1 and M_2 are $(N+P) \times (N+P)$ matrices. The M_1 and M_2 coefficients are defined from equations (4.34) and (4.37), and the *S*-matrix is simply obtained by calculating

$$S = -M_1^{-1} M_2, \tag{4.39}$$

the latter operation being numerically achieved by any routine dedicated to matrix inversion. Let us write down the coefficients of M_1, which are all obtained by inspection of equations (4.34) and (4.37). M_1 is a sparse coefficient matrix which can be subdivided into four parts M_{11}, M_{12}, M_{21} and M_{22}, as illustrated by Figure 4.8. Two of these submatrices are diagonal. We have

$$q \leq N, r \leq P \quad \Rightarrow \quad (M_1)_{q,r} = \left\langle \chi_q^{j+1} \middle| \chi_r^j \right\rangle \phi_r^j(-x_j)$$

$$q \leq N, r > P \quad \Rightarrow \quad (M_1)_{q,r} = \begin{cases} -\phi_q^{j+1}(x_j) & \text{if } r = P+q \\ 0 & \text{elsewhere} \end{cases}$$

$$q > N, r \leq P \quad \Rightarrow \quad (M_1)_{q,r} = \begin{cases} k_{q-N}^j \phi_{q-N}^j(-x_j) & \text{if } r = q-N \\ 0 & \text{elsewhere} \end{cases}$$

$$q > N, r > N \quad \Rightarrow \quad (M_1)_{q,r} = k_{r-P}^{j+1} \left\langle \chi_{q-N}^j \middle| \chi_{r-P}^{j+1} \right\rangle \phi_{r-P}^{j+1}(x_j) \quad,$$

(4.40)

where we introduced again additional indexes j or $j+1$ in order to characterize the transverse eigenfunctions and longitudinal wavevectors, so as to distinguish them from one slice to the next. Here we are going to give a trick to improve the efficiency of the inverting operation on M_1, which is useful if we want to write a numerical program. If $N=P$, the numerical precision and speed of the inverting operation can be greatly improved by noting that M_{12} and M_{21} are diagonal, and do not require a specific inversion routine to be inverted. Defining $A = M_1^{-1}$ and subdividing this matrix in A_{11}, A_{12}, A_{21} and A_{22} as for M_1, a straightforward calculation gives

$$A_{12} = \left[-M_{22} M_{12}^{-1} M_{11} + M_{21} \right]^{-1}$$
$$A_{11} = -A_{12} M_{22} M_{12}^{-1}$$
$$A_{21} = \left[M_{12} - M_{11} M_{21}^{-1} M_{22} \right]^{-1}$$
$$A_{22} = -A_{21} M_{11} M_{21}^{-1} \quad,$$

(4.41)

which reduces the inversion of a *2N×2N* matrix to inverting a set of *N×N* matrices. The coefficients of M_2 are also determined by a direct inspection of equations (4.34) and (4.37), and are given below:

$$q \leq N, r \leq P \quad \Rightarrow \quad (M_2)_{q,r} = \left\langle \chi_q^{j+1} \middle| \chi_r^j \right\rangle \phi_r^j(x_j)$$

$$q \leq N, r > P \quad \Rightarrow \quad (M_2)_{q,r} = \begin{cases} -\phi_q^{j+1}(-x_j) & \text{if } r = P+q \\ 0 & \text{elsewhere} \end{cases}$$

$$q > N, r \leq P \quad \Rightarrow \quad (M_2)_{q,r} = \begin{cases} -k_{q-N}^j \phi_{q-N}^j(x_j) & \text{if } r = q-N \\ 0 & \text{elsewhere} \end{cases}$$

$$q > N, r > N \quad \Rightarrow \quad (M_2)_{q,r} = -k_{r-P}^{j+1} \left\langle \chi_{q-N}^j \middle| \chi_{r-P}^{j+1} \right\rangle \phi_{r-P}^{j+1}(-x_j) \quad.$$

(4.42)

The procedure to calculate the *S*-matrix between two slices at a narrowing (see Figure 4.6) is very similar to the one described above in detail for the case of enlargement, and below we just summarize the corresponding equations. The continuity of the wave function and its derivative at the boundary between the two slices leads to the linear system

$$
\begin{cases}
a_p \phi_p(x_j) + b_p \phi_p(-x_j) = \sum_{n=1}^{N} \langle \chi_p | \chi_n \rangle \big(a_n \phi_n(x_j) + b_n \phi_n(-x_j) \big) \\
\sum_{p=1}^{P} k_p \langle \chi_n | \chi_p \rangle \big(a_p \phi_p(x_j) - b_p \phi_p(-x_j) \big) = k_n (a_n \phi_n(x_j) - b_n \phi_n(-x_j)).
\end{cases}
\tag{4.43}
$$

The coefficients of matrices M_1 and M_2 are given by

$$
q \le N, r \le P \implies
\begin{cases}
(M_1)_{q,r} = \begin{cases} \phi_q^j(-x_j) & if \quad q = r \\ 0 & elsewhere \end{cases} \\[2mm]
(M_2)_{q,r} = \begin{cases} \phi_q^j(x_j) & if \quad q = r \\ 0 & elsewhere \end{cases}
\end{cases}
$$

$$
q \le N, r > P \implies
\begin{cases}
(M_1)_{q,r} = -\langle \chi_q^j | \chi_{r-P}^{j+1} \rangle \phi_{r-P}^{j+1}(x_j) \\
(M_2)_{q,r} = -\langle \chi_q^j | \chi_{r-P}^{j+1} \rangle \phi_{r-P}^{j+1}(-x_j)
\end{cases}
$$

$$
q > N, r \le P \implies
\begin{cases}
(M_1)_{q,r} = k_r \langle \chi_{q-P}^{j+1} | \chi_r^j \rangle \phi_r^j(-x_j) \\
(M_2)_{q,r} = -k_r \langle \chi_{q-P}^{j+1} | \chi_r^j \rangle \phi_r^j(x_j)
\end{cases}
\tag{4.44}
$$

$$
q > N, r > N \implies
\begin{cases}
(M_1)_{q,r} = \begin{cases} k_{q-P} \phi_{q-P}^{j+1}(x_j) & if \quad q = r \\ 0 & elsewhere \end{cases} \\[2mm]
(M_2)_{q,r} = \begin{cases} -k_{q-P} \phi_{q-P}^{j+1}(-x_j) & if \quad q = r \\ 0 & elsewhere \end{cases}.
\end{cases}
$$

Submatrices M_{11} and M_{22} are diagonal, so that the inversion procedure of matrix M_I can be obtained by subdividing matrix $A = M_I^{-1}$ in four submatrices as in the case of enlargement. These submatrices are now given by the new set of relations

$$
\begin{aligned}
A_{11} &= \big[M_{11} - M_{12} M_{22}^{-1} M_{21} \big]^{-1} \\
A_{12} &= -A_{11} M_{12} M_{22}^{-1} \\
A_{22} &= \big[-M_{21} M_{11}^{-1} M_{12} + M_{22} \big]^{-1} \\
A_{21} &= -A_{22} M_{21} M_{11}^{-1} .
\end{aligned}
\tag{4.45}
$$

We have now almost finished describing the whole procedure to be followed in order to calculate the S-matrix between two slices. Let us now detail a useful point if we wish to write a numerical program. We must avoid any exponential growing of propagating numerical errors in the evanescent wave functions. For left-propagating states such functions are characterized by a purely imaginary, positive wavevector entering into equation (4.28). Thus, by replacing the form given in equation (4.28) by similarly-behaved wave functions of the type

$$
\begin{cases}
\phi_p^f(x) = \dfrac{1}{\sqrt{k_p}} \exp\!\big(ik_p(x - x_{j-1})\big) \\[2mm]
\phi_p^b(x) = \dfrac{1}{\sqrt{k_p}} \exp\!\big(-ik_p(x - x_j)\big)
\end{cases}
\quad and \quad
\begin{cases}
\phi_n^f(x) = \dfrac{1}{\sqrt{k_n}} \exp\!\big(ik_n(x - x_j)\big) \\[2mm]
\phi_n^b(x) = \dfrac{1}{\sqrt{k_n}} \exp\!\big(-ik_n(x - x_{j+1})\big)
\end{cases}
\tag{4.46}
$$

we ensure that the exponential terms of the evanescent states never exceed 1. Doing this, we must also replace the linear system represented by equations (4.34) and (4.37) by

$$
\begin{cases}
a_n \phi_n^f(x_j) + b_n \phi_n^b(x_j) = \displaystyle\sum_{p=1}^{P} \langle \chi_p | \chi_n \rangle \big(a_p \phi_p^f(x_j) + b_p \phi_p^b(x_j)\big) \\[4mm]
\displaystyle\sum_{n=1}^{N} k_n \langle \chi_n | \chi_p \rangle \big(a_n \phi_n^f(x_j) - b_n \phi_n^b(x_j)\big) = k_p \big(a_p \phi_p^f(x_j) - b_p \phi_p^b(x_j)\big).
\end{cases}
\tag{4.47}
$$

A strictly similar procedure can be applied to the case of narrowing and we do not reproduce it here. We now have all the equations needed to write down a numerical program, a task which can be rendered less time-consuming if we use a mathematical software with embedded numerical routines, and if we keep the same number of evanescent states in each slice. If we decide to write such a program, a good point to check the validity of the input parameters is to plot the transmission probabilities as a function of the number of evanescent states involved in the calculation. As we increase this number the transmission coefficients should converge towards constant values. Then just keep the minimum number required to achieve the desired accuracy.

4.5.2. Some numerical examples

Application of the formalism depicted in the previous section makes it possible to calculate the transmission probabilities and the wave function modulus inside arbitrary device geometries. Plotted in Figure 4.9 is the wave function modulus as a function of position in an InGaAs quantum ring. The plotted quantity is the sum of the square moduli over all occupied input channels. It is calculated for two different

Fermi energies, E_F=23 meV and 30 meV, respectively. Note that only the left-moving incoming states are represented, so that the figure should be properly symmetrized to give the local density of states at the Fermi energy. For such figures about 150 evanescent states per slice have been included.

A first point to note is that the electrons clearly do not follow anything like a linear trajectory along the path formed by each arm. Even with only one populated 1D channel in the entrance constriction, this would not necessarily be the case. Thus we realize that the simple-minded picture in which an electron follows two possible paths corresponding to the middle line of both arms will rarely give a correct dephasing value in realistic devices.

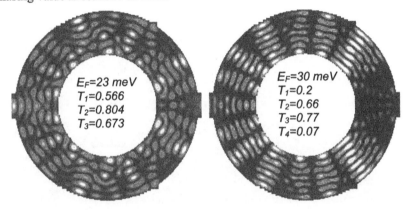

Figure 4.9. *Sum of the squared moduli of the wave functions in an InGaAs quantum ring, for two different E_F values. Effective mass m=0.04 m_0, input width: 70 nm, arm width: 125 nm. The transmission probabilities are given in the figures*

In addition, both the left and right figures illustrate the intrication between quantum coherence and semi-ballistic transport. Although the wave functions look like stationary states, exhibiting symmetric radial patterns with a size not directly related to the Fermi wavelength but instead linked to the device dimensions, those patterns with a higher probability density also exhibit short-range oscillations. Calculate half the Fermi wavelength; we should easily find $\lambda_F/2$=20.2 and 17.7 nm, respectively. If we divide the quantum arm width (*125 nm*) by those values we will find ratios of *6.18* and *7*, respectively. Now look more carefully at the figures: this is precisely the number of maxima that are observed in the radial patterns which are present in both cases. Thus, it is just as if the electron waves were preferentially bouncing back and forth on preferential trajectories, creating an interfering oscillation in the electron density with a period $\lambda_F/2$, as expected in a semi-classical approach. We have here some of the ingredients that we shall use in the next chapter, describing what is known as resonant tunneling: we can see this resonance either as a stationary state weakly coupled to the contacts through the constrictions, or as electrons passing a very long time bouncing back and forth in preferential areas inside the device before going out. We can also note that as mentioned in section 3.10, the electrons from the first input channel do not possess as many "transverse" kinetic energy, and are not transmitted as easily as the electrons in the next 1D channels.

Our second numerical example is that of a tapered constriction (Figure 4.10). Although the input lead is of limited width, the 25 occupied subbands and the tapered entrance are enough for the two 1D channels below the Fermi energy inside the constriction to be almost filled, so that the conductance is quite close to $4e^2/h$.

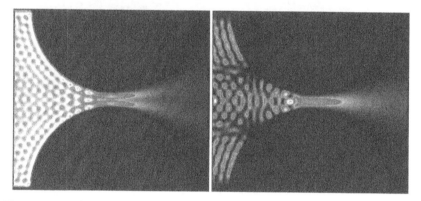

Figure 4.10. *Left: sum of the square moduli of the wave functions in an InGaAs tapered constriction; input width: 500 nm, EF=25 meV, minimum width: 45.5 nm. 25 input channels are occupied and only the left-moving incoming states are represented; G is extremely close to 4e2/h; right: third channel only*

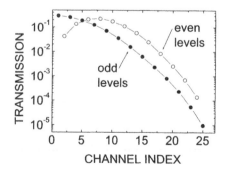

Figure 4.11. *Transmission versus input channel number for the device simulated in Figure 4.10; even and odd levels are plotted as two different curves*

It is instructive to study the variation of the transmission coefficient as a function of the 1D channel number in the input lead (Figure 4.11). The transmission strongly decreases with the index as the wave function spreads more and more over the whole width of the input lead, and channel transmissions with even and odd indexes fall on two different curves. It is quite clear that even for such a simple structure, the transmission eminently depends on the channel in the input lead. Reflection also leads to a strong wave mixing of the –*k* states in the leads. However, all these transmissions perfectly adjust so as not to exceed the maximum admissible occupation inside the constriction, and the conductance almost exactly obeys the Landauer formula for a perfect quantum wire. This might seem somewhat strange, because when we inject a wave from a given channel we might seem not to be concerned with what can happen when injecting something from another channel. Yet the way in which the *S*-matrix formalism has been formulated actually incorporates this feature.

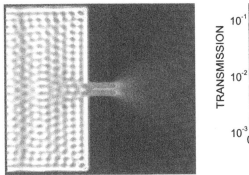 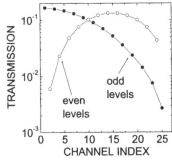

Figure 4.12. *Left: sum of the wave function square moduli in a square constriction with the same parameters as in Figure 4.10; right: transmission versus input channel number; even and odd levels are plotted as two different curves*

Furthermore, if we now examine the results obtained for a square constriction with identical dimensions, plotted in Figure 4.12, we can see that the distribution of the transmission across the various occupied input channels completely changes. However, the conductance is once again almost equal to $4e^2/h$.

4.6. Exercises

In this book most of the proposed exercises based on the use of *S*-matrices involve "tunneling" or "resonant-tunneling" matrices. Both phenomena are described in Chapter 5. Therefore, all exercises with *S*-matrices have been reported at the end of Chapter 5.

4.7. Bibliography

[FER 97] FERRY D.K., GOODNICK S.M., *Transport in Nanostructures*, Cambridge, Cambridge University Press, 1997.

[DAT 95] DATTA S., *Electronic Transport in Mesoscopic Systems*, Cambridge University Press, 1995.

Chapter 5

Tunneling and Detrapping

5.1. Introduction ●

In quantum mechanics the electrons can penetrate through energy barriers which may be higher than the electron kinetic energy, a phenomenon which is completely prohibited in classical mechanics. If the reader has taken an introductory quantum mechanical course they have probably already studied the tunneling phenomenon through a single barrier and derived the corresponding tunneling current. We will perform it once again to remind ourselves that in such devices the transmission coefficient can vary to a considerable extent as a function of the incident electron energy. This is somewhat different from the approach that we adopted in the previous chapter, where we made the assumption that this transmission coefficient did not vary so much, as the difference in chemical potential between the two terminals was kept small. In tunnel devices this is usually not the case, and the voltage applied to the sample can be much higher. In this section we shall focus on a specific tunneling phenomenon, known as resonant tunneling. Considering two single barrier tunneling devices in series, if the barriers are very far apart from one another, we expect the overall tunnel resistance to be multiplied by a factor of two. However, what happens if these two barriers come very close to one another, as for the device schematized in Figure 5.1? In fact the resistance is no longer the sum of the two tunneling resistances. We shall see that in such components the overall transmission is highly dependent on the distance separating the two barriers, and this can be easily understood if we consider this region as a quantum well coupled to the two contacts by tunnel barriers; we can readily imagine that something peculiar occurs whenever the energy of the electrons available for tunneling gets close to the quantized levels of the well.

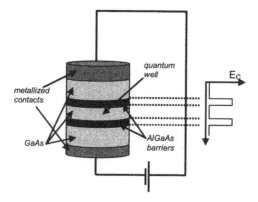

Figure 5.1. *A resonant tunneling device made from a double AlGaAs barrier enclosing a thin GAAs layer*

We shall demonstrate that in the *coherent* and *resonant* case, the double barrier device can be assimilated to the series combination of two transmission conductances, so that we can apply the Landauer formalism. To do this it must be understood that the electron phase is a key ingredient.

The operating principle of a double-barrier device can be understood from Figure 5.2. As long as a resonant level of the quantum well remains in an energy interval such that the corresponding levels in both contacts are filled (left scheme), there is no net current. Increasing the voltage bias leads to a situation for which the right levels are empty, and thus electrons can transit from one contact to the next (middle scheme). This is the resonant situation, which induces a sharply rising tunneling current. However, still enhancing the voltage leads to a situation in which either the resonant level is no longer aligned with left energy levels, so that we can observe a current decrease or, eventually, to a situation in which direct tunneling through the two barriers begins to prevail (right scheme), so that the overall current increases again.

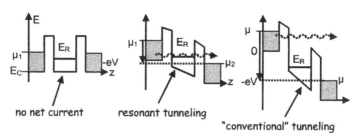

Figure 5.2. *Energy band diagram of a double-barrier tunneling diode for various bias voltages*

5.2. Single barrier tunneling ●

We consider a 1D energy barrier as in Figure 5.3. Looking back at the case of the finite quantum well discussed in section 2.5.1, it is readily seen that we can apply exactly the same principles to solve the tunneling case. In the parts where the potential is constant the solutions are plane or exponentially decaying waves, depending on the value of the electron energy with respect to that of the potential. With a single barrier and a single electron mode, a plane electron wave of the form e^{ikx} is partially transmitted and reflected so that the wave function is of the form

$$\begin{cases} \varphi(x) = e^{ikx} + re^{-ikx} & \text{before the barrier} \\ \varphi(x) = te^{ikx} & \text{after the barrier} \end{cases} \tag{5.1}$$

t and r are complex numbers. The squared modulus $|t|^2$ is the transmission probability T.

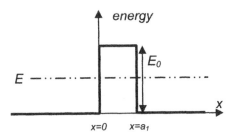

Figure 5.3. *Square potential barrier*

Keeping just a transmitted wave as in equation (5.1) and imposing the continuity of the wave function and its derivative at $x=0$ and $x=a$ provides a linear system whose solving gives the values of the transmission and reflection probabilities. This is what is proposed in most quantum-mechanical textbooks. However, this system is quite tedious to solve due to the number of intermediate unknowns to be eliminated. Here we propose a somewhat faster method [FER 66]. Consider a potential discontinuity at $x=0$ and inject a wave $\exp(ik_a x)$ on the left side (Figure 5.4). There will be a reflected part $r\exp(-ik_a x)$ and a transmitted part $t\exp(ik_b x)$. k_a and k_b are given by

$$k_a = \frac{\sqrt{2m(E - V_a)}}{\hbar} \qquad k_b = \frac{\sqrt{2m(E - V_b)}}{\hbar}. \tag{5.2}$$

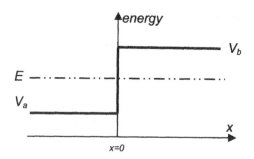

Figure 5.4. *Potential discontinuity*

Note that k_b can be real or imaginary, depending on the sign of E-V_b. Continuity of the wave and its derivative at $x=0$ straightforwardly gives

$$1 + r = t$$
$$k_a(1 - r) = tk_b \quad , \qquad (5.3)$$

from which we obtain

$$t_{ba} = \frac{2k_a}{k_a + k_b}, \qquad r_{ba} = \frac{k_a - k_b}{k_a + k_b} \quad . \qquad (5.4)$$

Be careful, as here t_{ab} is different from t_{ba}. In the case of a single barrier of magnitude E_0 as in Figure 5.3 we have two discontinuities and with the figure notations we can write

$$t_1 = \frac{2k_0}{k_0 + k_1}, \quad r_1 = \frac{k_0 - k_1}{k_0 + k_1}, \quad t_2 = \frac{2k_1}{k_0 + k_1}, \quad r_2 = \frac{k_1 - k_0}{k_0 + k_1}. \qquad (5.5)$$

A plane wave $exp(ik_0x)$ incident on the left barrier has a transmitted part $t_1exp(ik_1x)$ and a reflected part $r_1exp(ik_0x)$. The transmitted part impinges onto the second discontinuity, and a part $t_2t_1exp(ik_1a_1)exp(ik_0x)$ is transmitted. Here we included the phase shift $exp(ik_1a_1)$ acquired by going from $x=0$ to $x=a_1$. Part of the wave is reflected back to the first discontinuity. This last reflected part has in turn a fraction transmitted back through the first discontinuity, and a fraction which is reflected goes to the second discontinuity, etc. This "beam" division by each potential discontinuity can be schematically represented as in Figure 5.5.

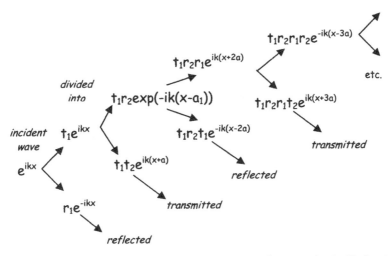

Figure 5.5. *The multiple division of a plane wave incident onto the double barrier*

It looks like the situation that we already faced when calculating Ohm's resistance from the Landauer formula, but here there is a crucial difference: we are not going to sum over the probabilities, which are real, but over the transmission and reflection coefficients, which are complex numbers. From Figure 5.9 it is clear that the part transmitted through the barrier is given by

$$
\begin{aligned}
t &= t_1 t_2 e^{ik_1 a_1} + t_1 t_2 r_2^2 e^{3ik_1 a_1} + t_1 t_2 r_2^4 e^{5ik_1 a_1} + \ldots + t_1 t_2 r_2^{2n} e^{ik_1(2n+1)a_1} + \cdots \\
&= t_1 t_2 e^{ik_1 a_1} \left(1 + \left(r_2^2 e^{2ik_1 a_1} \right)^1 + \left(r_2^2 e^{2ik_1 a_1} \right)^2 + \ldots \right)
\end{aligned}
\tag{5.6}
$$

which is a geometric series that can easily be summed as

$$
t = \frac{t_1 t_2 e^{ik_1 a_1}}{1 - r_2^2 e^{2ik_1 a_1}}
\tag{5.7}
$$

and turned into

$$
t = \frac{4 k_0 k_1 e^{ik_1 a_1}}{\left(k_0 + k_1 \right)^2 - \left(k_1 - k_0 \right)^2 e^{2ik_1 a_1}}
\tag{5.8}
$$

by substituting t_1, t_2 and r_2 by the expressions appearing in equation (5.5). The reflection coefficient can be calculated in the same way and is given by

$$r = \frac{(k_0^2 - k_1^2)(1 - e^{2ik_1a_1})}{(k_0 + k_1)^2 - (k_1 - k_0)^2 e^{2ik_1a_1}} .$$

(5.9)

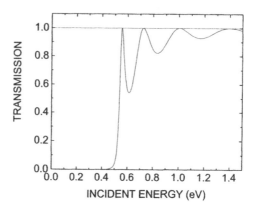

Figure 5.6. *Transmission through a square barrier of 0.5 eV with a width $a_1=10$ nm as a function of incident energy (effective mass $m=0.067\ m_0$)*

The transmission probability is therefore equal to

$$T = |t|^2 = \left| \frac{4k_0k_1}{(k_0 + k_1)^2 e^{-ik_1a_1} - (k_1 - k_0)^2 e^{ik_1a_1}} \right|^2$$

(5.10)

where k_0 is real and k_1 is real or pure imaginary. In Figure 5.6 this transmission probability is plotted as a function of energy for a particular barrier. Quantum-mechanical tunneling enables a non-zero transmission even when the energy of the incident electron is smaller than the barrier height. For E small in front of E_0, $k_1=i\beta$ is purely imaginary and $\exp(ik_1a_1)=\exp(-\beta a_1)$ is small, so that we can approximate the transmission by

$$T_{E \ll E_0} \cong \frac{16k_0^2\beta^2 e^{-2\beta a_1}}{(k_0^2 + \beta^2)^2} .$$

(5.11)

For $E>E_0$, k_1 is real and the oscillations are due to a constructive interference effect with maxima occurring whenever $k_1a_1=2n\pi$. This is readily seen by writing the transmission coefficient t as

$$t = \frac{4k_0k_1}{4k_0k_1\cos(k_1a_1) - 2i(k_0^2 + k_1^2)\sin(k_1a_1)} \qquad (5.12)$$

5.3. Two coherent devices in series: resonant tunneling ●

We shall calculate the conductance exhibited by two 1D mesoscopic devices put in series, assuming that there is phase coherence over the whole device (Figure 5.7).

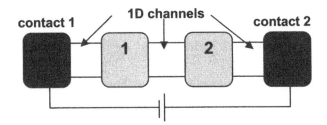

Figure 5.7. *Two coherent devices put in series*

We assume that we have two different barriers, each one being characterized by the coefficients t_1, r_1, t_2 and r_2, with a separation width a between the two barriers. Suppose that we put these two barriers close enough together so as to conserve phase coherence over the ensemble (Figure 5.8).

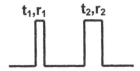

Figure 5.8. *The potential due to the two square barriers put in series*

A plane wave e^{ikx} incident on the left barrier has a transmitted part te^{ikx} and a reflected part re^{-ikx}. The transmitted part impinges on the second barrier, and part of it is transmitted ($t_2t_1e^{ik(x+a)}$), whereas part of it is reflected back to the first barrier ($r_2t_1e^{-ik(x-a)}$). This last reflected part has in turn a fraction transmitted back through the first barrier, and a fraction which is reflected by the first barrier and goes to the second one, etc. This "beam" division by each barrier can be treated exactly as in the case of a single barrier. It is schematically represented in Figure 5.9.

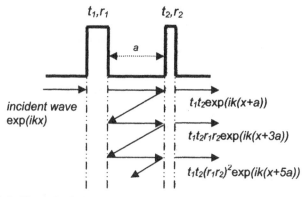

Figure 5.9. *The multiple division of a plane wave incident onto the double barrier*

The overall part transmitted through the double-barrier is given by

$$t = t_1 t_2 e^{ika} + t_1 t_2 r_1 r_2 e^{3ika} + t_1 t_2 (r_1 r_2)^2 e^{5ika} + t_1 t_2 (r_1 r_2)^3 e^{7ika} + ... \qquad (5.13)$$

This geometrical series is easily summed up and reduces to

$$t = \frac{t_1 t_2 e^{ika}}{1 - r_1 r_2 e^{2ika}} . \qquad (5.14)$$

Define the following quantities:

$$t_1 e^{ika/2} = \sqrt{T_1} e^{i\delta_1} \quad r_1 e^{ika} = \sqrt{R_1} e^{i\varphi_1}$$
$$t_2 e^{ika/2} = \sqrt{T_2} e^{i\delta_2} \quad r_2 e^{ika} = \sqrt{R_2} e^{i\varphi_2} \qquad (5.15)$$

If we replace the various terms in equation (5.14) by the expressions defined in equation (5.15), and if we write quantities such as e^{ikx} in the form $cos(kx)+i sin(kx)$, a few lines of trigonometric calculations lead you to

$$T = |t|^2 = \frac{T_1 T_2}{1 - 2\sqrt{R_1 R_2}\,\cos\varphi + R_1 R_2}, \tag{5.16}$$

where $\varphi = \varphi_1 + \varphi_2$. What does this angle represent? Suppose that a plane wave e^{ikx} propagates inside the two barriers. The part of this wave which bounces once onto the second barrier and once onto the first will pass again in the same position as the initial wave, but with a phase shift equal to $\varphi = \varphi_1 + \varphi_2$, since the wave is multiplied by a factor $r_1 r_2 \exp(2ika) = (R_1 R_2)^{1/2}\exp(i(\varphi_1 + \varphi_2))$. Thus, φ is the "phase shift acquired by bouncing once back and forth inside the barriers".

Now let us assume that we are close to a resonance, and that the transmission is quite weak (as is usually the case for a tunnel barrier). This means that $R_1 \approx R_2 \approx 1$. From equation (5.16), it is apparent that the transmission probability grows whenever $\cos\varphi$ gets close to 1. The resonance condition is thus

$$\varphi(E_L) = 2n\pi. \tag{5.17}$$

Around the resonance energy we can make a second order development of $\cos\varphi$

$$\cos\varphi \cong 1 - \frac{(\delta\varphi)^2}{2} \tag{5.18}$$

where $\delta\varphi = \varphi - 2n\pi$ is small, and a first order development of $\delta\varphi$ itself:

$$\delta\varphi \cong \frac{d\varphi}{dE_L}(E_L - E_R). \tag{5.19}$$

Close to a resonance, the denominator D in equation (5.16) can therefore be expressed as

$$D = 1 - 2\sqrt{R_1 R_2}\left(1 - \frac{1}{2}\left(\frac{d\varphi}{dE_L}\right)^2 (E_L - E_R)^2\right) + R_1 R_2. \tag{5.20}$$

$$= \left(1 - \sqrt{R_1 R_2}\right)^2 + \left(\frac{d\varphi}{dE_L}\right)^2 (E_L - E_R)^2 \sqrt{R_1 R_2} \quad .$$

Replace R_1 by $1-T_1$ and R_2 by $1-T_2$ in $\sqrt{R_1 R_2}$ and expand this quantity to first order:

$$\sqrt{R_1 R_2} = \sqrt{(1 - T_1)(1 - T_{2)}} \cong \sqrt{1 - T_1 - T_2} \cong 1 - \frac{T_1 + T_2}{2} \quad . \tag{5.21}$$

Now we define the two quantities Γ_1 and Γ_2 by

$$\Gamma_1 = T_1 \frac{dE_L}{d\varphi} \quad and \quad \Gamma_2 = T_2 \frac{dE_L}{d\varphi} \quad . \tag{5.22}$$

(of course this is not an arbitrary choice and the physical meaning of those two quantities will become clearer in the next section). Replace $\sqrt{R_1 R_2}$ by its first order expansion equation (5.21) in the denominator given by equation (5.20), and then into equation (5.16). Multiply the numerators and denominators of equation (5.16) by $(dE_L/d\varphi)^2$ so as to replace the terms T_1 or T_2 by Γ_1 and Γ_2. This leads to a formula giving the energy dependence of the transmission probability around the resonance energy:

$$T(E_L) \cong \frac{\Gamma_1 \Gamma_2}{\left(\frac{(\Gamma_1 + \Gamma_2)}{2}\right)^2 + (E_L - E_R)^2} . \tag{5.23}$$

Equation (5.23) is called the Breit-Wigner formula, and was initially derived in the context of nuclear physics. A mathematical function exhibiting a dependence such as equation (5.23) is also called a Lorentzian. Defining the Lorentzian of E as

$$Lor(E) = \frac{(\Gamma_1 + \Gamma_2)}{\left(\frac{\Gamma_1 + \Gamma_2}{2}\right)^2 + E^2} , \tag{5.24}$$

we can write

$$T(E_L) = \frac{\Gamma_1 \Gamma_2}{\Gamma_1 + \Gamma_2} Lor(E_L). \tag{5.25}$$

The transmission probability equation (5.25) is a simple function of the parameters Γ_1 and Γ_2, peaking around the resonance energy. The definition equation (5.22) of Γ_1 and Γ_2 does not imply that these parameters are constant in energy, but this is often assumed, and indeed *just around the resonance* energy E_R this is a rather well verified assumption. From equation (5.23) we can see that if the barriers are identical the resonance transmission reaches unity, but if the two barriers are different this is no longer the case. Note that even if the two barriers are identical, application of a non-negligible electric field changes their shape and increases the average height of one barrier with respect to the other, so that a unit transmission can no longer be obtained.

Now that we have derived the Breit-Wigner formula can we assign the resonance to a particular physical condition? If the reflection coefficients are close to 1 we can neglect the phase shift induced by the barrier when bouncing back and forth once, and the resonance condition is obviously given by $ka=n\pi$, so that the well width is equal to an integer number of half-electron wavelengths (see Figure 5.10). This is precisely the condition required for obtaining a bound state in an infinite well of the same width. We can also put it another way: let us assume that we are close to E_R: the resonance condition $\varphi(E_L)=2n\pi$ implies that the part of the wave which bounces back and forth is equal to the same wave multiplied by $(R_1R_2)^{1/2}\approx1$. Since the wave is the same after bouncing twice it is almost a stationary state of the quantum well considered alone, since it does not propagate, and Ψ is almost zero at the interfaces. E_R does correspond to a quantized well level, with $k\approx n\pi/W_{eff}$ (W_{eff} is an effective well width). For such a resonance condition we can of course expect the electrons to spend a lot of time inside the well. This is further detailed in the next section. For this energy the problem is also identical to the situation where the two barriers are now largely separated, but by an integer amount of the initial well width (see the scheme of Figure 5.10).

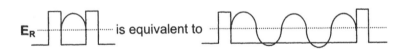

E_R is equivalent to

Figure 5.10. *Schematic illustration of the equivalence between shortly spaced and widely spaced double-barriers if the well width corresponds to an integer number of half wavelengths*

5.4. Physical meaning of the terms appearing in the resonant transmission probability ●

We are now ready to assign a physical meaning to the quantities Γ_1 and Γ_2 which appear in the Breit-Wigner formula equation (5.23). Although our two scatterers or energy barriers are not necessarily rectangular, we can assign an effective well width W_{eff} to the quantum well formed by the association of the two barriers, as schematized in Figure 5.11.

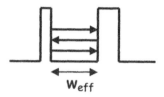

$$W_{eff}$$

Figure 5.11. *An electron wave impinging onto the second barrier bounces back and forth inside the well*

The phase shift acquired by a plane electron wave e^{ikx} bouncing back and forth inside the well (see Figure 5.11) is obviously equal to $\varphi \approx k \times 2w_{eff}$. Thus, from the definition of Γ_2 and with $E=\hbar^2k^2/2m$ we have

$$\Gamma_2 = T_2 \frac{\partial E_L}{\partial k} \frac{\partial k}{\partial \varphi} \approx T_2 \frac{\hbar v_G}{2w_{eff}} = \hbar T_2 v_{escape}. \tag{5.26}$$

The term $v_{escape}=v_G/2W_{eff}$ is the frequency at which the part of the wave bouncing back and forth impinges onto the second barrier so as to possibly transmit the associated electron towards the right terminal. This is called the attempt to escape frequency. We have thus obtained that the quantity $\hbar/\Gamma_2$ is equal to the reciprocal product of the transmission probability by the attempt to escape frequency:

$$\frac{\hbar}{\Gamma_2} = \frac{1}{T_2 v_{escape}}. \tag{5.27}$$

What is the physical meaning of this quantity? Suppose that we try to escape from a trap with a frequency v (Figure 5.12), fixed by the duration of each escape jump and time to fall back. Let us define the probability p of a jump being successful (it is a constant because we are still young and never tired, even if we do not put exactly the same energy into each jump). What is the expectation value $\langle \tau \rangle$ of the time spent in the trap?

Figure 5.12. *A person trapped in a pit and trying to escape*

We can either escape after the first attempt, or after the second, or after the third, etc. Each of these possibilities is incompatible with any other, and the average time spent in the trap is obviously equal to the expectation value

$$\langle \tau \rangle = \sum_i t_i p_i \qquad (5.28)$$

where p_i is the probability to escape after i attempts and t_i is the corresponding time spent inside the trap, equal to i/v. The probability of escaping after the first attempt is p_1. To escape after the second attempt, we must have been unsuccessful at the first attempt (with probability *(1-p)*) and successful at the second one (with probability *p*), so that the joint probability of those two events is *p(1-p)*, and the time spent in the well is *2/v*. The probability of succeeding after the third attempt is the joint probability of the events "fail at 1st attempt", "fail at the 2nd attempt" and "succeed at the 3rd attempt", which is equal to $p(1-p)^2$. The time spent in the well is thus equal to $3/v$. Equation (5.28) can thus be replaced by

$$\langle \tau \rangle = \frac{p}{v}\left(1 + 2(1-p) + 3(1-p)^2 + 4(1-p)^3 + ...\right). \qquad (5.29)$$

This last expression can be turned into

$$\langle \tau \rangle = \frac{p}{v}\left(1 + 2q + 3q^2 + 4q^3 + ...\right) \quad with \quad q = 1 - p. \qquad (5.30)$$

This is not exactly a geometric series, since we have the factors *1, 2, 3*, etc., but once we realize that this is nothing but the *derivative* of a geometric series, it becomes trivial to calculate it:

$$\langle \tau \rangle = \frac{p}{v}\frac{d}{dq}\left(1 + q + q^2 + q^3 + ...\right) = \frac{p}{v}\frac{d}{dq}\left(\frac{1}{1-q}\right) = \frac{p}{v}\frac{1}{(1-q)^2}, \qquad (5.31)$$

so that the average time spent inside the well is simply

$$\langle \tau \rangle = \frac{1}{vp}. \tag{5.32}$$

Compare this formula with equation (5.27). We have found that the physical meaning of Γ_2! $\hbar/\Gamma_2$ *is the electron lifetime in the well before escaping through barrier 2*. Reciprocally, $\Gamma_2/\hbar$ is the escape rate. Of course the same explanation is also applicable to Γ_1. Thus these two quantities, which were introduced in a seemingly arbitrary way, do indeed have a deep physical signification. From the way a double barrier is physically modified, we can now infer how these two coefficients are going to vary, and thus how the resonant transmission evolves.

Let us check all these results on an example corresponding to an exact calculation in the case of a double-square barrier. In Figure 5.13 we can find a good summary of all the features developed in this section:

– the transmission through a 1D resonant tunneling device with identical barriers exhibits resonances with unit transmission whenever the incident energy gets close to a resonant level of the quantum well;

– an asymmetry in barrier height or width reduces the value of these resonances;

– the width of the resonance depends on the strength with which the quantum well level is coupled to the continuum: as the incident energy increases, the barrier which prevents electron escape from the well is reduced, and the resonant state is more strongly coupled to the continuum, so that the lifetime decreases. From the uncertainty relation $\Delta E \Delta t \geq h$ this induces a spreading in energy of the resonance, as clearly observable in Figure 5.13. This point is further discussed in the next section (section 5.5) and discussed in a quantitative context in section 5.7.

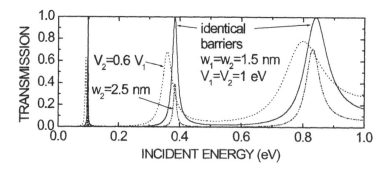

Figure 5.13. *Transmission through a double square barrier calculated with a mass $m=0.067\ m_0$ and a width between the barriers $a=6$ nm; V_1 and V_2 are the barrier heights*

Note that the case of the double square barrier is also more thoroughly expounded using exercises 5.11.2 and 5.11.3.

5.5. Tunneling current ●

In order to calculate the overall tunneling current at $T=0$ we have to sum over energy using the transmission probability of equation (5.23). In one channel labeled m, with quantized transverse energy E_T and longitudinal energy E_L ($E=E_T+E_L$), the current is given by

$$I_m = \frac{2e}{h} \int_{E_T}^{\mu_1} T_L(E) dE \tag{5.33}$$

and using equation (5.23) we obtain

$$I_m = \frac{2e}{h} \frac{\Gamma_1 \Gamma_2}{\Gamma_1 + \Gamma_2} \int_{E_T}^{\mu_1} Lor(E - E_R) dE . \tag{5.34}$$

Note that the lower bound of the integration interval is E_T if we face a situation as in Figure 5.2, but it may occur that μ_2 is higher than E_T. In such a case the lower bound becomes μ_2. We are led to integrate something similar to the scheme of Figure 5.14. Close to the resonance we can consider that the significant part of the transmission peak is included inside the integration interval, and the current is the same as that which we would obtain by integrating from minus to plus infinity. The primitive of a Lorentzian such as equation (5.24) is

$$\int \frac{\Gamma_1 + \Gamma_2}{\left(\frac{\Gamma_1 + \Gamma_2}{2}\right)^2 + E^2} dE = 2A \tan\left(\frac{2E}{\Gamma_1 + \Gamma_2}\right), \tag{5.35}$$

so that at the resonance (i.e. if the transmission peak is located within the integration interval) the tunneling current is equal to

$$I_m^{peak} = \frac{2e}{\hbar} \frac{\Gamma_1 \Gamma_2}{\Gamma_1 + \Gamma_2} . \tag{5.36}$$

If the voltage is not high enough we are in a situation such as the right schema of Figure 5.14, and then it is clear that the tunneling current will be lower than that given by equation (5.36). In the same way, if the voltage is too high, the lower bound E_T will pass the transmission peak and the current will start to decrease. So we expect the current to rise with the voltage, to reach a maximum I_{peak} and then to decrease. Of course, if another resonant level is reached we can expect a new peak, or when the voltage becomes very high the current will continuously increase because we shall have a "direct" carrier tunneling through the two barriers.

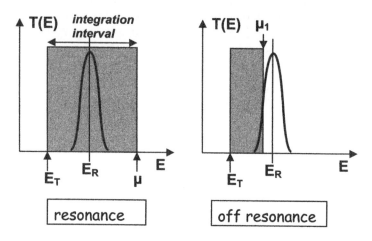

Figure 5.14. *Energy interval over which the transmission must be integrated to give the overall current*

Now consider carefully equation (5.36), and for the sake of simplicity let us assume that $\Gamma_1=\Gamma_2=\Gamma$, so that we have $I_m^{peak}=e\Gamma/\hbar$. Suppose that we continuously increase the coupling between the resonant level and the contacts (e.g. by reducing the barrier height or the barrier width), so that we continuously increase Γ as well. From the peak current formula, proportional to Γ, it would seem that we can increase the current to arbitrarily high values, in contradiction to the limited conductance of a point contact. The reader should pause before reading the lines below for at least a couple of seconds and try to find the reason for this paradoxical result.

Certainly the reader will have already found it by themself but let us write the explanation down: we can increase the coupling, but then *we increase the spreading in energy of the transmission* (see the Lorentzian formula and Figure 5.15). Eventually this spreading becomes larger than the separation between μ_1 and E_T, or μ_1 and μ_2. Thus, we cannot integrate over the total Lorentzian and *at max* (i.e. if the integration interval is around the transmission maximum, see Figure 5.15), we clearly have:

$$I_{max} = \frac{2e}{h} \max(\int_{\mu_1}^{\mu_2} T(E)dE) \cong \frac{2e}{h} T_{max}(\mu_1 - \mu_2) = \frac{2e^2}{h} T_{max}V , \qquad (5.37)$$

in agreement with Landauer's formula! There is no longer any paradox. It is worth noting that this is not a phenomenon specific to resonant tunneling, but a very general quantum-mechanical feature: from the uncertainty relationship $\Delta E \Delta t \geq h$, we can see that if we increase the coupling, i.e. if we decrease the lifetime, we must increase the spreading in energy.

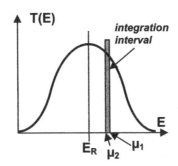

Figure 5.15. *Energy interval over which the transmission must be integrated to give the overall current; case of a quantum well strongly coupled to the leads*

5.6. Resonant tunneling in the real world ●

Figure 5.16 is an illustration of the phenomena which we described theoretically in the previous sections, which lead to the apparition of a negative dynamic conductance in the I-V characteristics of resonant-tunneling devices. Such curves are usually obtained from double-barrier AlAs/GaAs/AlAs tunneling diodes, possibly operated at low temperature. What can such a device be used for? If we have a negative resistance part, we know that if we insert such a component in an RC circuit we can obtain oscillations, and this is a typical resonant tunneling diode application, which aims at producing fast oscillators. Furthermore, resonant tunneling diodes are in fact among the fastest electronic components fabricated so far, and can be operated in the terahertz range. In Figure 5.16 the characteristic is roughly symmetric, and if we concentrate on the positive voltage part we can verify that the current increases up to the resonance point, and then decreases. The final increase is due to direct tunneling through the two barriers as the electric field reaches high enough values and to thermal excitation over the barriers.

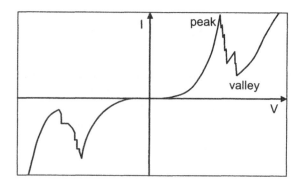

Figure 5.16. *Typical shape exhibited by the I(V) characteristics which can be measured from a double barrier formed by, e.g., an AlGaAs/GaAs/AlGaAs stack*

What we did not predict theoretically is the somewhat unstable part just after the resonance peak. In addition, the valley current is always much higher than expected from a simple coherent model. In fact it would be difficult to obtain an accurate fit of the I-V curve without taking into account two factors that we neglected in the previous sections: first, the electrons which are transmitted through the quantum well are not always coherent, and this induces scattering processes which are responsible for a higher valley current than expected in a simple model. Interface roughness is also quite important because in general application devices do not consist of simple 1D occupied channels at the contacts, but are 3D. Secondly, the electrons which are located inside the double-barrier create a potential which in the independent electron approximation would be given by the 1D Poisson equation

$$\frac{\partial^2 V}{\partial x^2} = -\frac{\rho(x)}{\varepsilon}. \tag{5.38}$$

This accordingly modifies a band diagram such as in Figure 5.2. In practice, the device engineer uses a self-consistent approach which can be summarized as follows: first he takes a simple band picture such as in Figure 5.2, and then he calculates the quantized levels by numerically solving the Schrödinger equation. Once he obtains all the energy levels, he is able to populate them with the relevant distribution function and to calculate the tunneling current. With the electron distribution he solves the Poisson equation, which gives him a new potential profile. With this new potential profile he solves the Schrödinger equation again, and repeats the same procedure as in the previous step. The whole scheme is repeated until a satisfying convergence has been reached, and this method is called a self-consistent calculation. In the case of a resonant-tunneling diode such calculations can even predict an intrinsic bistability in the negative conductance part: a diode may exhibit two stable states for a given applied bias, corresponding to two different charge states inside the well. Thirdly, instabilities are also induced by the oscillations created by the measurement circuit in the negative dynamic conductance part, and this is the reason for which the decreasing current part of the I-V characteristics represented in Figure 5.16 is somewhat jagged and noisy.

5.7. Discrete state coupled to a continuum ●

In section 5.4 we presented a quite simple treatment for obtaining the lifetime inside a quantum well coupled to a reservoir. However, the ease with which it can be understood has a number of negative counterparts. First, we have no idea about how to calculate this lifetime if we are given a potential and second, the picture of this semi-classical electron which bounces back and forth inside a small well is not always adapted to practical cases and more rigorous calculations. In addition, in nanostructures, electron transport is not always adequately described by a ballistic picture, and we may also be interested in studying devices in which not only does an electron spend a non-negligible time inside the device, but can in fact be trapped for a considerable amount of time before escaping. We need to know how to estimate that, if an electron stays too long a time in a device, as small as it may be, it will lose phase coherence and this must be taken into account to assess the transport and quantum interference properties. Both "memory-type" nanostructures and undesirable parasitic traps require such an adapted modeling. The tunneling lifetime also governs the properties of key electronic components such as quantum well-based optoelectronic devices, from optical modulators to lasers (for the latter class of devices the electron-hole recombination time must clearly not exceed the tunneling lifetime).

In this section we present a simple but more rigorous derivation of the tunneling lifetime and of the probability of finding an electron in a discrete state coupled to a continuum as a function of time. The formalism is applied to a simplistic system, in which we can take full advantage of treating only linear combinations of stationary states, but it can easily be generalized[1]. The calculation will first enable us to establish in a rather cheap way the physical expression universally used to calculate the transition probability per unit time between a discrete state and a continuum, better known under the name of Fermi's golden rule. Eventually, it will also allow us to introduce a remarkable effect known as Fano resonance (section 5.8). The latter occurs when the initial state is not only coupled to a continuum, but also to a bound state which is itself coupled to the same continuum. The overall coupling

---

1 In this section we develop a model originally proposed by U. Fano [FAN 35]; a quite similar but more detailed presentation of this model can be found in [COH 92], in the context of radiation-matter interaction, and from which this introduction is directly inspired.

between the initial state and the continuum is the result of the combined interference between the two escape paths which can be followed by the particle.

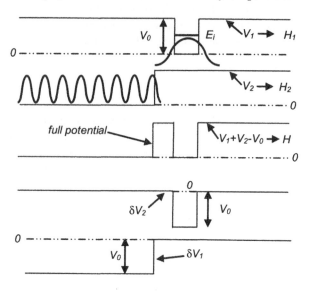

Figure 5.17. *A discrete state coupled to a continuum, with the potentials used to dissociate the full Hamiltonian into two solvable Hamiltonians corresponding to a discrete state and to the continuum, respectively*

Consider an electron initially prepared in a discrete quantum state $|\varphi_i\rangle$ with energy E_i, for instance in a quantum well as in Figure 5.17, for which we know how to calculate the energy level from the knowledge of the Hamiltonian H_1 with potential V_1. Consider also the continuum of quantum states which correspond to the Hamiltonian H_2 and potential V_2 appearing in Figure 5,17. We wish to study the stationary states of the Hamiltonian H described by Figure 5.17, in which our initial well is now coupled to a continuum, the potential of which is equal to $V=V_1+V_2-V_0$ (here V_0 is obviously a constant but this is not required by the demonstration). The point is that H can be expressed either as $H=H_1+\delta V_1$, where $\delta V_1=V_2-V_0$ takes values different from V_1 only on the interval where the wave function of the discrete state vanishes, or as $H=H_2+\delta V_2$, where δV_2 takes values different from zero only on the interval over which the wave functions of the continuum states vanish[2].

---

2 In textbooks this problem is most often treated with a full Hamiltonian $H=H_0+V$, where *both* the discrete state and the continuum states are common eigenstates of H_0, and a coupling potential V is adiabatically imposed to couple these states. It is in fact quite difficult to identify such a Hamiltonian H_0 and perturbation V in a practical situation such as illustrated by Figure 5.17. This is the reason why we use a slightly different mode of presentation. The overall treatment and the final results remain the same anyway.

The continuum is represented by equally distant energy states with energy $E_k = k\delta E$ (as in a harmonic potential), which we design as $|k\rangle$ and where k is an integer. To alleviate the derivations we can assume that the matrix elements between the discrete state and any of the continuum states are all equal, without too serious a loss of generality. Thus, we have

$$\langle k | H_2 | k \rangle = E_k = k\delta E$$
$$\langle \varphi_i | H_1 | \varphi_i \rangle = E_i$$
$$\langle k | H | \varphi_i \rangle = E_i \langle k | \varphi_i \rangle + \langle k | \delta V_1 | \varphi_i \rangle = v = \langle \varphi_i | H | k \rangle$$
$$\langle \varphi_i | H | \varphi_i \rangle = E_i + \langle \varphi_i | \delta V_1 | \varphi_i \rangle \cong E_i$$
$$\langle k | H | k \rangle = E_k + \langle k | \delta V_2 | k \rangle \cong E_k \qquad (5.39)$$
$$\langle k | H | k' \rangle = \langle k | \delta V_2 | k' \rangle \cong 0$$

where the simplifications made consist of neglecting the terms where waves corresponding to only one initial potential (V_1 or V_2) are implied in a matrix element and both waves take small values on the fraction of space where the perturbation potential is appreciably different from zero (see Figure 5.17). The only matrix element with cross products between the continuum and the discrete state which survives to this approximation is $\langle k | \delta V_1 | \varphi_i \rangle$, since in this case only the discrete state wave function takes small values when V_1 is substantial (see Figure 5.17). This is our coupling term. Design by $|\psi_f\rangle$ and E_f the stationary states and the energies of the full Hamiltonian, respectively[3]. We thus have

$$H | \psi_f \rangle = E_f | \psi_f \rangle \qquad (5.40)$$

Assume that one such state can still be developed on the basis formed by the $|k\rangle$s and $|\varphi_i\rangle$:

$$| \psi_f \rangle = \sum_k \langle k | \psi_f \rangle | k \rangle + \langle \varphi_i | \psi_f \rangle | \varphi_i \rangle \qquad (5.41)$$

---

3 Be careful where "f" stands for "final". E_f is the "final" energy value and is *not* the Fermi energy E_F.

Projecting equation (5.40) onto the state $|\varphi_i\rangle$ or a state $|k\rangle$ gives the relations

$$\left\langle k\left|H\right|\psi_f\right\rangle = E_k\left\langle k\left|\psi_f\right.\right\rangle + v\left\langle\varphi_i\left|\psi_f\right.\right\rangle = E_f\left\langle k\left|\psi_f\right.\right\rangle$$

$$\left\langle\varphi_i\left|H\right|\psi_f\right\rangle = \sum_k v\left\langle k\left|\psi_f\right.\right\rangle + E_i\left\langle\varphi_i\left|\psi_f\right.\right\rangle = E_f\left\langle\varphi_i\left|\psi_f\right.\right\rangle \qquad (5.42)$$

Assuming that E_k is always different from E_f (a property which we will verify at the end of the calculations), we first obtain from the upper equality in equation (5.42)

$$\left\langle k\left|\psi_f\right.\right\rangle = \frac{v}{E_f - E_k}\left\langle\varphi_i\left|\psi_f\right.\right\rangle. \qquad (5.43)$$

Then, by replacing the corresponding matrix element by the expression above in the bottom equality of equation (5.42) we can write the eigenvalue equation

$$E_f - E_i - \sum_k \frac{v^2}{E_f - E_k} = 0. \qquad (5.44)$$

Here we note (and admit without demonstration) the mathematical result

$$\sum_{k=-\infty}^{k=+\infty} \frac{1}{x - k} = \pi \cot(\pi x) , \qquad (5.45)$$

from which we can approximate the eigenvalue equation (5.44) by

$$E_f - E_i - \frac{\pi v^2}{\delta E}\cot\left(\pi \frac{E_f}{\delta E}\right) = 0. \qquad (5.46)$$

The energy values are given by the intersection points between the straight line $E_f\text{-}E_i$ defined by the first term of the left-hand side in equation (5.46), and the curve defined by the last term of the left-hand side in equation (5.46), as illustrated by Figure 5.18. From this figure it is clear that no E_f can be equal to any E_k: there is exactly one E_f value between each adjacent E_k value. It is also clear from the figure that when initially the E_k value is very far from E_i, the corresponding E_f value is extremely close to E_k, so that far from the initial discrete level energy the continuum states are not appreciably perturbed.

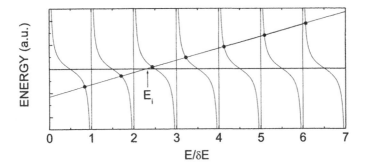

Figure 5.18. *Plot of the left-hand side and right-hand side of equation (5.46) according to energy*

If we define the frequency $\Omega=\Gamma/\hbar$ by

$$\Omega = \frac{\Gamma}{\hbar} = \frac{2\pi v^2}{\hbar\,\delta E} \ ,$$ (5.47)

the argument inside the Atan is $\Gamma/2(E_f\text{-}E_i)$ and we see that the difference between E_k and E_f is appreciable if this argument is close to zero, so that $E_f\text{-}E_i\gg\Gamma$. From the fact that the wave function ψ_f is normed to unity we can write that

$$\left\langle \varphi_i \middle| \psi_f \right\rangle^2 + \sum_k \left\langle k \middle| \psi_f \right\rangle^2 = 1$$ (5.48)

which along with equation (5.43) leads to

$$\left\langle \varphi_i \middle| \psi_f \right\rangle = \frac{1}{\sqrt{1+\sum_k \dfrac{v^2}{\left(E_f - E_k\right)^2}}}$$ (5.49)

and to

$$\left\langle k \middle| \psi_f \right\rangle = \frac{v}{E_f - E_k} \times \frac{1}{\sqrt{1+\sum_k \dfrac{v^2}{\left(E_f - E_k\right)^2}}}$$ (5.50)

By differentiating equation (5.44) we obtain

$$1 + \sum_k \frac{v^2}{\left(E_f - E_k\right)^2} = 1 + \frac{\pi^2 v^2}{\delta E^2}\left(1 + \cot g^2\left(\frac{\pi E_f}{\delta E}\right)\right),$$ (5.51)

Replacing the last term in the parenthesis of the right hand side by expressing it in a simpler form as a function of E_f which can be obtained through the use of equation (5.46), we can write the matrix element in equation (5.49) as

$$\left\langle \varphi_i \middle| \psi_f \right\rangle^2 = \frac{v^2}{v^2 + \frac{\Gamma^2}{4} + \left(E_f - E_i\right)^2}.$$ (5.52)

Here we note that the ratio between the first and second terms in the denominator of equation (5.52),

$$\frac{4v^2}{\Gamma^2} = \frac{1}{\pi^2}\left(\frac{\delta E}{v}\right)^2$$ (5.53)

vanishes when the level spacing δE is much smaller than the coupling term v, so that we can neglect the first term in front of the second and write

$$\left\langle \varphi_i \middle| \psi_f \right\rangle^2 = \frac{v^2}{\frac{\Gamma^2}{4} + \left(E_f - E_i\right)^2}.$$ (5.54)

In an energy interval ΔE large in front of δE, the number of discrete energy values E_f is equal to $\Delta E/\delta E$, since there is one such value in between each E_k value. If ΔE is small in front of Γ the probability of finding the electron in the state $|\varphi_i\rangle$ is given by

$$\Delta N = \frac{\Delta E}{\delta E}\left\langle \varphi_i \middle| \psi_f \right\rangle^2$$ (5.55)

so that in the continuum the density of states which results from the coupling with the discrete state is given by the Lorentzian

$$\frac{dN}{dE} = \frac{\Gamma/2\pi}{\frac{\Gamma^2}{4} + \left(E_f - E_i\right)^2}.$$ (5.56)

The discrete state is diluted into the continuum over an energy interval of order Γ, as anticipated in our previous analysis of resonant tunneling. Let us demonstrate the physical meaning of Γ by showing its connection with the electron lifetime in the discrete state. If at $t=0$ we prepare the electron in the discrete state we can expand it with respect to the stationary eigenstates of the full Hamiltonian as

$$|\psi(t = 0)\rangle = |\varphi_i\rangle = \sum_f \frac{v}{\sqrt{\dfrac{\Gamma^2}{4} + (E_f - E_i)^2}}|\psi_f\rangle \tag{5.57}$$

with the help of equation (5.54). At any time t, the state is a linear combination of the stationary states which evolves with time due to the interference between the complex energy terms (see equation (2.19)), and is given by

$$|\psi(t)\rangle = \sum_f \frac{ve^{-i\dfrac{E_f t}{\hbar}}}{\sqrt{\dfrac{\Gamma^2}{4} + (E_f - E_i)^2}}|\psi_f\rangle \tag{5.58}$$

By combining equations (5.57) and (5.58), we can write the probability amplitude to find the system in the initial state as

$$\langle\varphi_i|\psi(t)\rangle = \delta E \frac{\Gamma}{2\pi}\sum_f \frac{e^{-i\dfrac{E_f t}{\hbar}}}{\dfrac{\Gamma^2}{4} + (E_f - E_i)^2} \tag{5.59}$$

where we used equation (5.47) to replace v in the numerator, and put to good use the orthogonality of the basis states. With a vanishing energy spacing this expression can be advantageously replaced by a continuous integral:

$$\langle\varphi_i|\psi(t)\rangle = \frac{\Gamma}{2\pi}\int_{-\infty}^{+\infty} \frac{e^{-i\dfrac{E_f t}{\hbar}}}{\dfrac{\Gamma^2}{4} + (E_f - E_i)^2}dE_f \tag{5.60}$$

To calculate this, consider the complex integral

$$I = \oint_C \frac{e^{-izt}}{z_0^2 + z^2} dz \tag{5.61}$$

where the integral contour is defined in Figure 5.19 and $t>0$.

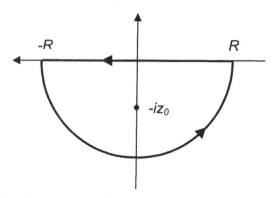

Figure 5.19. *Complex contour used to calculate the integral given by equation (5.61)*

The two poles are given by $z=\pm iz_0$ and to calculate I application of the residue theorem is straightforward. Only the pole $-iz_0$ is contained inside the contour, and the residue is easily obtained as the corresponding coefficient in its Laurent expansion, for the integrand can be written as

$$\frac{e^{-izt}}{z_0^2 + z^2} = \frac{e^{-izt}}{2iz_0} \left(\frac{1}{z - iz_0} - \frac{1}{z + iz_0} \right)$$

$$= \frac{e^{-izt}}{2iz_0} \frac{1}{z - iz_0} - \frac{e^{-z_0 t}}{2iz_0} \left(\frac{e^{-i(z + iz_0)t}}{z + iz_0} \right) \tag{5.62}$$

Expanding the exponential term inside the parenthesis leads to a coefficient (and thus a residue) equal to $-exp(-z_0t)/2iz_0$ in front of $1/(z+iz_0)$. For R tending to infinity we thus have

$$I = -2i\pi \times \frac{e^{-z_0 t}}{2iz_0} = \int_{arc} - \int_{-\infty}^{+\infty} \frac{e^{-izt}}{z_0^2 + z^2} dz \tag{5.63}$$

The sum over the curved arc tends towards zero when the radius R tends towards infinity because expressing the integral in polar coordinates leads to a prevailing exponential term $exp(Rtsin\theta)$ which is always negative, because θ varies between $-\pi$ and zero and thus the sine function can only take negative values. We finally obtain

$$\int_{-\infty}^{+\infty} \frac{e^{-izt}}{z_0^2 + z^2} dz = \frac{\pi}{z_0} e^{-z_0 t} \tag{5.64}$$

By comparing equation (5.61) to equation (5.60) and to its result equation (5.64) we obtain

$$\langle \varphi_i | \psi(t) \rangle = \exp\left(-i\frac{E_i t}{\hbar}\right) \exp\left(-\frac{\Gamma t}{2\hbar}\right) \tag{5.65}$$

and the probability of finding the electron in the initial state after a time t is equal to

$$\left| \langle \varphi_i | \psi(t) \rangle \right|^2 = \exp\left(-\Omega t\right) \tag{5.66}$$

This is the well known exponential decay which characterizes the disintegration of the initial discrete state into the continuum. The lifetime is $\hbar/\Gamma$ and $\Gamma/\hbar$ is the escape rate. Here we can extend our result to an arbitrary density of states to know how to calculate Γ in the general (and practical) case: clearly in our particular case $1/\delta E$ is the (constant) density of states per unit energy in the continuum, and v has been defined as the matrix element between the discrete state and the continuum, so that generalizing equation (5.47) leads to the relation

$$\Omega = \frac{2\pi}{\hbar} \rho(E) \left| \langle k | \delta V_1 | \varphi_i \rangle \right|^2 , \tag{5.67}$$

where $\rho(E)$ is an arbitrary density of states in the continuum. Ω is the transition probability per unit time as defined in the previous section. Equation (5.67) enables its quantitative evaluation, indicating that it is proportional both to the square matrix element between the initial and final state and to the density of final states. This relation is better known as *Fermi's golden rule*, and is one of the most useful quantum-mechanical formulae ever produced, for if we derived it in a particular case, its applicability is in fact quite general.

5.8. Fano resonance ●

Consider an electron prepared in a discrete state coupled to a continuum as in the previous section. Add to this system a bound state to which are coupled both the continuum and the discrete state. It turns out that the excitation spectrum resulting from the discrete state dilution into the continuum is the product of the interference between the two escape paths that can be followed by the electron, either through direct coupling or through an indirect coupling mediated by the bound state. The analytical shape of the spectrum is specific to the effect and thus constitutes a clear signature of its occurrence. For this reason Fano interference is without doubt one of the few essential landmarks with respect to quantum interference experiments ([FAN 35] and [FAN 61]), to be put on the same footing as the double-slit interference set-up, the Aharonov-Bohm effect or the weak localization phenomenon. In the double slit experiment, interference is evidenced through the spatially-dependent collected signal. In the Fano effect, interference is evidenced through the energy-dependent collected signal. It is worth noting that the discrete state need not be bound, so that the overall transmission through a device which incorporates in parallel both an open channel and a channel resulting from a resonant-tunneling state is a typical Fano resonance example. This explains why Fano profiles have in fact been obtained in a number of mesoscopic physics experiments, ranging from Aharonov-Bohm rings with a quantum dot in one arm to carbon nanotubes.

Consider a system similar to the discrete state coupled to a continuum as in section 5.7, but add to this another discrete state, which can be coupled either to the first discrete state with a matrix element

$$\left\langle \xi \middle| W \middle| \varphi_i \right\rangle = w_i \tag{5.68}$$

and to the continuum with a matrix element

$$\left\langle \xi \middle| W \middle| k \right\rangle = w \tag{5.69}$$

The problem can in fact be reduced to studying the transition probability between the discrete state $|\xi\rangle$ and the "new" continuum formed by the states resulting from the dilution of the discrete state $|\varphi_i\rangle$ into the continuum $|k\rangle$. Thus, we have to apply Fermi's golden rule between the initial state $|\xi\rangle$ and the final state $|\psi_f\rangle$. From equation (5.67) the transition probability from the discrete state $|\xi\rangle$ to the continuum is proportional to the squared modulus of the matrix element

$$\left\langle \xi \middle| W \middle| \psi_F \right\rangle = \sum_k \left\langle k \middle| \psi_f \right\rangle \left\langle \xi \middle| W \middle| k \right\rangle + \left\langle \varphi_i \middle| \psi_f \right\rangle \left\langle \xi \middle| W \middle| \varphi_i \right\rangle \tag{5.70}$$

where the right-hand side has been obtained by using the decomposition of this final state over $|\varphi_i\rangle$, the $|k\rangle$s and $|\xi\rangle$, and where we assumed as before that $\langle\xi|W|\xi\rangle=0$. Equation (5.70) shows that the probability amplitude is now the sum of two complex amplitudes corresponding to the continuum and to the discrete state, respectively. By inserting the expressions given by equations (5.50) and (5.52) into equation (5.70), and taking equation (5.44) into account, we obtain

$$\langle\xi|W|\psi_F\rangle = \frac{w_i v + w\left(E_f - E_i\right)}{\sqrt{v^2 + \Gamma^2/4 + \left(E_f - E_i\right)^2}} \qquad (5.71)$$

It is common practice to re-write this formula with reduced variables defined as

$$\varepsilon = \frac{2\left(E_f - E_i\right)}{\Gamma} \quad and \quad q = \frac{\delta E}{\pi v}\frac{w_i}{w} \qquad (5.72)$$

where q is a term which measures the asymmetry degree between the coupling of the discrete state $|\xi\rangle$ to the evanescent bound state $|\varphi_i\rangle$ and to the continuum $|k\rangle$, respectively. This enables us to turn the square of equation (5.71) into

$$\left|\langle\xi|W|\psi_F\rangle\right|^2 \cong \frac{|\varepsilon + q|^2}{1 + \varepsilon^2} , \qquad (5.73)$$

in which we neglected the term v^2 in the denominator of equation (5.71). Let us now examine this last expression in further detail: the magnitude of the numerator depends on the respective signs of the level energy and q, which may add or susbtract, depending on whether the energy is below or above the resonance and the sign of q is positive or negative. The excitation profiles into the continuum are thus expected to reflect this asymmetry, and some particular cases are illustrated in Figure 5.20.

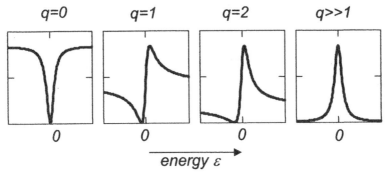

Figure 5.20. *Fano profiles for different values of the coupling parameter q*

For some limiting cases the shape of the spectrum can be either inverted (anti-resonance for $q=0$) or rendered symmetric ($q>>1$). This physical situation thus leads to a much larger variety of resonance spectra than the simpler phenomenon of resonant tunneling studied in section 5.5. It must also be stressed that an evanescent bound state can have an appreciable lifetime, so that observing such an effect in mesoscopic devices usually requires very low temperatures, to keep the wave function coherent over times longer than this lifetime.

5.9. Fano resonance in a quantum-coherent device ●

Here we are going to illustrate by a simple analytical example how a Fano resonance may occur in a quantum device. To make the analysis easier we shall restrict ourselves to the case of one input channel. A first ingredient is of course to define a structure in which the electron waves can follow two differentiated paths, and a second one is to include a resonant tunneling part. Consider a structure as in Figure 5.21, which incorporates both aspects. The electrons issued from the left contact can be transferred to the output either through the upper arm (which gives us a resonance equivalent to that of the bound state communicating with the continuum in section 5.8), or through the lower arm. First we have to define the S-matrix of the tunneling resonant structure. A typical matrix is[4]

$$S_u = \frac{1}{E + i\Gamma} \begin{bmatrix} E & i\Gamma \\ i\Gamma & E \end{bmatrix}.$$
(5.74)

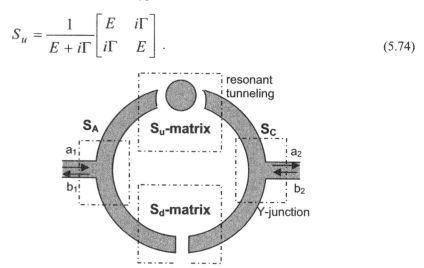

Figure 5.21. *Quantum ring exhibiting a Fano-like resonance*

---

4 Each time we introduce a new type of S-matrix in this section we should check that it is unitary and convince ourselves that it does correspond to what we are looking for. For instance here check that the transmission probability is formally similar to equation (5.23).

Note that here we use the symbol Γ to define an energy rather than an emission rate, so as to simplify the notations. The lower arm could be just a line inducing wave dephasing, the corresponding S-matrix being obviously of the form

$$S_d = \begin{bmatrix} 1 & e^{-i\theta} \\ e^{i\theta} & 1 \end{bmatrix}$$

(5.75)

However, it is more reasonable to also expect some wave attenuation in the lower arm, so that a possible S-matrix is

$$S_d = \begin{bmatrix} \sqrt{1-t_d^2} & it_d \\ it_d & \sqrt{1-t_d^2} \end{bmatrix} = \begin{bmatrix} r_d & it_d \\ it_d & r_d \end{bmatrix}.$$

(5.76)

where t_d is a number that is positive, real and lower than 1 (note that we did not choose an arbitrary phase as in equation (5.75), because the final result critically depends on it; this point is discussed later on). The S-matrices of the Y-junctions are given by equation (4.20). To calculate the overall transmission we can just apply the results that are thoroughly expounded in section 4.4, and in particular use the analytical transmission coefficients given by equation (4.25) or equation (4.26). First we voluntarily choose a Y-junction with a small transmission t, so that the full transmission can possibly correspond to the sum over two different paths, as explained in section 4.4, so that it is conceptually close to the principle of the Fano resonance. The various factors which enter into equation (4.24) are easily deduced from the matrix coefficients appearing in equations (5.74) and (5.76):

$$\alpha_u = \frac{4E}{E + i\Gamma}, \quad s_u = \frac{E - i\Gamma}{E + i\Gamma}, \quad \alpha_d = 2\left(1 + r_d\right), \quad s_d = 1.$$

(5.77)

However, we see that factor α_u defined by equation (4.24) gives us a quantity which vanishes at the resonance. Thus the zero[th] order approximation cannot be used, and we have to calculate the transmission using equation (4.26) instead of the simpler formula equation (4.25)! After inserting the expressions above into equation (4.26) and a few lines of algebra, we find an overall transmission coefficient

$$t_T = \frac{it^2 t_d}{2(1+r_d)(1-t^2)} \frac{E + \dfrac{\Gamma(1+r_d)}{2t_d}}{E + i\dfrac{\Gamma t^2}{4(1-t^2)}}, \qquad (5.78)$$

so that we obtain a transmission probability

$$|t_T|^2 = \lambda^2 \frac{|\varepsilon + q|^2}{1+\varepsilon^2}, \qquad (5.79)$$

where we have defined the reduced energy ε, the constant λ and the factor q as

$$\varepsilon = E\left(\frac{\Gamma t^2}{4(1-t^2)}\right)^{-1}, \quad \lambda = \frac{t^2 t_d}{2(1+r_d)(1-t^2)}, \quad q = \frac{2(1+r_d)(1-t^2)}{t_d t^2}. \qquad (5.80)$$

Equation (5.79) has the form of a Fano resonance. However, with small transmission coefficients t and t_d, the factor q is clearly much larger than 1, and we do not expect a clear Fano-like resonance shape (see the case $q \gg 1$ in Figure 5.20). For this example it is therefore preferable to make the calculation for a large transmission factor t. This is however achieved at the price of more involved analytical results, and a higher conceptual difficulty, because we know that for a large t value the final transmission is the result of multiple scattering inside the two arms, which are not clearly separated. However, in this case we are going to see that it is indeed easier to obtain a clearly asymmetric resonance line shape. Inserting the parameters given by equation (5.77) into the general formula equation (4.23) leads after some tedious but easy algebra to

$$t_T = \frac{2it^2\left(2t_d E + \Gamma(1+r_d)\right)}{\left((1-t^2)(5+3r_d)+3\sqrt{1-2t^2}\right)E + \Gamma\left(t_d\left(1-t^2-\sqrt{1-2t^2}\right)+i\times 2t^2(1+r_d)\right)} \qquad (5.81)$$

Since there is an energy offset between the real parts appearing in the numerator and the denominator, we do expect a Fano resonance. This is readily observable in Figure 5.22, for which the transmission is plotted versus energy, with the transmission in the lower arm as a parameter, and taking for the Y-junction transmission the maximum admissible value $t=1/2^{1/2}$.

Although it exhibits a Fano-like resonance, the case illustrated by Figure 5.22 is somewhat more complicated than the archetype expounded in section 5.8, due to the multiple scattering which occurs between both arms. Let us examine if by choosing a more reasonable matrix, i.e. putting some attenuation in the upper arm as well, we can clearly separate the transmission through both arms before realizing a quantum interference at the output and still obtain an outcome which exhibits a Fano-like resonance, even for small transmissions. Let us combine in series a resonant tunneling matrix as given by equation (5.74) and an "attenuation" matrix as given by equation (5.76). After some easy algebra, the usual combination rules (equation (4.15)) lead us to a new upper S-matrix of the form

$$S_u = \begin{bmatrix} \dfrac{E + i\gamma r_a}{E + i\gamma} & \dfrac{i\gamma t_a}{E - i\gamma} \\ \dfrac{i\gamma t_a}{E + i\gamma} & \dfrac{E - i\gamma r_a}{E - i\gamma} \end{bmatrix}, \tag{5.82}$$

where r_a and t_a are the reflection and transmission coefficients corresponding to an attenuation, and γ is equal to

$$\gamma = \frac{\Gamma}{1 - r_a}. \tag{5.83}$$

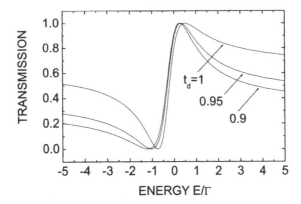

Figure 5.22. *Fano-like resonance in a quantum ring with resonant tunneling in the upper arm, calculated according to equation (5.82) and with $t=2^{-1/2}$*

The transmission is still resonant, but now the resonance magnitude is limited to t_a, instead of being equal to 1. If we calculate the factor α_u defined by equation (4.24) from the coefficients appearing in S_u (equation (5.82)), we now find that it is equal to

$$\alpha_u = 2\frac{2E^2 + \gamma^2\left(1 + r_a\right)}{E^2 + \gamma^2}$$

(5.84)

Provided that the attenuation is important (*i.e.* r_a is close to 1), we see that $\alpha_u \cong 4$. Thus, the inverse quantity never exhibits a pathological divergence, and we can apply the zero[th] order approximation found for the quantum ring transmission coefficient, to find

$$t_T \cong t^2\left(\frac{t_u}{4} + \frac{t_d}{\alpha_d}\right)$$

(5.85)

Choose for the lower arm an attenuation S-matrix as given by equation (5.76). If the lower arm transmission is weak we have $\alpha_d \cong 4$ and we obtain a transmission

$$t_T \cong \frac{t^2}{4}\left(\frac{i(\gamma t_a + Et_d) - \gamma t_d}{E + i\gamma}\right).$$

(5.86)

Since in the numerator the imaginary part is shifted in energy we do find a Fano-like resonance component, to be added to a conventional resonant tunneling part. This can be re-written as

$$\left|t_T\right|^2 \cong \frac{t^4 t_d^2}{16}\left(\frac{1 + \left|q + \varepsilon\right|^2}{1 + \varepsilon^2}\right).$$

(5.87)

where the energy $\varepsilon = E/\gamma$ is normalized and the coefficient q is defined as $q = t_a/t_d$. This form is quite close to equation (5.73). It is instructive to see that here the q factor is really physically equivalent to the one found in the derivation given in section 5.8, since it is the ratio between the transmissions corresponding to the two interfering paths that can be followed by the electrons. This factor thus reflects the asymmetry between these two paths. In addition, by adjusting this transmission ratio, Fano resonance line shapes are clearly no longer restricted to large transmission values. For this particular example, making $q=0$ leads to a constant transmission and not to an anti-resonance.

Note that the dephasing terms appearing in the S-matrix coefficients are essential to obtain a Fano-like resonance, which is of course not surprising since this phenomenon results from an interference effect. For instance, just consider a lower arm S-matrix S_d of the kind

$$
S_d = \begin{bmatrix} \sqrt{1 - t_d{}^2} & -t_d \\ t_d & \sqrt{1 - t_d{}^2} \end{bmatrix} = \begin{bmatrix} r_d & -t_d \\ t_d & r_d \end{bmatrix},
\tag{5.88}
$$

which only differs by transmission phase factors from equation (5.76), along with the upper matrix S_u defined by equation (5.74). As in the first case, we have to calculate the transmission coefficient using equation (4.26), and here we just quote the final outcome,

$$
|t_T|^2 = t_u^2 \beta^2 \frac{\Gamma^2 + \alpha^2 E^2}{\Gamma^2 \beta^2 + E^2},
\tag{5.89}
$$

where parameters α and β are defined as

$$
\alpha = 1 + \frac{2t_d}{t_u(1 + r_d)} \qquad \beta = \frac{t^2}{4(1 - t^2)}.
\tag{5.90}
$$

This transmission is always an even function of energy, and there is simply no Fano resonance at all! This explains why experimental data of Fano resonance are often obtained by varying a magnetic field through the ring, because the magnetic field value allows us to adjust the dephasing between the two arms.

5.10. Fano resonance in the real world ⓪

Fano resonance has now been observed in a number of quantum devices (see, e.g., [GOR 00], [KOB 02], [KOB 03], [KOB 04]), and here we reproduce data obtained from quantum rings incorporating a dot in one arm and purposely designed to exhibit and control a Fano resonance effect.

A scanning electron micrograph of the full device is shown in Figure 5.23. The conductance curves reproduced in the same figure very clearly exhibit asymmetric resonance peaks, and a Fano-like signature is obtained only when the upper arm transmission is not reduced down to zero, demonstrating thereby that an asymmetric shape is induced by the interference between the two arms. The shape of these peaks can be simply fitted by adjusting the value of the q factor. This shape varies qualitatively by passing from small to high q values. In addition, Fano resonance only occurs at the lowest temperatures, which are required to produce a large enough coherence length (see the conductance curves at the bottom of equation (5.28)). Note that the oscillation periodicity is due to the Coulomb blockade effect, which is the subject of a full chapter of this book.

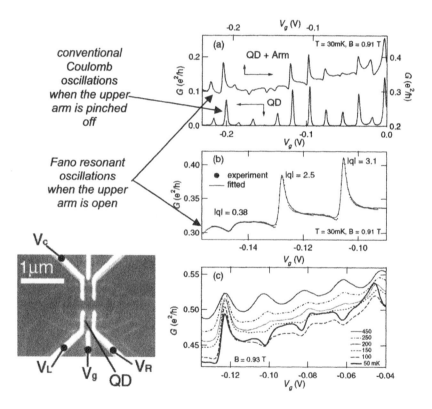

Figure 5.23. *Quantum ring with a dot exhibiting a Fano-like resonance.*
V_c is used to control the upper arm transmission; reproduced with permission
from K. Kobayashi et al., Phys. Rev. Lett. 88, 256801 (2002), copyright (2002) by the
American Physical Society

5.11. Exercises

5.11.1. *Exercise*

Consider a 1D quantum dot connected to two reservoirs, as in Figure 5.24.

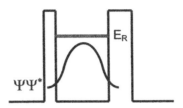

Figure 5.24. *A 1D quantum dot*

1) Assume that there is only one stationary state in the quantum dot with associated wave function ψ and energy E_R. Depict the variation of the current as a function of voltage and express this current as a function of the main dot parameters.

2) By approaching a metallized AFM tip close to the sample, we are able to locally perturb the potential as illustrated in Figure 5.25. The small perturbation V can be moved at will above any part of the sample, and its maximum is V_{max}. Assume that the potential is local and only slightly perturbative, so that first order perturbation theory applies. Explain which parameter is mainly affected and express its modification as a function of V.

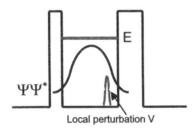

Local perturbation V

Figure 5.25. *A 1D quantum dot affected by some local perturbation potential*

3) Give the resistance change as a function of the perturbing potential.

4) Devise a method to deduce in an experimental way the wave function inside the dot.

5) Imagine the problems which can prevent a simple procedure such as the one devised in the previous question being straightforwardly applied.

6) What are the advantages and disadvantages brought by this procedure with respect to scanning tunneling microscopy?

5.11.2. Exercise

Consider a 1D resonant structure formed by a thin GaAs layer in between two AlGaAs layers, so as to obtain a conduction band profile as in Figure 5.26. The barrier offset is equal to 0.36 eV, and the isotropic effective mass is equal to $m=0.067\ m_0$. We assume that the effective mass is the same in both GaAs and AlGaAs (this exercise is intended to show that the resonant tunneling energy is given by a simple transcendent equation).

1) Calculate the dephasing φ acquired by bouncing back and forth once inside the well. Show that the resonance condition is given by

$$n\pi = kL - A\tan\left(\frac{2k\beta}{(k^2 - \beta^2)\tanh(\beta\, x_B)}\right)$$

where n is an integer.

2) Take for w the same value as the well width considered in exercise 2.19.2 (8 nm), and a=1.5 nm. Calculate the first two resonant levels of the 1D resonant structure (we can plot the transmission-energy curve with any appropriate software and find the resonance values on the graph, or use a numerical solver). Compare them with the values to be found in the finite well case.

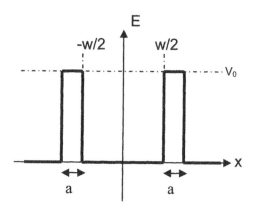

Figure 5.26. *A 1D resonant structure*

5.11.3. *Exercise*

The purpose of this exercise is to calculate the lifetime and the resonance width of the same double-barrier structure as in exercise 5.11.2 using Fermi's golden rule (equation (5.67)), and to assess its accuracy by comparing the result found with an exact numerical calculation. To calculate the lifetime we first calculate the escape rate from the left barrier by using the energy profile below, composed of a large quantum well separated from a second well by the same barrier as in the case of the resonant tunneling device, and with the same width (Figure 5.27).

1) Use Fermi's golden rule to calculate the escape rate Ω of an electron initially located in the right quantum well, and make L tend towards infinity to obtain the rate corresponding to the escape through the left barrier of the resonant tunneling structure (this calculation requires some time).

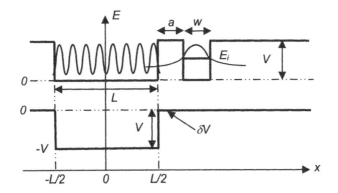

Figure 5.27. *A narrow quantum well coupled to a wide quantum well*

2) Plot the exact value of the transmission as a function of energy (use equation (5.14)). Below is what we should find for the first resonance peak (Figure 5.28).

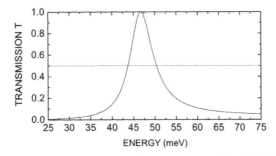

Figure 5.28. *Analytical plot showing the first resonance of the double barrier device*

3) Give the half transmission width of a resonant peak as a function of Γ.

4) Compare the numerical value of the first resonant level width, as calculated from Fermi's golden rule, with the width obtained from the plot (use the figure above or your own plot).

5.11.4. *Exercise*

In this exercise we are going to check that Y-junctions defined by equation (4.21) are in fact quite singular. Consider a quantum ring formed by putting in series two Y-junctions defined by equation (4.21), as illustrated by Figure 5.29. Assume that there is only one populated input channel and that the Fermi energy is adjusted so as to obtain no dephasing in each arm.

1) Show that using equation (4.23) to calculate the transmission results in an undetermined form.

2) Write the two S-matrix relations and calculate the overall transmission coefficient t_T.

3) What is the ratio between the amplitude of the wave in one arm and the input amplitude as a function of the input transmission t of the Y-junction? Discuss.

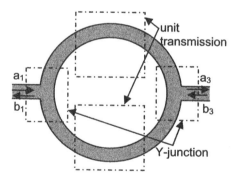

Figure 5.29. *A simple quantum ring*

5.11.5. *Exercise*

Consider a quantum ring with two resonant tunneling dots in each arm, as shown in Figure 5.30. Assume that there is only one occupied channel in the input lead, that the dephasing in each arm is the same and that they possess the same resonant energy, so that we can model each arm using S-matrices of the form

$$S_1 = \frac{1}{E + i\Gamma_1}\begin{pmatrix} E & i\Gamma_1 \\ i\Gamma_1 & E \end{pmatrix}, \qquad S_2 = \frac{1}{E + i\Gamma_1}\begin{pmatrix} E & i\Gamma_2 \\ i\Gamma_2 & E \end{pmatrix} .$$

The input and output Y-junctions are described by a matrix of the form given by equation (4.21), characterized by a single transmission coefficient t.

1) Calculate the transmission coefficient t_T of the whole device.

2) What is the behavior of the ring?

3) Give the electron lifetime in the ring. As the input transmission t of the Y-junction tends towards zero, what is the evolution of the lifetime? What is the variation of the resonance width? Give a physical comment.

4) We consider the largest input transmission approximation, i.e. $t = 1/\sqrt{2}$. Give the value of the overall transmission. What is the electron lifetime at the resonance?

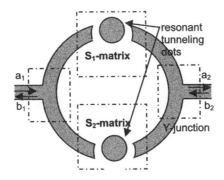

Figure 5.29. *A quantum ring with a dot in each arm*

5.11.6. *Exercise*

Consider the same ring as in exercise 5.11.4. In this exercise we are going to take into account the dephasing induced in each arm as a function of its length L and of the wavevector k.

1) What is the S-matrix of a straight wire of length L with one occupied input channel?

2) Calculate the transmission coefficient of the ring by assimilating propagation through one arm to that in a straight wire.

3) What is the resonance condition?

4) Plot the transmission probability as a function of wavevector or energy for some values of the input transmission t of the Y-junction.

5.12. Bibliography

[COH 92] COHEN-TANNOUDJI C., DUPONT-ROC J., GRYNBERG G., *Atom-photon Interactions: Basic Processes and Applications*, New York, Wiley, 1992.

[FAN 35] FANO U., "Absorption spectrum of the noble gases near the limit of the arc spectrum", *Nuovo Cimento*, vol. 12, 1935, p. 154-161.

[FAN 61] FANO U., "Effects of configuration interaction on intensities and phase shifts", *Physical Review*, vol. 124, no. 6, 1961, p. 1866-1878.

[FER 66] FERMOR J.H., "Quantum-mechanical tunneling", *American Journal of Physics*, 1966, p. 1168-1170.

[GOR 00] GORES J., GOLDHABER-GORDON D., HEEMEYER S., KASTNER M.A., "Fano resonances in electronic transport through a single-electron transistor", *Physical Review B*, vol. 62, no. 3, 2000, p. 2188-2194.

[KEL 95] KELLY M.J., *Low-dimensional Semiconductors*, Oxford, Oxford University Press, 1995.

[KOB 02] KOBAYASHI K., AIKAWA H., KATSUMOTO K., IYE Y., "Tuning of the Fano effect through a quantum dot in an Aharonov-Bohm interferometer", *Physical Review Letters*, vol. 88, no. 256806, 2002, p. 1-4.

[KOB 03] KOBAYASHI K., AIKAWA H., KATSUMOTO S., IYE Y., "Mesoscopic Fano effect in a quantum dot embedded in an Aharonov-Bohm ring", *Physical Review B*, vol. 68, no. 235304, 2003, p. 1-8.

[KOB 04] KOBAYASHI K., AIKAWA H., SANO A., KATSUMOTO S., IYE Y., "Fano resonance in a quantum wire with a side-coupled quantum dot", *Physical Review B*, vol. 70, no. 035319, 2004, p. 1-6.

Chapter 6

An Introduction to Current Noise in Mesoscopic Devices

6.1. Introduction ●

In the general introduction we stated that a mesoscopic device is a wonderful material support to probe the dual particle-wave nature of the electrons. However, if we look back at all that we did in the previous chapters only one of the two aspects is really necessary to obtain the results: up to now we just invoked the wave nature of the electrons to calculate the current or any other property. To illustrate that point remember the double-slit experiment with electrons, or a Geiger counter which records the radioactive disintegration of some instable materials. If we just measure the time-averaged signal at a given detector plate position or from the Geiger counter we only have access to the wave properties. However, of course we well know that we are looking at a collection of scattered points appearing on the plate detector, or hearing those "click-clicks" from the Geiger counter. Thus, the average properties are definitely given by the wave formalism, but the measurement sequence is formed by this succession of discrete events. Whenever we measure something, so that a macroscopic display at the end of a measurement line gives us a number, we must be sure that even if the average value does not reflect the particle nature of what we measure, when looking carefully at the fluctuations we are most probably going to find some "click-clicks" which are definitely a particle signature. We are not going to investigate in detail either where the quantum measurement processes take place, or if we need a human observer in front of the amperemeter for the quantum measurement to happen, but remember that if we measure a current we actually achieve a quantum measurement as defined in the section devoted to quantum mechanics reminders. Thus, the average value and its variation as a function of a macroscopic or microscopic parameter reflects the wave nature, but if we become interested in the noise then we can obtain some experimental evidence for the particle nature of the electrons. Once again we shall see that in this respect quantum mesoscopic devices may differ in a noticeable way from their macroscopic counterparts. In this chapter we shall first remind ourselves how the granular nature of the current-carrying charges creates shot noise in a macroscopic conductor with a barrier, and then we shall show that mesoscopic conductors exhibit appreciable deviations from the conventional behavior, due to the electron correlations which appear thanks to their transit through 1D channels.

6.2. Ergodicity and stationarity ◎

Imagine a game in which we have a huge number of dice, with the usual numbers $i=1$ to 6 on each face. If you take just one of those dice and throw it, the probability to get i is $p_i=1/6$. If we take the same dice and repeat the experiment many times (say N), we can calculate the relative frequency f_i at which we get i, equal to n_i/N, where n_i is the number of successful events. f_i tends to p_i as N tends to infinity. Now consider the full dice ensemble and ask one person per die to take one each. If they throw all dice simultaneously and if we calculate the relative frequencies for each n value, we will obviously find the same probability values as in the case with one die, and thus the same averages. The fact that the time sequence averages of the first case and the statistical averages of the second case are equal is called *ergodicity*. A random signal $x(t)$ is something which is measured as a function of time, but just as in the second case of the dice game we can imagine many realizations of it occurring simultaneously, each different but with the same statistical properties (see Figure 6.1, for which the N values are spanned over the s axis). Then if this random signal is assumed to be ergodic (which in most cases is a quite reasonable assumption) we can identify the statistical averages obtained by summing over the s axis and the time averages obtained for one of the signal realizations. For any time t_n we can thus define a random variable x_n with given statistical properties such as a probability density $p(x_n, t_n)$ to obtain the particular value x_n (p is obtained by averaging over the s axis). We can also calculate things such as the joint probability density $p(x_1, x_2, t_1, t_2)$, which represents the probability density of obtaining the particular value x_1 at t_1 and x_2 at t_2.

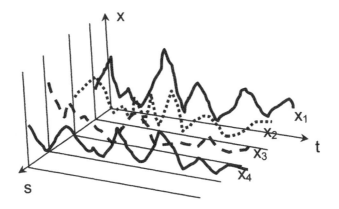

Figure 6.1. *Several realizations of a random ergodic signal x(t); statistical averages along the s-axis are the same as averages along the time axis for just one x_i*

A second useful assumption is that those statistical properties do not change with time, so that $p(x_n, t_n)$ remains the same at any time t_n (of course, in reality, it is not possible to obtain something which really is a permanent current regime, with stable average and higher order moments, because at some time in the past there has been a transient when we switched on the voltage, and our device has not been here for ever but was fabricated, so that this assumption is just a good approximation). This second property is called *stationarity*. If the signal is stationary, we can write that a joint probability just depends on the time interval $\tau = t_1 - t_2$, and not on the particular values of t_1 and t_2, i.e. $p(x_1, x_2, t_1, t_2) = p(x_1, x_2, \tau)$.

Ergodicity is particularly useful to calculate an important quantity known as the autocorrelation. The latter is defined by

$$R_x(\tau) = \lim_{T \to +\infty} \frac{1}{T} \int_{-T/2}^{+T/2} x(t) x(t + \tau) dt \quad . \tag{6.1}$$

The *statistical* autocorrelation can be written in the form

$$R_x(t_1, t_2) = R_x(\tau) = \iint_{-\infty}^{+\infty} x_1 x_2 \, p(x1, x2, \tau) dx_1 dx_2 \tag{6.2}$$

if the x_is are continuous or as

$$R_x(t_1, t_2) = R_x(\tau) = \sum_{x_i, x_j} x_i x_j P(x_i, x_j, \tau)$$

(6.3)

if the possible values are discrete. Then, *from the ergodicity property, we can identify the time autocorrelation defined by equation (6.1) to the statistical autocorrelation defined by equation (6.2) or equation (6.3).* This will be quite useful in the following.

6.3. Spectral noise density and Wiener-Khintchine theorem ○

Consider a random signal $x(t)$. Observe one of its realizations x_i on a time interval of width T between $-T/2$ and $T/2$, and define $x_i^{obs}(t)$ as equal to $x_i(t)$ in this interval and zero everywhere else ($x_i(t)$ is obtained as the limit of $x_i^{obs}(t)$ when $T \to +\infty$). $x_i^{obs}(t)$ can be Fourier transformed into a function $X_i^{obs}(f,T)$ given by

$$X_i^{obs}(f,T) = \int_{-T/2}^{+T/2} x_i^{obs}(t) e^{-2i\pi f t} dt$$

(6.4)

and of course if we want to investigate all the frequency components which form the overall signal this is the type of quantity that we have to calculate. The power spectral density of $x_i^{obs}(t)$ is defined as

$$\Phi_i^{obs}(f,T) = \frac{1}{T}\left|X_i^{obs}(f,T)\right|^2 = \frac{X_i^{obs}(X_i^{obs})^*}{T}$$

(6.5)

The average power of this signal is given by

$$P_i^{obs}(T) = \frac{1}{T}\int_{-\infty}^{+\infty}\left|X_i^{obs}(f,T)\right|^2 df .$$

(6.6)

The power spectral density $\Phi(f)$ of $x(t)$ is obtained by averaging over all realizations and passing to the limit:

$$\Phi(f) = \lim_{T \to +\infty}\left\langle \Phi_i^{obs}(f,T)\right\rangle.$$

(6.7)

This gives us the frequency distribution of the signal energy. The Wiener-Khintchine theorem states that the power spectral density $\Phi(f)$ of a stationary random signal $x(t)$ is the Fourier transform of the autocorrelation function:

$$\Phi(f) = \int_{-\infty}^{+\infty} R_x(\tau) e^{-2i\pi f \tau} d\tau \tag{6.8}$$

(for those interested by the demonstration see section 10.7). To assess noise properties it is most often more useful to investigate its characteristics in the frequency domain rather than in the time domain, and to do so we have to assess (either experimentally or theoretically) the noise power spectral density. This is just what we shall do in the next sections. The Wiener-Khintchine theorem enables us to use the same expression for calculating the power spectral density of deterministic and random signals. As we shall work it out in all the following examples, the correlation function and the power spectral density of random signals are in fact truly deterministic.

First of all, from equation (6.1) we can note that the autocorrelation value at $\tau=0$ is equal to the mean square of x:

$$R_x(0) = \lim_{T \to +\infty} \frac{1}{T} \int_{-T/2}^{+T/2} x^2(t) dt = \left\langle x^2 \right\rangle, \tag{6.9}$$

Secondly, when τ tends towards infinity it frequently occurs that the autocorrelation function of a random signal tends towards zero (there is no longer any causal relation between infinitely separated times).

An interesting property can be deduced if the appreciable values taken by the autocorrelation function are restricted to a finite time interval. We shall work out this property through the description of a simple example: consider an autocorrelation function with a triangular shape

$$R_x(\tau) = \left\langle x^2 \right\rangle \left(1 - \frac{\tau}{\tau_0} \right) \tag{6.10}$$

when $-\tau_0 < x < \tau_0$, and equal to zero elsewhere. This function takes appreciable values on the finite time interval $2\tau_0$. A straightforward calculation of its Fourier transform will give us

$$\Phi_x(f) = \left\langle x^2 \right\rangle \tau_0 \left(\frac{\sin(\pi \tau_0 f)}{\pi \tau_0 f} \right)^2, \tag{6.11}$$

from which we can deduce that if the frequency is much smaller than $1/\tau_0$ then the spectral density can be approximated as a constant and is given by

$$\Phi_x(f) = \langle x^2 \rangle \tau_0. \tag{6.12}$$

This formula can in fact be generalized to roughly approximate the noise intensity when the autocorrelation function takes appreciable values over a finite interval of width $2\tau_0$. When the power spectral density is constant over the considered frequency interval, the spectrum (or the noise) is said to be white.

6.4. Measured power spectral density ◉

In section 6.3 we obtained a noise formula for which the frequency can be negative, and these negative frequencies will also show up in the shot noise calculations that we are going to carry out. Here we are going to shed some light on this aspect. In fact the *measured* power spectral density is not the quantity given by equations (6.7) or (6.8), but is equal to exactly twice this value. The reason is as follows: first, when we measure a signal we always use some type of linear filter with a finite bandwidth B. Output component $y(f)$ at a frequency f is related to the input component $x(f)$ by an expression of the type $y(f)=H(f)x(f)$, where $H(f)$ is the gain of the filter. If the observed signal $x(t)$ is real, by writing equation (6.4) with $-f$ instead of f it is seen that its Fourier transform necessarily verifies $x(-f)=x^*(f)$, where the asterisk denotes complex conjugation. The filter output is of course also real so that we can write $y(-f)=y^*(f)$. Thus we also have $H(-f)x(-f)=(H(f)x(f))^*=H^*(f)x^*(f)$. We therefore conclude that a real pass-band filter exhibits two symmetric parts as in Figure 6.2.

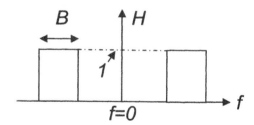

Figure 6.2. *Gain of a real pass-band filter versus frequency*

Measuring the power spectral density at a given frequency is equivalent to measuring the power output of an extremely narrow pass-band filter, and the real signal power measured at the filter output is the sum of the power corresponding to the negative and positive frequency bands[1]. The measured spectral density is thus given by

$$\Phi_{meas}(f) = 2\Phi(f),$$

(6.13)

The power spectral density $\Phi_{meas}(f)$ of a physical signal, measured at a positive frequency, corresponds to the sum of the densities calculated at f and $-f$. Thus at the end of each of our calculations we have to multiply our result by a factor of two.

6.5. Shot noise in the classical case ⓪

Consider an energy barrier in a conducting device, for example an old vacuum tube. Carriers thermally excited above the barrier at the cathode are randomly emitted and *ballistically* travel to the next electrode through the action of the electric field. Here we assimilate the electrons to point particles, and we also assume that the current through a section is a sum of Dirac peaks, each one corresponding to one electron crossing the section:

$$I(t) = \sum_n e\delta(t - t_n),$$

(6.14)

The t_ns are the random times at which one electron is crossing the section. For independent and uncorrelated electrons the t_ns are *independent* random variables. This is the case if electrons are not too numerous so that we can neglect electron interactions inside the barrier, and if the electron emission processes above or through the barrier are independent as well.

To calculate the power spectral density of this signal we shall assume that the Dirac peaks can be approximated as rectangular functions of width ε and magnitude $1/\varepsilon$, and eventually we shall calculate the result for the limit $\varepsilon \to 0$. The variation $x(t) = I(t)/e$ looks the same as in Figure 6.3. We define λ as the average number of peaks per unit time. Over a time T we have N peaks with $N = \lambda T$. Each t_n is uniformly distributed over the interval T. The probability that the time interval over which a given peak occurs includes a given time t is obviously equal to the ratio $p = \varepsilon/T$.

---

1 Here we implicitly assumed that the output signals of two filters operating at two different frequency bands with the same input signal are uncorrelated, and extended this result to separate frequency bands of a single filter; this can in fact be demonstrated.

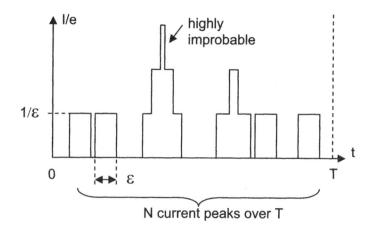

Figure 6.3. *Time variation of the current over T*

At a given time t the signal can take the discrete values 0, $1/\varepsilon$, $2/\varepsilon$, etc., depending on the number of peaks which have begun at times between t-ε and t. Thus, the probability $P(k/\varepsilon)$ that at a given time t the signal is equal to k/ε is given by the binomial law (we must place the peaks one by one between 0 and T and we must assess the probability of obtaining k successes and N-k failures; the binomial law which describes such a process is demonstrated in section 10.8):

$$P(k/\varepsilon) = C_N^k \left(\frac{\varepsilon}{T}\right)^k \left(1 - \frac{\varepsilon}{T}\right)^{N-k},$$
(6.15)

The ratio $P(k+1/\varepsilon)/P(k/\varepsilon)$ is given by

$$\frac{P(k+1/\varepsilon)}{P(k/\varepsilon)} = \frac{N-k}{N} \times \frac{1}{k+1} \times \left(1 - \frac{\varepsilon\lambda}{N}\right)^{-1} \times \varepsilon\lambda,$$
(6.16)

The two first terms of the right-hand side are obviously smaller than 1, so that when ε tends towards zero the ratio is always smaller than a quantity which tends to $\varepsilon\lambda$ and thus towards zero. Thus, when ε becomes very small the probability of obtaining k peaks present at the same time becomes negligible considering the probability of obtaining one, and the probability of just obtaining one is thus

$$P(1/\varepsilon) \cong \varepsilon\lambda$$
(6.17)

This probability is characteristic of a random mechanism better known as a Poisson process. This is in fact a very important process, occurring in many areas of physics, and is described in further detail in section 10.9. We are now ready to calculate the autocorrelation $R_I(\tau)$. If τ is larger than width ε, any peak occurring at t is different from a peak occurring at $t+\tau$, so that the I_1 and I_2 values at t and $t+\tau$ are uncorrelated and $P(i_1,i_2)=P(i_1)P(i_2)$. We can also neglect the possible values corresponding to $k>1$ since their probability is negligible considering the case $k=1$ for a small ε. Thus the prevailing term corresponds to the possible values $i_1=i_2=1/\varepsilon$. Using the statistical autocorrelation equation (6.14) we obtain

$$R_I(\tau) = \sum P(i_i)P(i_j)i_i i_j = e^2 P(1/\varepsilon)^2 \left(\frac{1}{\varepsilon}\right)^2 = e^2 \lambda^2 \quad \text{if } \tau > \varepsilon. \tag{6.18}$$

If τ is smaller than ε, and neglecting once again the possibility of obtaining more than one peak in such a small time interval, the only possibility of obtaining a term different from zero in the autocorrelation corresponds to a single pulse which spreads over τ. The probability of one peak occurring at t is $P(x_1=1/\varepsilon)=\varepsilon\lambda$, and this corresponds to middle peak positions uniformly distributed between $t-\varepsilon/2$ and $t+\varepsilon/2$, i.e over a width ε (see Figure 6.4).

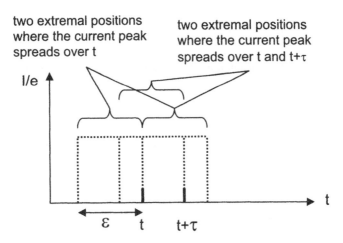

Figure 6.4. *Extremal positions of a single current peak to spread over t or t and t+τ*

We wish to obtain the conditional probability that this peak also spreads over $t+\tau$. With a middle peak position uniformly distributed between $t-\varepsilon/2$ and $t+\varepsilon/2$, the favorable events are the one for which this middle peak position is located between $t-\varepsilon/2+\tau$ and $t+\varepsilon/2$, i.e. over an interval $\varepsilon-\tau$ (see Figure 6.4), so that the conditional probability is $P(x_2=1/\varepsilon$ if $x_1=1/\varepsilon)=(\varepsilon-\tau)/\varepsilon$. The probability that $x_1=x_2=1/\varepsilon$ is thus equal to $P(x_1=1/\varepsilon)\times P(x_2=1/\varepsilon$ if $x_1=1/\varepsilon)=\lambda(\varepsilon-\tau)$ and the autocorrelation is

$$R_I(\tau)=\frac{e^2\lambda}{\varepsilon^2}(\varepsilon-\tau) \quad \text{if } \tau<\varepsilon. \tag{6.19}$$

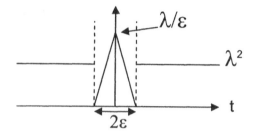

Figure 6.5. *Autocorrelation of a signal with random and independent square pulses*

From equations (6.18) and (6.19) we can plot the autocorrelation function (Figure 6.5) and by taking the limit when ε tends to zero we obtain

$$R_I(\tau)=e^2(\lambda\delta(\tau)+\lambda^2). \tag{6.20}$$

Since $\lambda=N/T=I/e$ the current noise power spectral density is equal to

$$\Phi_I(f)=eI+I^2\delta(f). \tag{6.21}$$

where I is the average current. The second term is the power spectral component of the continuous part, and the first term is due to the current fluctuations. The measured power spectral density is

$$\Phi_I^{meas}(f)=2eI, \tag{6.22}$$

where we omitted the Dirac peak component. From equation (6.22) we see that the obtained density represents a white noise (i.e. it does not vary with frequency) and its usual name is "shot noise". Note that if we straightforwardly apply the low frequency approximation equation (6.12) using the standard deviation of a Poisson process, also equal to its mean (see section 10.9), we also obtain the same result. This formula was first derived by Schottky. If we wish to have a good idea of what such a noise looks like, we could just listen to the water droplets falling on the roof of a house during a rainy day. What we must retain is that a crucial ingredient for obtaining this formula is the statistical independence of the individual current pulses. As we shall see below in quantum mesoscopic conductors we can also expect shot noise, since they are of a ballistic nature, but with a crucial difference, which lies in the fact that both the spatial electron confinement and the Pauli exclusion principle may substantially attenuate and even destroy this statistical independence. We might also ask why shot noise is not always observed, because a simple interpretation of what we wrote above might lead us to think that shot noise is obtained whenever a current is measured in any solid-state device. This is in fact not the case, and although answering this question does not exclusively pertain to the field of mesoscopic physics, for the sake of clarity it is discussed in the next section.

6.6. Why the shot noise formula is not valid in a macroscopic conductor ⑩

6.6.1. *Current pulse shape*

First of all we need to explain, at least in a qualitative way, the shape of the pulses responsible for the apparition of shot noise. Just consider a point charge e comprised in between two infinite capacitor plates separated by a distance a, as in Figure 6.6. It is not very easy to calculate this, but this point charge induces charges at both plates, which are given by

$$q_1 = -\left(1 - \frac{x}{a}\right)e \tag{6.23}$$

and

$$q_2 = -(x/a)e, \tag{6.24}$$

at the bottom and top plates, respectively, and where x is the distance which separates the charge from the bottom plate (if we assume that the charge in between the plates is uniformly distributed on a plane parallel to the plate then a simple application of Gauss theorem also leads to equations (6.23) and (6.24)). If there is some applied voltage this does not qualitatively modify the situation, but now we also have to deal with additional and opposite accumulated charges $+Q$ and $-Q$ at both plates due to the action of the electric field between the plates, and given by the usual $Q=CV$ law, where C is the plate capacitance.

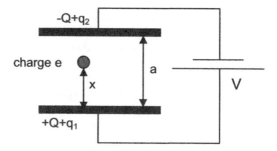

Figure 6.6. *An electron between two metallized plates induces charges on both plates*

As the emitted charge is pushed by the electric field from one plate to the next this induces in turn a time variation of the charges accumulated at both plates following equations (6.23) and (6.24). These charges are provided by the voltage generator, so that each single electron transit between the two plates induces a continuous current pulse from the generator as illustrated by Figure 6.7.

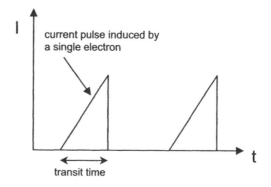

Figure 6.7. *Current pulses induced by electrons moving from one plate to another*

Note that the generator current is not limited by charge granularity since the plate charges can be obtained by a continuous displacement of the free electrons with respect to the fixed positive ion background. Of course the pulse duration is limited to the electron transit time, and provided that it is short enough, this does not put in danger the conclusions drawn from our Dirac peak model in the previous section. Each pulse carries exactly one elementary charge when integrated over the transit time. What we must keep in mind is that the ballistic electron transfer induces a transfer of charges between the generator and the sample. It is these latter charges that are measured in a current noise experiment; we do not directly measure the ballistic electron transfer.

6.6.2. *Non-ballistic conductor*

Consider the same idealized ballistic structure as in the previous section, but introducing now randomizing collisions inside the capacitor plates, as illustrated in Figure 6.8, so that the momentum mean free path is much smaller than the distance between the plates $\lambda_m << a$.

Each electron traveling from one plate to the next experiences at least a number of randomizing events of order a/λ_m. Thus, with the same approach as in the previous section it is clear that each electron now contributes to many independent current pulses in the generator, but *each one carrying an overall charge of order $q=e\lambda_m/a$.* Thus, if the ratio λ_m/a is small this induces in turn a vanishing of the shot noise with respect to the other common noise contributions.

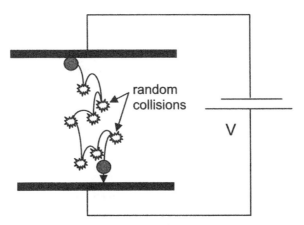

Figure 6.8. *An electron transiting between two metallized plates and experiencing randomizing collisions*

This argument, which can be simply developed using our somewhat simple model, is in fact also valid for solid-state macroscopic and even mesoscopic conductors, when they are operated in a diffusive rather than a ballistic regime. In such a case each electron induces random current pulses corresponding to charges of order $q=e\lambda_m/L$, where a has been replaced by the device length L. As in our idealized example, shot noise is reduced much below the level of the other noise sources.

6.7. Classical example 1: a game with cannon balls ⑩

Consider a classical apparatus with cannon balls as illustrated by Figure 6.9. A man feeds an inclined tube with small cannon balls at a pace faster than the cannon balls can roll and leave the tube. There is some friction so that the balls do not accelerate but move at a constant velocity. The cannon ball "current" at the way out is measured as a function of time. If the balls are provided at a fast enough pace, all balls are touching one another and their positions are perfectly correlated. All balls have the same velocity, and in such a case there is absolutely no current noise. The current peaks are periodically spaced in time.

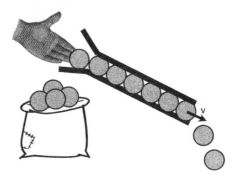

Figure 6.9. *A classical game with cannon balls*

In contrast with the case of the previous section, we can see that perfect correlations totally suppress the shot noise. Noise can arise only if the operator is distracted, so that the pace at which he feeds the tube is random and slow enough for the balls to remain separated from one another inside the tube. Thus we can see from this first simple example that shot noise is destroyed by the introduction of correlations. This can also be the case in a vacuum tube if the electron density is high enough between the electrodes: the creation of a space charge correlates the electrons and partially reduces the shot noise intensity. We shall see in section 6.9 that for quantum-coherent 1D channels fed by macroscopic reservoirs the Fermi exclusion principle may introduce correlations sufficient in order to partially or totally suppress the shot noise, depending on the value of the transmission coefficient.

6.8. Classical example 2: cars and anti-cars ⑩

Consider an "infinitely long" train with a velocity v_{car} and its passage on a line of length L. The train is formed by cars all with the same length L_{car}. Each car can contain just one passenger. The maximum admissible passenger traffic is just like in Figure 6.10, where all cars are occupied. In such a case it is clear that all passenger positions are completely correlated and then the passenger flow is just a noiseless succession of perfectly separated peaks, with an average flow given by $F_{max}=v_{car}/L_{car}$.

Figure 6.10. *A fully occupied train composed of small cars*

If the number of passengers is very small with regards to the number of cars and their spatial distribution in the cars is random, the passenger positions are independent and we again find the same shot noise situation as in section 6.5, but playing with passengers rather than with electrons. Thus, the average flow is $F=F_{max}\times f_{car}$ where f_{car} is the average occupation of the cars by the passengers. The shot noise power spectral density is $S_{car}(f)=2F=2F_{max}\times f_{car}$.

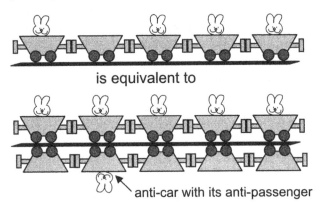

is equivalent to

anti-car with its anti-passenger

Figure 6.11. *An almost full train and its science-fiction analog with anti-cars*

Now imagine that the train is almost full. The situation now looks like that in Figure 6.11. Most passenger positions are almost completely correlated, and to calculate the power spectral density it is easier to replace the real flow by introducing an additional lane with "anti-passengers". Replace each unoccupied car by one occupied car and one occupied anti-car. The flow properties obviously remain the same, but we have decomposed the flow F into two parts: one is the maximum flow, fully correlated and noise free, and we add to it the anti-passenger flow with the anti-passengers now largely separated and independent. Thus, the average flow part due to the anti-passengers is equal to $F_{max}(1-f_{car})$, and the noise intensity is $S_{car}(f)=2F_{max}(1-f_{car})$, *different from and smaller than* $2F=2F_{max}f_{car}$. When $f_{car}=1$ we recover the noiseless maximum flow. We thus see that with a passenger occupation number from 0 to 1, the noise power is not always equal to $2F$, but increases, passes through a maximum, and then decreases down to zero. Thus, we see how the correlations can progressively reduce the shot noise. The simplest interpolation formula would be something like $S_{car}(f)=2F_{max}f_{car}(1-f_{car})=2F(1-f_{car})$. This formula is in fact an exact classical analog of the quantum shot noise expression, which includes the factor $T(1-T)$ as heuristically derived in the next section.

6.9. Quantum shot noise ●

6.9.1. *Fluctuations and Pauli exclusion principle*

Consider a system of fermions. Due to the exclusion principle each quantum state can be occupied at most by one electron. At non-zero temperatures the average statistical occupation of one state $\langle n \rangle = f$ is given by the value of the Fermi-Dirac distribution function at the corresponding energy. The fluctuations, and more precisely the mean square of the fluctuations, are given by

$$\sigma_n^2 = \left\langle \left(n - \langle n \rangle \right)^2 \right\rangle = n^2 - 2\langle n \rangle n + \langle n \rangle^2. \tag{6.25}$$

Since n is equal to zero or 1, we have $n^2=n$ and from equation (6.25) we obtain

$$\sigma_n^2 = \langle n \rangle \left(1 - \langle n \rangle \right), \tag{6.26}$$

from which we can see that the fluctuations vanish at zero temperature, and are determined in a simple way by the occupation number at non-zero temperatures. These thermal fluctuations in the occupation number are responsible for fluctuations in the current, and thus for the thermal noise which occurs even in the absence of average current.

In the case of shot noise, we have to investigate how the exclusion principle rules the electron transfer between reservoirs which are not at the same chemical potential, and thus give rise to an out-of-equilibrium current. The simplest system we can think of consists of a particle incident onto a barrier, with a transmission probability T and a reflection probability R, with $R+T=1$. First we shall assume that the state which describes the incident electron is in contact with the left reservoir and is always occupied, so that we can write $\langle n_{in}\rangle=1$. Obviously we also have $\langle n_T\rangle=T$ and $\langle n_R\rangle=R$. Here the fluctuations in the occupation number are not due to thermal fluctuations, but to the fact that an incident electron is either transmitted or reflected, so that it is often called partition noise. However, we can use exactly the same argument as for thermal fluctuations, and from the Pauli exclusion principle we immediately find that

$$\sigma_{n_T}^2 = \left\langle \left(n_T - \langle n_T\rangle\right)^2 \right\rangle = \left\langle \left(n_R - \langle n_R\rangle\right)^2 \right\rangle = T(1-T) = TR \ . \tag{6.27}$$

This is the standard deviation that we are going to use for calculating the current noise. It is maximum when $T=1/2$, and rises from zero to $1/2$ and then decreases from $1/2$ to zero as T runs from zero to 1, just as in the train example.

6.9.2. Shot noise power spectrum at T=0

There is one strong conceptual difficulty in tackling the quantum noise problem in small, open structures, and although we will not solve it we will not let it be buried in a text clever enough so as to keep all the appearance of logical deduction, but ignoring the major difficulty. In the classical shot-noise case, there is no conceptual problem to define the basic stochastic process which leads to the current and to the noise fluctuations: an electron is randomly emitted from the cathode and travels towards the anode. It induces a well-defined, time-varying charge transfer from the voltage supply to the capacitance plate. This is clearly not so for our mesoscopic devices. We have derived all our transport formulae by assuming that electrons are lying in states which are plane waves, possibly normalized by dividing them by the square root of the device length. This could be conceptually correct with a ring, and *this leads us to expressions which are well verified by the experiment.* Nevertheless, it is not fully consistent to describe a state which extends over a finite length by a wave which describes a continuous probability current. An electron which continuously moves forward cannot stay in a finite area forever, unless this area forms a closed orbit. In other words, the very same electron cannot stay in a spatially finite quantum state exhibiting a non-zero probability current forever, because such a state is clearly non-stationary. It must be transferred to an "outside" state and then replaced by a new electron coming from the left. Our formalism does not tell us either the rate at which this occurs, or how it may affect the overall system fluctuations. A more comfortable view would consist of replacing our plane waves by wave packets. However, although a formal mathematical procedure can be applied to independent electrons so as to transform our set of orthogonalized plane waves into a set of orthogonalized wave packets, there is no demonstration that the chosen set actually describes the reality, and the whole picture becomes much more complicated as the states no longer have a well defined energy. More precisely, considering a ballistic wire, if we make use of wave packets we can readily imagine that their space and energy density will correspond to that of our plane waves, but there is no simple demonstration that their successive departures from the left side and their arrival at the right side should be perfectly time-ordered (and no such attempt has been made in the papers which derive the quantum shot noise in such a way; the authors just use a set of perfectly time-ordered wave packets, so that the demonstration is achieved by putting the result in the set of assumptions). Even in a second quantification formalism, with all due rigor we should not use creation and annihilation operators operating on current-carrying states characterized by a finite spatial development. We shall thus have to assume (and this is confirmed by the experiment) that if the left reservoir is such that it immediately fills everything that we call electron states inside the device, the Pauli exclusion principle acts so as to completely suppress any possibility of fluctuations, just as in the fully occupied train case. This will then allow us to derive the noise formula when the transmission is no longer equal to unity. A more rigorous treatment would imply developing a quantum-mechanical formalism adapted to account for electron correlations, involving creation and annihilation operators, and able to tackle all the subtleties of the measurement problem. For instance, even if the electron is not yet "measured", it influences the wave functions of the electrons in the electrode through its Coulomb potential, and all wave functions must indeed be replaced by a many-body wave function which incorporates exchange effects – electrons are undistinguishable particles – and quantum correlations. This many-body formalism is well beyond the scope of this introductory book. What we shall retain is that the same ingredients which make the cannon ball tube noiseless operate in our quantum devices: the left reservoir is able to feed the wire at a pace such that it is always full, and all occupied electron states inside the wire are thus perfectly correlated.

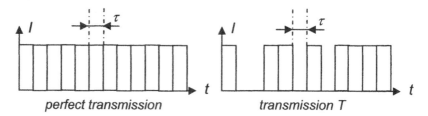

perfect transmission   transmission T

Figure 6.12. *Current due to the propagating states in an energy interval dE of the output lead with perfect (left) or imperfect (right) transmission through the 1D conductor; τ is the electron transit time in a propagating state*

To put some figures on the considerations expounded above, we now go into the heart of the matter, and approximate a perfect 1D device using the rough model which follows: first, we assume that $T=0$ and consider that there is no transition between the different energy channels; we also assume that just one subband is opened. Secondly, we take into account the fact that an electron does not stay in a propagating state of the perfect output lead of a 1D conductor forever, by assuming that it occupies it only during a transit time $\tau_T(E)=L/v(E)$, where L is the length of the output lead and v is the energy-dependent velocity. Thirdly, we attribute to the electron occupying a propagating state in the output lead the usual value of the probability current associated with this state. Current fluctuations are thus ascribed to occupation number fluctuations. No two electrons can occupy the same state at the same time, to account for the Pauli exclusion principle, and an electron leaving the channel is immediately replaced by a new one coming from the contact if the conductor is perfect. Thus, as in the cannon ball example the electrons are perfectly correlated, and the time behavior of the current in this perfect channel can be represented as in Figure 6.12. This is constant, formed by the exact time superposition of the rectangular current pulses issued from each transiting electron. Noise is equal to zero. The current fraction corresponding to an energy slice dE is given by

$$dI_{perfect\,lead} = \frac{2e}{h}dE \cdot$$

(6.28)

and the transit time is obviously equal to

$$\tau_T = \frac{e}{dI_{perfect\,lead}} = \frac{h}{2dE}.$$

(6.29)

Note that in our simple model this transit time has no reason to change if not all carriers are transmitted. Then if we introduce a scatterer such that only a fraction T of the electrons is transferred into the supposedly perfect output lead, we arrive at the situation depicted by Figure 6.12. The current consists of a binary signal with two levels, and the probability of obtaining the high level is equal to the transmission probability T. The current carried by an energy fraction dE is now given by

$$dI = \frac{2e}{h}n_T(E)dE$$

(6.30)

where $n_T(E)$ is the occupation number, whose average value is equal to that given by the Fermi-Dirac distribution, but which fluctuates in time according to the process described in section 6.9.1. Clearly the current values of a signal such as in Figure 6.12 are uncorrelated if they are separated by a time interval larger than τ_T, and in this case the autocorrelation is simply equal to $\langle dI \rangle^2$. If $|t| < \tau_T$, we can use equation (6.3). Six possible time sequences must be considered. They are enumerated and schematically represented in Figure 6.13. First we can note that for two realizations of a signal such as in Figure 6.12 the two train of pulses are randomly shifted from one another by an amount t_r which is uniformly distributed over the interval τ_T., i.e. with a probability density $p=1/\tau_T$. The probability of obtaining a transition over the time τ is equal to $P_{trans}=P(t_r<|\tau|)=|\tau|/\tau_T$, and the probability of having none is $P_{none}=1-|\tau|/\tau_T$.

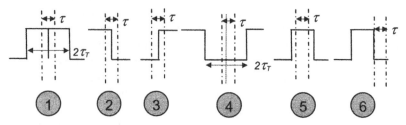

Figure 6.13. *The six possible current time sequences that can occur for a time τ smaller than the transit time τ_T*

Defining dI_u and dI_d as the higher and lower current levels (where, in our particular case, $dI_d=0$), from equation (6.3) and Figure 6.13 we can write

$$R_{dI}(\tau) = P_{trans}(p_u^2 dI_u^2 + 2p_u p_d dI_u dI_d + p_d^2 dI_d^2) + P_{none}(p_u dI_u^2 + p_d dI_d^2) \qquad (6.31)$$

where the first term inside the first bracket corresponds to case 1, the second to cases 2 and 3 and the third to case 4, after the numbering defined in Figure 6.13. The first term inside the second bracket is for case 5 and the second for case 6. The above expression obviously simplifies to

$$R_{dI}(\tau) = P_{trans}\left(\sum p_i dI_i\right)^2 + P_{none}\sum p_i dI_i^2 = P_{trans}\langle dI\rangle^2 + P_{none}\langle dI^2\rangle \qquad (6.32)$$

which can be re-arranged as

$$R_{dI}(\tau) = \langle dI\rangle^2 + P_{none}\,\sigma_{dI}^2 \qquad (6.33)$$

and thus as

$$R_{dI}(\tau) = \langle dI\rangle^2 + \sigma_{dI}^2\left(1-\frac{|\tau|}{\tau_T}\right) \qquad (|\tau|<\tau_T). \qquad (6.34)$$

Omitting the contribution of the average $\langle dI\rangle^2$ (which is the same for $|\tau|<\tau_T$ and $|\tau|>\tau_T$), we again find a triangular function as that of equation (6.10). We can thus apply the low frequency approximation equation (6.30) of the power spectral density given by equation (6.12), and we obtain

$$S_{dI} = 2\left(\frac{2edE}{h}\right)^2 \sigma_{n_T}^2 \tau_T \cdot \tag{6.35}$$

where we have already included a factor of two to obtain the measured power spectral density, and used equation (6.30) to replace σ_{dI}^2 by $\sigma_{n_T}^2$. Substituting $\sigma_{n_T}^2$ and τ_T by equations (6.27) and (6.29) in equation (6.35), we thus obtain the noise intensity

$$S_{dI} = 2e \times \frac{2e}{h} T(1-T) dE \cdot \tag{6.36}$$

Summing over energy gives the full noise intensity at $T=0$:

$$S_I = 2e \times \frac{2e}{h} \int T(1-T) dE , \tag{6.37}$$

from which we can see that if the transmission is weak, the noise power is equal to $2eI$ as in the case of classical shot noise, but reduces to zero if the transmission is perfect. From this simplified model we thus deduce that the Pauli exclusion principle induces quantum correlations which are able to reduce and even totally suppress the shot noise, depending on the value of the transmission coefficient. Neglecting the variation of the transmission probability with energy, the zero-temperature shot noise of a two-terminal coherent device with one open channel is thus given by

$$S_I = \frac{4e^3}{h} T(1-T) V = 2eI(1-T) \tag{6.38}$$

where V is the applied voltage. When the transmission T is equal to unity, noise is totally suppressed, and is maximum when $T=1/2$.

In order to be measured, such a noise must exceed the equilibrium thermal noise contribution. This contribution is given by the Nyquist formula $S_I=4k_B\Theta G$, where we use the symbol Θ for the temperature in order to avoid any confusion with the transmission probability. To compare the shot noise contribution to the Nyquist formula, noise is often expressed as an equivalent noise temperature, defined by

$$\Theta^* = \frac{S_I}{4k_B G} \tag{6.39}$$

where G is the device conductance. From equation (6.38) we expect this equivalent noise temperature to be given by

$$\Theta^* = \frac{eV}{2k_B}(1-T)$$

(6.40)

Note that treating the non-zero temperature case implies considering energy intervals over which both input and output states are unoccupied, which somewhat complicates the matter and requires some changes in our set of assumptions. However, the same kind of analysis can still be carried out if the input states are identified to wave packets, with the additional hypothesis that wave packets propagating in opposite directions are perfectly time-ordered and impinge at the same time on the potential barrier [MAR 92]. The analytical results obtained with such an *ad hoc* argument are nevertheless comforted by a more rigorous treatment [BLA 00]. Here we shall just mention an analytical formula which gives the overall noise temperature for a multi-channel, two-terminal sample, and which includes the equilibrium and shot noise contributions:

$$\Theta^* = \Theta\left(1 + \left(\left(\frac{eV}{2k_BT}\right)\left(\tanh\left(\frac{eV}{2k_BT}\right)\right)^{-1} - 1\right)\frac{\sum_i T_i(1-T_i)}{\sum_i T_i}\right).$$

(6.41)

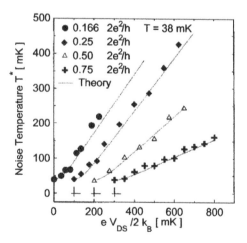

Figure 6.14. *Experimental noise temperature measured at different values of the transmission T and applied voltage V_{DS}, in a constriction with a single open channel; reproduced with permission from Kumar A. et al., Phys. Rev. Lett. 15, p. 2778 (1996), copyright (1996) by the American Physical Society*

We refer the reader in search of a rigorous derivation of equation (6.41) to the review article by Blanter and Büttiker [BLA 00], a very detailed account of quantum noise in mesoscopic devices. It is worth noting that equation (6.41) is well approximated by equation (6.40) if the applied voltage V substantially exceeds the thermal voltage $k_B\Theta/e$.

The validity of equation (6.40) has been checked in a particularly clear experiment [KUM 96], the main result of which is reproduced in Figure 6.14. In Figure 6.14 the noise temperature is plotted as a function of the voltage applied to a point contact. For each curve the transmission T is adjusted by tuning the extent of the lateral depletion areas which define the physical constriction. This depletion width is controlled by the voltage applied on the side gates. All curves are obtained for gate biases corresponding to the opening of the first 1D channel (i.e. in the rising edge before the first quantized conductance plateau). The lines are obtained using equation (6.41), just entering the measured transmission value. For high enough voltages the equilibrium contribution is small with regards to the shot noise, and thus the experimental curves very closely follow equation (6.40). The quantum shot noise value gradually falls below the classical value as the transmission rises.

6.10. Bibliography

[BLA 00], BLANTER Ya.M., BUTTIKER M., "Shot noise in mesoscopic conductors", *Physics Reports*, vol. 336, 2000, p. 1-166.

[KUM 96] KUMAR A., SAMINADAYAR L., GLATTLI D.C., "Experimental test of the quantum shot noise reduction theory", *Physical Review Letters*, vol. 76, no. 15, 1996, p. 2778-2781.

[MAR 92] MARTIN Th., LANDAUER R., "Wave-packet approach to noise in multi-channel mesoscopic systems", *Physical Review B*, vol. 45, no. 4, 1992, p. 1742-1755.

Chapter 7

Coulomb Blockade Effect

7.1. Introduction ●

In a sample with tunneling barriers two different physical factors determine the current: on the one hand, a double-barrier device which forms a quantum well exhibits quantized energy levels, and conduction can occur only when those levels can exchange carriers with the device terminals. On the other hand, the electrons moving through the quantum well or inside a dot generate an electrostatic potential and thus modify the potential energy: the electrostatic potential due to one electron in the well acts on all other electrons and enhances their potential energy, since the electric force exerted by one electron on another one is repulsive. In the chapter dealing with resonant tunneling we only took into account the first factor, and we completely neglected the second factor. This gives us a correct picture as long as the potential energy of any electron which is due to the action of all other electrons inside the well remains small against the initial separation between the quantized energy levels. However, suppose that just the opposite is true, and even that the potential energy modification brought by adding just one electron is much larger than the initial quantized energy level spacing. In such a case the current is dominated by classical charge effects. Of course this depends on the device dimensions but in practice this regime can actually be achieved. Thus, in this chapter we shall take the point of view opposite from the one which we adopted in the previous section. We shall neglect the quantum-mechanical quantization of the energy levels, and we shall study cases in which the effect of the electron charge prevails. This will lead us to introduce an important phenomenon known as "Coulomb blockade", in which tunneling is completely forbidden at certain voltages due to the electron charge effect.

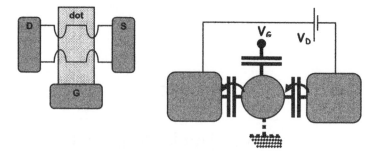

Figure 7.1. *A gated dot connected to drain and source by constrictions*

The Coulomb blockade effect appears in small devices which are separated from the contacts by tunneling barriers. A typical device configuration is shown in Figure 7.1, which is called a single-electron transistor (SET) for reasons that will become clear later on. It consists for instance of a MOS field effect transistor of small dimensions, where drain and source are separated from the gate-controlled dot by tunneling barriers. An alternative way to represent such a device is that of the right scheme in Figure 7.1. Each tunneling barrier can be described both by its conductance and its capacitance, and in some senses it can be assimilated to a "leaky" capacitor (but the gate is not leaky). For the Coulomb blockade to be observed the various capacitances have to be very small. From the energetic point of view we might draw a scheme like the one in Figure 7.2. The device is composed of a dot, in which electrons can be injected, this dot being separated from the contacts by potential barriers. Injection of additional electrons modifies the potential inside the dot.

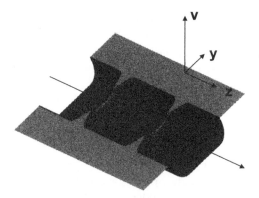

Figure 7.2. *The potential landscape seen by the electrons in a single-electron transistor*

What is the origin of the Coulomb blockade? If two classical capacitors are put in series and if there is no leakage, the conductor which links them is globally neutral and the total charge between them is zero (see Figure 7.3). The charge Q can take any real value, depending on V, because the charges at the electrodes (>0 or <0) are obtained by a displacement of the free electrons with respect to the (fixed) positive background charge. Since this displacement can be arbitrarily small and continuous, there is no limitation imposed on the charge value except the one imposed by the charge-capacitance-voltage relationships. However, if the capacitors are "leaky", we can inject electrons into the cell (Figure 7.3). However, in such a case the net charge is obviously an integer number N of elementary charges, and it cannot be continuously varied, even if it can be shared in non-integer parts lying at each capacitor plate. In addition, this uncompensated charge changes the cell potential. We shall see that if the capacitance is small this may result in observable phenomena, referred to as Coulomb blockade, and which result from the granularity of the charge. Thus, even if the carriers are injected or emitted from the dot through tunneling barriers, which is a quantum-mechanical feature, the Coulomb blockade itself should instead be viewed as a classical effect, which is primarily due to the quantification of an uncompensated charge in a partially isolated part, and not to quantum-mechanical effects. Here the "quantum" does not result from the Schrödinger equation, but is just the elementary electron charge, which cannot be divided into smaller parts.

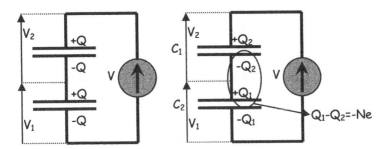

Figure 7.3. *Two capacitors put in series; in the left figure they are perfect and no charge can be injected through the plates; in the right figure it is possible to inject some net charge inside the cell formed by the two capacitors*

7.2. Energy balance when charging capacitors ●

Consider a situation as in Figure 7.4, where a circuit is initially formed by two unconnected capacitors, the left one being uncharged and the right one being charged with charge Q_0 and resulting voltage V_0. If we connect the two capacitors, the right one will be partially discharged into the second one, until we get the same potential V at both capacitor plates (Figure 7.4).

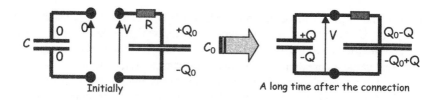

Figure 7.4. *Two capacitors are connected in series so that the initially charged one is partially discharged into the initially uncharged one*

In addition, whatever a given charge distribution ρ, if we know it and if we know the electrostatic potential V, we can calculate the electrostatic potential energy, given by

$$E = \frac{1}{2} \int_{all\ space} \rho(\vec{r}) V(\vec{r}) d^3 \vec{r} \ .$$ (7.1)

Let us define E_{C0} as the electrostatic energy contained in the right capacitor plates and E_C as the energy of the left capacitor. In the initial situation (unconnected capacitors), a direct application of equation (7.1) and of the relationship $Q=CV$ gives us

$$Q_0 = C_0 V_0, \ E_C = 0 \ \text{and} \ E_{C0} = \frac{1}{2} \int \rho V dV = \frac{Q_0^2}{2C_0} = \frac{1}{2} C_0 V_0^2$$ (7.2)

and the total energy is obviously equal to $E_{tot}^{init} = E_{C0}$. A long time after the connection we have the capacitive relationships

$$Q = CV \ \text{and} \ Q_0 - Q = C_0 V \ ,$$ (7.3)

from which we readily obtain

$$V = \frac{C_0}{C + C_0} V_0 \ .$$ (7.4)

The total energy is thus equal to

$$E_{tot}^{fin} = \frac{1}{2}(C + C_0)V^2 = \frac{C_0^2 V_0^2}{2(C_0 + C)} \ .$$ (7.5)

The energy variations in the left capacitor, in the right capacitor and in the overall circuit which result from the connection of the two capacitors are given by

$$\Delta E_C = + \frac{CC_0^2}{2(C+C_0)^2} V_0^2,$$

(7.6)

$$\Delta E_{C0} = - \frac{C_0 C(C+2C_0)}{2(C_0+C)^2} V_0^2$$

(7.7)

and

$$\Delta E_{tot} = - \frac{CC_0}{2(C+C_0)} V_0^2,$$

(7.8)

respectively. The charges have been re-distributed so as to minimize the total energy accumulated in the circuit. The energy lost in the charging process is dissipated in the resistive part of the circuit, and the reason for which such an energy transfer is favorable lies in the entropy increase which results from it. If we assume that C_0 is much larger than C, we can simplify these energy variations as

$$\Delta E_C = \frac{1}{2} CV_0^2 \qquad \Delta E_{C0} = -CV_0^2 = -QV_0 \qquad \Delta E_{tot} = -\frac{1}{2} CV_0^2.$$

(7.9)

This is equivalent to a situation in which the right capacitor is assimilated to a voltage generator (C_0 is so big that it is an inexhaustible source of charges and the voltage $V \approx V_0$ appearing at its plates cannot be appreciably varied). Thus, from this simple example we found that *a voltage generator from which +Q is withdrawn from the anode and –Q from the cathode loses an energy –QV, and a capacitor charged with +Q and –Q acquires an energy $Q^2/2C$*. This is a result that we shall use without restraint in the next sections. To understand and derive the Coulomb blockade effect we do not need anything else!

7.3. Coulomb blockade in a two-terminal device ●

Consider two "leaky" capacitors put in series as in Figure 7.5, and suppose that we have injected a number N of electrons inside the "dot" formed by those two capacitors.

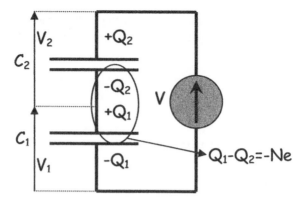

Figure 7.5. *A two-terminal Coulomb blockade device viewed as the series combination of two leaky capacitors*

We have the following charge-voltage and charge conservation relationships:

$$\begin{cases} Q_2 = C_2 V_2 \\ Q_1 = C_1 V_1 \\ Q_1 - Q_2 = -Ne \\ V = V_1 + V_2 \end{cases} \tag{7.10}$$

From these relations it is easy to find that we have

$$V_1 = \frac{1}{C_1 + C_2}(C_2 V - Ne) \tag{7.11}$$

and

$$V_2 = \frac{1}{C_1 + C_2}(C_1 V + Ne) \tag{7.12}$$

Note that if $V>0$, V_2 is always positive. Thus, if C_2 is leaky, electrons can only pass through C_2 going from the cell to the anode. From equation (7.11) V_1 can be either positive or negative, because the uncompensated electrons in the cell contribute to the potential and tend to raise the potential energy. Notwithstanding the leakage through C_2, to inject $1,2,3,...N$ electrons from the cathode requires that V_1 is positive before each injection (otherwise the electric field in C_1 opposes their injection). However, this is not a sufficient condition, because it may happen that injecting the Nth electron renders V_1 negative, so that this electron should pass back (and forth!) to the cathode! Clearly, the equilibrium state cannot be found with so simple a reasoning, and it must be found by making a detailed energy balance, including both the cell charging energy and the generator energy loss. We cannot content ourselves with calculating just the energy change inside the cell, we also have to consider the work achieved by the generators to provide the charges to the structure.

The electrostatic energy in the cell can be calculated by straightforwardly applying equation (7.1). A few lines of calculations applied to the circuit of Figure 7.5 gives

$$E = \frac{1}{2}\frac{C_1 C_2}{C_1 + C_2}V^2 + \frac{1}{2}\frac{(Ne)^2}{C_1 + C_2}.$$
(7.13)

The second term of the right-hand side is the charging energy E_C. It is positive since putting more electrons together has an energy cost, as they exert repulsive forces on one another. Now let us calculate the energy lost by the voltage generator. The cathode gives an additional charge $-Ne$ inside the cell and $-(Q_1-Q_0)$ at the bottom plate of C_1 (Q_0 is the charge with $N=0$). Using equation (7.10) leads to

$$Q_{by\ cathode}^{given} = \frac{C_2}{C_1 + C_2}Ne .$$
(7.14)

The anode gives an additional charge Q_2-Q_0 at the top plate of C_1. Using equation (7.10) we find

$$Q_{by\ anode}^{given} = +\frac{C_2}{C_1 + C_2}Ne .$$
(7.15)

Of course the two quantities above have the same absolute value and opposite signs. Now we can apply the basic result demonstrated in section 7.2 to state that the generator loses an energy equal to

$$\Delta E_G = -\frac{C_2}{C_1 + C_2}NeV .$$
(7.16)

Adding N electrons to the cell thus gives a total energy variation

$$\Delta E = \Delta E_C + \Delta E_G = \frac{(Ne)^2}{2(C_1 + C_2)} - \frac{C_2}{C_1 + C_2} NeV \tag{7.17}$$

The charging process actually realized for a given applied voltage V is the one which maximizes the energy loss through dissipation (i.e. ΔE must be as negative as possible). From equation (7.17) we can calculate that passing from N to $(N+1)$ electrons gives a change in ΔE equal to

$$\Delta E(N+1) - \Delta E(N) = \frac{e}{(C_1 + C_2)}\left(\left(N + \frac{1}{2}\right)e - C_2 V\right) \ . \tag{7.18}$$

It is an unfavorable process if the quantity above is positive, i.e. if

$$V < \left(N + \frac{1}{2}\right)\frac{e}{C_2} \ , \tag{7.19}$$

and if the equality

$$V = \frac{e}{C_2}\left(N + \frac{1}{2}\right) \tag{7.20}$$

is verified, two charge states are equally acceptable. We can now determine the variation of the net charge contained in the cell or the dot as a function of the voltage. Initially we have a net charge in the cell $Q=0$, and if we increase the voltage from zero, it is not possible to inject any electron as long as $V<e/2C_2$ (make $N=0$ in equation (7.19)). Above $V=e/2C_2$ one electron can be added, but to add a second one V must exceed $3e/2C_2$ (make $N=1$ in equation (7.19)), etc. The overall variation is plotted in Figure 7.6.

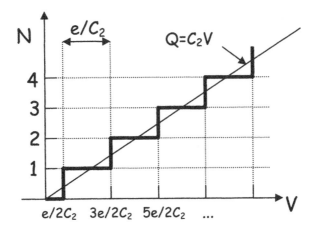

Figure 7.6. *Variation of the charge inside the two-terminal dot as a function of the applied voltage*

In contrast to the case of ideal, non-leaky capacitors, the possibility of injecting electrons inside the dot leads to a discontinuous charge *versus* voltage relation. The granularity of the charge gives a staircase variation. The impossibility of adding a new electron if a high enough voltage is not applied, is not due to the tunneling capability of the barriers, but to a pure charge effect. This is called a Coulomb blockade.

Suppose that the tunneling current is controlled by a much thicker barrier 2, so that the current passing through it is controlled by the number of electrons which can be stored in the dot according to the Coulomb blockade effect and by their transmission probability through barrier 2. Such a sample will exhibit a staircase form as the charge in Figure 7.6 ("Coulomb" staircase).

Why was this effect not observed before the advent of small structures? In our calculations we did not make any assumption about the dot size, and we should expect to obtain Coulomb blockade phenomena provided that the "capacitors" are leaky. As a matter of fact this is not so, because to have a measurable impact the energies which come into the play must exceed the thermal energy k_BT, so as not to be wiped out by thermal fluctuations. Consider a voltage interval over which N electrons are contained in the cell, i.e. such that $(N-1/2)e/C_2<V<(N+1/2)e/C_2$ (from equation (7.19)). It is easy to see that the maximum energy difference which separates either $E(N)$ and $E(N+1)$ or $E(N)$ and $E(N-1)$ is obtained at the middle of the interval, and is equal to $\pm e^2/2(C_1+C_2)$. Thus, for the effect to be observable $e^2/2(C_1+C_2)$ must be larger than k_BT. If we work out some numerical examples we will realize at once that the capacitances must be *very* small, especially at room temperature. This is the reason why the Coulomb blockade is observable only in nanostructures, and the lower the temperature, the better it is. Device engineers are currently struggling to obtain it at room temperature in order to use it for electronic applications.

7.4. Coulomb blockade in a single-electron transistor ●

Consider a gate-controlled dot separated from drain and source by tunneling barriers formed by physical constrictions, such as in Figure 7.7, and the corresponding electrical diagram. Suppose that N electrons have been injected into the dot.

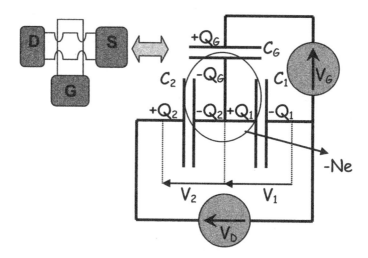

Figure 7.7. *A single-electron transistor and the corresponding electric circuit*

The charge-voltage relationships in each of the capacitors and the charge conservation relation give

$$\begin{cases} Q_G = C_G(V_G - V_1) \\ Q_2 = C_2 V_2 \\ Q_1 = C_1 V_1 \\ Q_G + Q_2 - Q_1 = Ne \\ V_D = V_1 + V_2 \end{cases} \qquad (7.21)$$

Solving this system leads after a few manipulations to

$$V_1 = \frac{1}{C_1 + C_2 + C_G}\left(C_2 V_D + C_G V_G - Ne\right)$$
(7.22)

and

$$V_2 = \frac{1}{C_1 + C_2 + C_G}\left(Ne + (C_1 + C_G)V_D - C_G V_G\right).$$
(7.23)

The electrostatic energy of the dot can be easily calculated using equation (7.1), and is equal to

$$E = \frac{1}{2}\left(Q_G V_G - Ne V_1 + Q_2 V_D\right),$$
(7.24)

which can be re-written as follows using equation (7.22):

$$E = \frac{1}{2(C_1 + C_2 + C_G)}\left(C_1 C_2 V_D^2 + C_1 C_G V_G^2 + C_2 C_G (V_G - V_D)^2 + (Ne)^2\right).$$
(7.25)

The last term inside the parenthesis corresponds to the charging energy term. If $V_D > 0$, it is easier to inject an electron into the cell from the source than from the drain. Let us look for the condition required to inject an additional electron from the source into the cell if there are already N uncompensated electrons in the dot ($N \rightarrow N+1$). This is the process illustrated by Figure 7.8.

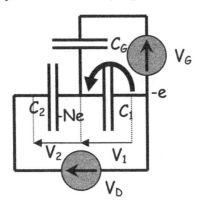

Figure 7.8. *Injecting an additional electron from the source into the dot*

If we add an electron with charge e from the source into the structure the charge changes at the gate and at the drain are given by

$$\delta Q_G = \frac{C_G}{C_G + C_1 + C_2}e$$

(7.26)

and

$$\delta Q_2 = \frac{C_2}{C_1 + C_2 + C_G}e \; ,$$

(7.27)

respectively (to find these expressions just manipulate equation (7.21)). The energy change in the generator is given by

$$\Delta E_G = -\delta Q_G V_G - \delta Q_2 V_D = \frac{-e(C_G V_G + C_2 V_D)}{C_1 + C_2 + C_G}$$

(7.28)

and the energy change inside the dot is equal to .

$$\Delta E_C = \left(N + \frac{1}{2}\right)\frac{e^2}{C_1 + C_2 + C_G} \; .$$

(7.29)

Thus, we find that the total energy change when adding an electron into the dot from the source is equal to

$$\Delta E = \frac{e}{C_1 + C_2 + C_G}\left(\left(N + \frac{1}{2}\right)e - C_G V_G - C_2 V_D\right).$$

(7.30)

We can still add an electron if the quantity above is negative, so that *the condition for injecting an electron from the source if N uncompensated electrons are already present inside the dot and $V_D>0$ is*

$$V_D > \left(N + \frac{1}{2}\right)\frac{e}{C_2} - \frac{C_G}{C_2}V_G \; .$$

(7.31)

If $V_D > 0$, it is easier to withdraw an electron from the cell to inject it into the drain than into the source. Let us find the condition for injecting an electron from the dot into the drain if there are already N uncompensated electrons in the dot ($N \rightarrow N-1$). We have already almost performed the same job in the lines above, and this situation is schematically illustrated by Figure 7.9.

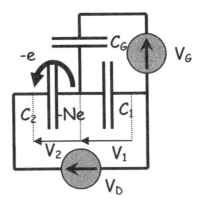

Figure 7.9. *An electron escapes from the dot to the drain*

The charge changes at the gate and at the source are given by

$$\delta Q_G = \frac{-C_G}{C_G + C_1 + C_2} e$$

(7.32)

and

$$-\delta Q_1 = \frac{-C_1}{C_1 + C_2 + C_G} e \; ,$$

(7.33)

respectively (note that the source is the common cathode of both generators, and δQ_1 is a charge which originates in both generators). The charge change in the drain generator is equal to

$$\delta Q_{drain} = -\delta Q_1 + \delta Q_G = \frac{-(C_1 + C_G)}{C_1 + C_2 + C_G} e \; .$$

(7.34)

From the relations above we can calculate the energy change in both generators,

$$\Delta E_G = -\delta Q_G V_G + \delta Q_{Drain} V_D = \frac{e(C_G V_G - (C_1 + C_G)V_D)}{C_1 + C_2 + C_G} \quad , \qquad (7.35)$$

and the energy change inside the dot is

$$\Delta E_C = -\left(N - \frac{1}{2}\right)\frac{e^2}{C_1 + C_2 + C_G} . \qquad (7.36)$$

Thus, the entire system energy change when injecting an electron from the dot to the drain is given by

$$\Delta E = \frac{e}{C_1 + C_2 + C_G}\left(-\left(N - \frac{1}{2}\right)e + C_G V_G - (C_1 + C_G)V_D\right) . \qquad (7.37)$$

To take out an electron this energy change must be negative. *Thus, with N uncompensated electrons already present in the dot and $V_D > 0$, we can still withdraw an electron from the dot into the drain if the condition*

$$V_D > -\left(N - \frac{1}{2}\right)\frac{e}{C_1 + C_G} + \frac{C_G}{C_1 + C_G}V_G \qquad (7.38)$$

is fulfilled.

With a discussion quite similar to the case $V_D > 0$, we find that if $V_D < 0$, it is easier to inject an electron into the cell from the drain than from the source, and *the condition for injecting an electron from the drain into the cell if there are already N uncompensated electrons in the dot ($N \rightarrow N+1$) and $V_D < 0$ is*

$$V_D < -\left(N + \frac{1}{2}\right)\frac{e}{C_1 + C_G} + \frac{C_G}{C_1 + C_G}V_G . \qquad (7.39)$$

If $V_D < 0$, it is easier to withdraw an electron from the cell to inject it into the source than into the drain. *The condition for injecting an electron from the dot into the source if there are already N uncompensated electrons in the dot ($N \rightarrow N-1$) and if $V_D < 0$ is*

$$V_D < \left(N - \frac{1}{2}\right)\frac{e}{C_2} - \frac{C_G}{C_2}V_G. \tag{7.40}$$

The four previous equations allow us to map the Coulomb blockade domains in the (V_D, V_G) plane, which are represented in Figure 7.10. The conditions above represent straight lines, each of them addressing a case in which there are already N uncompensated electrons in the dot. Each of these lines separates the plane into two domains, one in which it is for example favorable to add an electron from the source if there are already N in the dot, and one in which it is impossible to do it. It is worth noting that N is not the total number of electrons inside the dot, but the number of *"uncompensated"* electrons. For example, a neutral dot can already contain electrons, the charge of which is counterbalanced by impurities. Thus, we have to take into account that even if the net charge is equal to zero, this does not necessarily mean that we cannot withdraw one electron, so that the charge state of the dot becomes $+e$. N can be negative, up to a value which is fixed by the initial number of compensated electrons in the neutral dot.

If we examine Figure 7.10 it is not difficult to see that in the domains which are colored gray, the number of electrons is fixed and it is impossible that electrons move through the device. These are the Coulomb blockade domains or "Coulomb diamonds", in which electron transport from source to drain is completely forbidden. For example, consider the diamond which includes the origin, and the part with $V_D>0$. When we bias the sample initially it contains no additional electron in the dot, so that we start from zero uncompensated electrons. The gray line corresponds to equation (7.31) with $N=0$. Below this line it is impossible to add an electron if we have less than one in the cell. Since we started from zero in the considered Coulomb diamond it is clearly impossible to add a new one. Is it possible to withdraw one? The black, dotted line is given by equation (7.38) with $N=0$: below it it is impossible to withdraw one electron if there are zero uncompensated electrons inside the dot. Thus, it is clear that the charge state of this Coulomb diamond is equal to zero in its upper part. In the lower part ($V_D<0$), above the black, dashed line it is not possible to add an electron if we start from zero (equation (7.39) with $N=0$), and above the dotted, gray line it is not possible to withdraw one if we start from zero (equation (7.40) with $N=0$). Thus, in the lower part it is not possible to either add or withdraw an electron, and the charge state is also equal to zero.

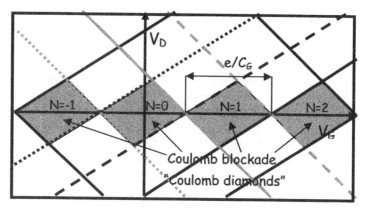

Figure 7.10. *The SET Coulomb blockade areas in the (V_D, V_G) plane*

We can also consider the diamond labeled "*N=1*" in Figure 7.10. In the upper part with $V_D>0$, initially there is no uncompensated electron in the dot, but as above the gray line we can add a new one if we have *N=0*, the dot will be charged with *N=1*. Once this electron has been injected is it possible to add a new one? No, because below the dashed gray line (equation (7.31) with *N=1*), it is impossible to add a new electron if we have only one in the dot. Now, is it possible to withdraw an electron if we have one in the dot? To do so we should be above the dashed, black line (equation (7.38) with N=1), which is not the case of our diamond. Thus after charging one electron, no new electron can be injected and no electron can be withdrawn. The same kind of discussion can be applied to the lower part, and once again, we obtain a Coulomb blockade but now with a net charge state *N=1*. Is it possible to have electron transport, e.g., in the diamond defined by the solid gray, dashed gray, dotted black and dashed black lines? We start from zero but since we are above the gray line we can add one electron. Since we are above the dashed black line we can also withdraw this electron from the dot to the drain. In such a case we can again add an electron from the source, and then take it to the drain, etc. Current is allowed, because different charge states are equally favorable.

Of course once we are in a regime where passing a current is allowed this current depends on the charge state and on the tunneling barrier properties. However, choose a drain voltage such that its value remains smaller than the top of the Coulomb diamonds. If we sweep the gate voltage we periodically enter and go out of the Coulomb diamonds. Thus, we expect the current to exhibit peaks with a periodicity equal to e/C_G.

7.5. Single-electron turnstile ○

Using appropriate frequencies and signal level it is possible to obtain a regime in which a single-electron transistor exhibits a current directly proportional to the frequency. Suppose that we apply a squared voltage signal such as in Figure 7.11, with a drain voltage which does not exceed the top of the Coulomb diamonds in Figure 7.10.

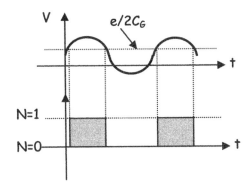

Figure 7.11. *Single-electron turnstile; the upper curve is the gate voltage versus time, and the bottom curve is the net charge number inside the dot versus time*

If the gate signal is chosen so as to put the device in the diamond with $N=1$ when the rising voltage exceeds a specific value, the dot is charged and discharged periodically, whenever the voltage passes from the domain N=0 to $N=1$ and from $N=1$ to $N=0$ during the falling part. Thus at each period exactly one electron has been transferred from the source into the dot and then from the dot into the drain. In such a configuration the current is equal to $I=|e|f$. We obtain a frequency current converter. In practice this has been realized in a number of Coulomb blockade devices, in which several tunnel junctions are put in series (see, e.g., [GEE 90]).

7.6. Coulomb blockade in the real world ●

A current regime dominated by Coulomb blockade effects such as the one described in the previous sections is nowadays routinely obtained in laboratories where nanostructures are fabricated and tested. A typical example is given in Figure 7.12, where the Coulomb diamonds and the Coulomb Blockade oscillations are plotted for a small, silicon-based single-electron transistor [TAK 02]. We can recognize the periodic conductance oscillations due to the Coulomb blockade diamonds that we studied in section 7.4. The oscillation period allows us to extract the gate capacitance, which is of course pretty small. The peak values are different, and this is due to the fact that for each peak the tunneling double barrier exhibits a different shape, which depends on the voltage (note that the actual current is also a function of a factor that we neglected throughout: in such devices the dot dimensions may result in the existence of levels whose separation is quite small in comparison with the Coulomb gap, but which nevertheless have an impact on the overall transmission once the charge effect does not prevent electrons being transmitted).

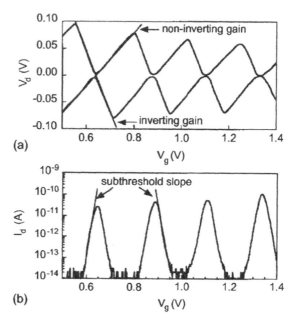

Figure 7.12. *Electrical characteristics of a silicon SET operated at T=27K. In (a) the Coulomb diamonds are obtained by measuring the output drain voltage for a fixed current threshold of ±10 pA. In (b) the periodic current oscillations are measured for $V_D=10$ mV. Reproduced with permission from [TAK 02], copyright (2002) by IOP Publishing Ltd*

Let us mention a few points about the Coulomb diamonds that we derived theoretically in the previous section, and that you can see in Figure 7.12. It is worth noting that in order to fabricate an inverter logic gate with a SET, the output gain must exceed unity, so as to allow the voltage output to be used as the input of a new logic gate, and the output of this new logic gate to be used in turn as the input of another gate, etc., without completely losing the distinction between the low and high logic levels after a few steps. This is a necessary condition to build logic circuits. From equations (7.38) and (7.40) we can deduce that the gain $G=dV_D/dV_G$

is given by $G=C_G/C_G+C_j$ without inversion, and $G_{inv}=C_G/C_2$ with inversion (see Figure 7.12). Thus, only the latter can be made higher than unity. In addition, to increase this gain requires us to increase the gate capacitance over that of the junction, but then a trade-off must be found since increasing this capacitance reduces the possibility of using the device at high temperatures. This makes the optimization of such devices a somewhat difficult task [TAK 02].

What can such devices be useful for, if the researchers succeed in operating them at room temperature? They can be used in logic circuits to further miniaturize future logic circuits, to drastically reduce power consumption and to increase device speed. A detailed account of the efforts made to achieve this goal in Si devices can be found in [TAK 02]. Although obtained at very low temperature, here I wish to report a nice experimental demonstration that basic logic gates can be realized using the Coulomb blockade effect. These particular results were from a research team from Delft University of Technology. The SET inverter is fabricated using electron beam lithography and a scanning electron microscope image of the circuit is shown in Figure 7.13.

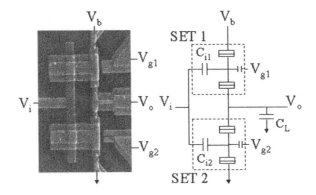

Figure 7.13. *SEM image of a logic inverter using two SETs and scheme of the logic circuit. Reproduced with permission from C.P. Heij et al., Appl. Phys. Lett. 78, p. 1140 (2001) [HEI 01], copyright (2001) by the American Institute of Physics*

The devices consist of Al stripes deposited on an SiO_2 layer thermally grown on an Si substrate. The gate insulator is an 8 nm thick Al_xO_y layer. The capacitance values are $C_j=280aF$, $C_{i1}\approx C_{i2}\approx810aF$, $C_{G1}\approx C_{G2}\approx45aF$. The logic gate circuit is that of Figure 7.13. V_{G1} and V_{G2} are "tuning gates". They are adjusted so that 1/ SET1 is in the Coulomb blockade regime and SET2 conducts when V_i is high and 2/ SET1 conducts and SET2 is in the Coulomb blockade regime when V_i is low. The gate is thus an inverter working much like a conventional CMOS inverter. Transfer characteristics are shown in Figure 7.14.

It is worth noting that the inverter gain (the negative slope of the curves in Figure 7.14) is equal to *G=2.6*, thus allowing the voltage out to be used as the input of a new logic gate and the output of this new logic gate to be used in turn as the input of another gate. Although these results were obtained at quite a low temperature (30mK!), researchers involved in this field of activity obtain steadily improving results as the years pass on. For instance, a Silicon-On-Insulator (SOI) SET inverter operating with *G=1.3* at 27 K has already been realized by the NTT basic research labs in Japan, using the same technology as that used in Figure 7.12 [TAK 02].

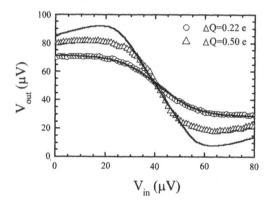

Figure 7.14. *Transfer characteristics of an SET inverter such as in the previous figure. Reproduced with permission from C.P. Heij et al., Appl. Phys. Lett. 78, p. 1140 (2001) [HEI 01], copyright (2001) by the American Institute of Physics*

7.7. Exercises

7.7.1. *Exercise*

Consider two metallic dots labeled 1 and 2. They are put in series and linked to macroscopic metal pads as in Figure 7.15. The equivalent capacitors are labeled as in Figure 4. C_1 and C_2 are "leaky" and also behave as tunnel resistances, but C_3 is perfectly insulating. It is assumed that N_1 electrons have been injected into dot 1 and N_2 electrons into dot 2. The dots are submitted to a potential drop V applied by a voltage generator.

1) Explain in a few lines the Coulomb blockade effect.

2) Express the potential drops V_1, V_2 and V_3 across each capacitor as a function of V, N_1, N_2 and the various capacitances.

3) From now on assume that $C_1=C_2=C_3=C$. Simplify the formulae found in the previous question.

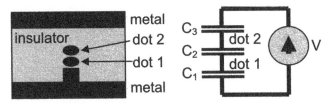

Figure 7.15. *Double dot structure and equivalent circuit*

4) Calculate the electrostatic energy of the two cells.

5) What is the change in the energy generator if we inject an additional electron from the cathode into dot 1.

6) What is the change in the electrostatic energy if we inject an additional electron from the cathode into dot 1?

7) What is the overall energy change if we inject an additional electron from the cathode into dot 1?

8) What is the voltage condition required to inject an additional electron from the cathode into dot 1?

9) What is the generator energy change if one electron is transferred from dot 1 into dot 2?

10) What is the electrostatic energy change if one electron is transferred from dot 1 into dot 2?

11) What is the voltage condition required to transfer one electron from dot 1 into dot 2?

12) From the answers given to the questions above deduce the charge present in the two dots as a function of the applied voltage. A detailed reasoning and careful explanations are required.

13) What is the difference between this system and the single dot case studied during the course?

7.7.2. Exercise

Consider a silicon-on-insulator (SOI) device as in Figure 7.16. All insulating and passivation layers are formed with SiO_2 (dielectric constant $\varepsilon_{SiO2}=3.9$). The active silicon film thickness is $t_{Si}=15$ nm. The buried SiO_2 film which forms the gate has a thickness $t_{SiO2}=100$ nm. The electron transverse effective mass is $m_T=0.19$ m_0 and the longitudinal mass is $m_L=0.98$ m_0. The central Si island is separated from source and drain by a 2 nm thick insulator. $W=50$ nm and $L=100$ nm. The Si work function is supposed to be the same in all Si parts (a phenomenon which occurs only in textbooks).

1) The temperature T is close to zero. We apply $V_G=0$. What is the value that V_{DS} must exceed to overcome the Coulomb blockade?

2) We apply $V_{DS}=0.5$ mV. What is the approximate width of the current peaks in the $I_D(V_G)$ characteristics?

3) All dot dimensions are reduced down to 5 nm. Do you expect the Coulomb blockade to remain qualitatively the same as in the previous structure? If the answer is no, justify this and describe the change in a qualitative way.

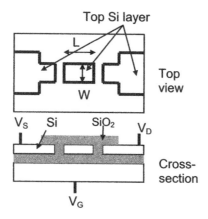

Figure 7.16. *SOI coulomb blockade island*

7.7.3. *Exercise*

Consider a single-electron transistor as defined in the course. Take the same notations as in the course as well. Let $V_D=0$.

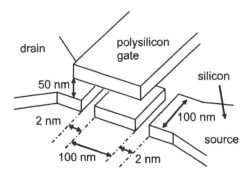

Figure 7.17. *Single-electron transistor*

1) Consider a Coulomb diamond with N electrons added inside the cell. What is the gate voltage interval over which it is defined?

2) Over this gate voltage interval, plot the absolute value of the energy gain or loss for passing from N to (N+1) electrons inside the dot by injecting one more electron from the source.

3) Over this interval, plot the absolute value of the energy gain or loss for passing from (N-1) to N electrons inside the dot by injecting one more electron from the source.

4) Give the voltage for which the minimum energy difference which separates the state with N electrons from any other state is maximized. Calculate this energy

5) Give an approximate temperature value below which a Coulomb blockade effect can be expected with an SET.

6) Numerical application: calculate this temperature with the device defined by Figure 7.17. The active layers are made from silicon and the insulating parts from silicon dioxide (dielectric constant 3.9).

7.8. Bibliography

[GEE 90] GEERLIGS L.J., ANDEREGG V.F., HOLWEG P.A., MOOIJ J.E., "Frequency-locked turnstile device for single electrons", *Physical Review Letters*, vol. 64, no. 22, 1990, p. 2691-2694.

[HEI 01] HEIJ C.P., HADLEY P., MOOIJ J.E., "Single-electron inverter", *Applied Physics Letters*, vol. 78, no. 8, 2001, p. 1140-1142.

[TAK 01] TAKAHASHI Y., ONO Y., FUJIWARA A., INOKAWA H., "Silicon single-electron devices", *Journal of Physics: Condensed Matter*, vol. 14, 2002, p. R995-R1033.

Chapter 8

Specific Interference Effects

8.1. Classical Lagrangian with a magnetic field ⦿

In this chapter we shall not be very rigorous when deriving the equations and introducing the formalism useful to describe the interference effects that we want to study. For those interested by a more serious approach the reader can consult a number of books or articles, a list of which is given in the bibliography at the end of the chapter (in particular see [COH 75] for a brief introduction, [FEY 48] for the "original paper", [FEY 65] for the "original book", [SCH 81] for a more advanced analysis and [DAT 95] for an introduction to Green's function transport formalism in the case of mesoscopic physics).

The idea lying behind the path integral version of quantum mechanics is to replace the standard postulates of QM, with the Schrödinger equation and wave function, by the amplitude probability $K(\vec{r}_2, t_2, \vec{r}_1, t_1) = K(2,1)$ for a particle to go from a space-time point $(\vec{r}_1, t_1)$ to $(\vec{r}_2, t_2)$, and adequate postulates which allow us to calculate K. Then, if at a given time t_1 the wave function is known, we can apply something which looks like the Huygens' principle in optics, i.e. to sum over all possible probability amplitudes to go from all possible points $(\vec{r}_1, t_1)$, which are considered as "sources", to the point $(\vec{r}_2, t_2)$, and thus to calculate the full wave function at a time t_2 if this is repeated for all points of space. In other words, for each possible position at a time t_2 we have to sum over all the possible paths followed by the particle, which can have started from any point $\vec{r}_1$ at time t_1 with a probability amplitude $\psi_1(\vec{r}_1, t_1)$. With the definition of the K function we can write

$$\psi(\vec{r}_2,t_2)= \int K(\vec{r}_2,t_2;\vec{r}_1,t_1)\psi(\vec{r}_1,t_1)\,d^3\vec{r}_1 \tag{8.1}$$

where $K(\vec{r}_2,t_2,\vec{r}_1,t_1) = K(2,1)$ is called the propagator of the Schrödinger equation. It can be demonstrated (easily enough) that K obeys the equation

$$\left(i\hbar\frac{\partial}{\partial t_2} - H(\vec{r}_2,\vec{p}_2)\right)K(\vec{r}_2,t_2;\vec{r}_1,t_1)= i\hbar\delta(t_2 - t_1)\delta(\vec{r}_2 - \vec{r}_1), \tag{8.2}$$

and the solutions of such an equation are also called "Green's functions". It can also be demonstrated (but this is quite difficult) that *K(2,1) is the sum of all partial amplitudes corresponding to all space-time paths from 1 to 2, and a partial amplitude along a given space-time path is given by*

$$K(2,1)= \sum_{\text{all space-time paths } \Gamma} A\exp\left(\frac{iS_\Gamma}{\hbar}\right) \tag{8.3}$$

where S_Γ is the *classical action* along that path,

$$S_\Gamma = \int L(\vec{r},\vec{p},t)\,dt \tag{8.4}$$

and L is the *classical particle Lagrangian* (in the absence of magnetic field $L=T-V$ is the difference between the kinetic and potential energies). A is a normalization constant.

Reciprocally, this result can be taken as a postulate and then from this we can demonstrate the Schrödinger equation. This is what Feynman did, and is the "Lagrangian" version of quantum mechanics.

Of course this can produce many possible paths ... So what can be done with such a formalism? As a matter of fact it is quite difficult to handle if the possible trajectories are numerous, as in empty space, because we must sum over an infinity of spacetime paths. However, it becomes quite interesting to use to understand qualitatively interference effects when the number of possible paths is reduced, as, e.g., in some quantum nano-devices, or when only a restricted number of paths interfere constructively, as, e.g., in the case of weak localization, or to understand some spectacular phenomena induced by a vector potential, as, e.g., with the Aharonov-Bohm effect in mesoscopic rings. This is precisely the aim of this chapter, and although we shall not make any detailed (and most often very complicated) path integral calculations, we shall use the formalism to deduce or explain phenomena which would be more difficult to handle in another approach.

8.2. Classical Lagrangian without a magnetic field ◉

In the absence of a magnetic field, the classical equation of movement of a point particle is given by the second Newton's law:

$$m\frac{d^2\vec{r}}{dt^2} = -\vec{\nabla}U \quad (\Leftrightarrow m\ddot{x} = -\frac{\partial U}{\partial x} \quad m\ddot{y} = -\frac{\partial U}{\partial y} \quad m\ddot{z} = -\frac{\partial U}{\partial z}) \tag{8.5}$$

The Lagrangian is defined as $L=T-U$, where $T=mv^2/2$ is the kinetic energy and U is the potential energy. We easily verify that the laws of motion written above are identical to ($q_i=x,y,z$):

$$\frac{d}{dt}\frac{\partial L}{\partial \dot{q}_i} - \frac{\partial L}{\partial q_i} = 0 , \tag{8.6}$$

which are the Lagrangian equations of motion. With a magnetic field B the law of motion for a particle with charge q is

$$m\frac{d^2\vec{r}}{dt^2} = q\left(\vec{E} + \frac{d\vec{r}}{dt} \wedge \vec{B}\right). \tag{8.7}$$

It is somewhat tedious but not conceptually difficult to show that the Lagrangian law of motion equation (8.6) is still identical to the equation above if the Lagrangian function is now [COH 77]

$$L = \frac{1}{2}m\left(\frac{d\vec{r}}{dt}\right)^2 + q\frac{d\vec{r}}{dt}\vec{A} - qV(\vec{r}) \tag{8.8}$$

where $\vec{A}$ is the vector potential ($\vec{B} = \vec{\nabla} \times \vec{A}$).

8.3. Phase shift due to a magnetic field ●

With a magnetic field the contribution of the vector potential to the action integrated over a particular path is that of the second term in equation (8.8). From equations (8.4) and (8.8) the time can be eliminated and we immediately obtain

$$\Delta S_A = -e \int_0^t \frac{d\vec{r}}{d\tau} \vec{A} \, d\tau = -e \int_\Gamma \vec{A} \, d\vec{l} \cdot \qquad (8.9)$$

The contribution of the vector potential to the action along a path Γ thus has a particularly simple form, since it is proportional to the circulation of the vector potential along that path. It is worth noting that in the case of a closed orbit, we can turn this circulation into the magnetic field flux by applying Stoke's theorem, which states that the circulation of a vector around a closed orbit is equal to the flux of the rotational through the corresponding oriented surface. Thus, the probability amplitude phase shift due to a given trajectory around a *closed* orbit is (see equation (8.9))

$$\Delta \varphi = -\frac{e}{\hbar} \oint \vec{A} d\vec{l} = -\frac{e}{\hbar} \oint_S \vec{B} d\vec{S}. \qquad (8.10)$$

This is a quite simple formula, which means that the phase shift experienced by electron waves traveling over closed orbits is simply proportional to the magnetic field flux through this orbit. Note that it is not even necessary that the electrons themselves be submitted to the magnetic field: imagine that the field is produced by an infinitely long coil located perpendicularly to and inside a ring where some electrons are constrained to move. There is no magnetic field outside the coil, yet the electrons are submitted to a vector potential and they experience a phase shift as given by equation (8.10). This thought experiment was originally proposed by Aharonov and Bohm to demonstrate the physical reality of the vector potential independently of that of the magnetic field, something which is impossible to show in classical mechanics, but which should occur in quantum mechanics.

8.4. Aharonov-Bohm effect in mesoscopic rings ●

8.4.1. *Theory*

Although in a mesoscopic structure the magnetic field is applied to the whole substrate containing the mescoscopic samples, so that such devices cannot really be used to demonstrate the action of a vector potential in areas where the magnetic field is absent, a phenomenon like the Aharonov-Bohm effect can be observed. Submit a mesoscopic, coherent semiconductor ring to a magnetic field B applied perpendicularly to the plane of the ring, as illustrated by Figure 8.1.

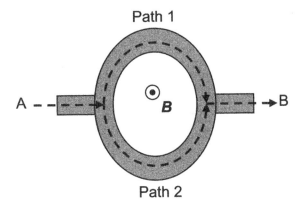

Figure 8.1. *A coherent semiconductor or metal ring submitted to a magnetic field*

The probability amplitudes passing through both arms of one 1D channel are complex numbers which can be written as

$$t_1 = \sqrt{T_1}\, e^{i\left(\varphi_1^0 + \varphi_1^B\right)} \quad t_2 = \sqrt{T_2}\, e^{i\left(\varphi_2^0 + \varphi_2^B\right)}, \tag{8.11}$$

where φ_1^0 and φ_2^0 are the phase shifts which would be obtained without a magnetic field (resulting, e.g., from the action of scatterers in the two arms and from wave propagation along paths 1 and B), and T_1 and T_2 are the transmission probabilities from the two arms, respectively. φ_1^B and φ_2^B are the phase shifts induced by the vector potential as expressed by equation (8.9).

According to the results obtained in the previous section, the phase shift difference between the two arms which is due to the action of the vector potential is simply

$$\varphi_1^B - \varphi_2^B = \frac{e}{\hbar} \oint_{ring} \vec{A}\, d\vec{l} = \frac{eBS}{\hbar} = 2\pi \frac{\Phi}{\Phi_0}, \tag{8.12}$$

where Φ_0 is the elementary flux quantum

$$\Phi_0 = \frac{h}{e}. \tag{8.13}$$

As justified in Chapter 5, we shall assume that the conditions are such that the overall transmission probability is proportional to the modulus of the sum of the complex probability amplitudes in each of the two arms, and to simplify the matter even further we shall only take the sum without explicitly considering any proportionality factor. Therefore, transmission is equal to

$$T = |t_1 + t_2|^2 = T_1 + T_2 + 2\sqrt{T_1 T_2} \cos\left(\varphi^0 + 2\pi \frac{\Phi}{\Phi_0}\right). \qquad (8.14)$$

Applying the Landauer formula to the ring thus gives a conductance

$$G = \frac{2e^2}{h}\left(T_0 + T_B \cos\left(\varphi^0 + 2\pi \frac{\Phi}{\Phi_0}\right)\right). \qquad (8.15)$$

where $\varphi_0 = \varphi_1^0 - \varphi_2^0$. The oscillatory term is due to the magnetic-field induced interference between the two arms; it can be periodically varied simply by sweeping the magnetic field. Thus, with such a solid-state device we can realize a quantum-mechanical interference experiment "just" by using the device at low temperature and measuring the current as a function of magnetic field. We must obtain oscillations with a universal period, depending only on the elementary flux quantum. Note that we only calculated the contribution arising from probability amplitudes corresponding to a direct transmission, but rigorously we should also take into account the possibility for an electron to travel over one full orbit before being transmitted, or to travel over two orbits, etc. The higher the overall transmission, the larger multiple scattering inside the ring and the more pronounced this effect will be. If the coherence length λ_Φ is large enough, we can thus observe additional oscillations with periodicity Φ_0/N, with N integer, due to paths where the particle travels over more than one circular orbit.

Yet another surprising point: from equation (8.15) it would seem that the phase φ_0 can take any arbitrary value. However, *this is in fact not the case*. Since we have a two-terminal device in the linear operation regime, it is obvious that if we reverse the sign of the small voltage that we apply to measure the conductance, the current magnitude is conserved even with an applied magnetic field, so that the squared modulus of the transmission probability obeys $|t_{12}(B)|^2 = |t_{21}(B)|^2$. However, if we now apply the reciprocity relation we can write for the second term $t_{21}(B) = t_{12}(-B)$, so that $G(B) = G(-B)$. Thus, the conductance of a two-terminal device is an even function of the magnetic field. However, if we take this into account, from equation (8.15) it is easy to find that this implies the striking result $\varphi_0 = n\pi$. The phase is "rigid", equal to zero or π, not depending on device geometry! A more correct two-terminal formula is thus

$$G = \frac{2e^2}{h} \left(T_0 + T_B \cos \left(n\pi + 2\pi \frac{\Phi}{\Phi_0} \right) \right).$$
(8.16)

The only phase changes that can be observed, e.g., by varying the Fermi level position are abrupt and equal to π [YAC 96]. Note that this reasoning does not apply to more than two contacts, because if we have more terminals then the overall current conservation does not imply that $t_{pq}(B)=t_{qp}(B)$ (consider a three-terminal device such as in Figure 3.22 in the quantum Hall effect regime: we obviously have $T_{31}(B)\neq0$ whereas $T_{13}(B)=0$).

8.4.2. Aharonov-Bohm effect in the real world

The Aharonov-Bohm effect has now been observed by a large number of experimental scientists involved in mesoscopic physics, and below we reproduce one of these experiments, which very nicely demonstrates the wave and quantum-coherent nature of the electron.

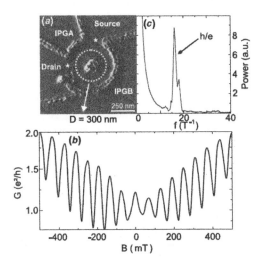

Figure 8.2. *(a)AFM image of an GaAs-GaAlAs heterostructure used to measure (b) Aharanov-Bohm oscillations; the dashed circle in (a) is calculated from the power spectrum (c) (T=25 mK); reprinted with permission from [KEY 02], copyright (2002) by IOP Publishing Ltd*

In Figure 8.2 we can observe the resistance oscillations measured from a GaAs-GaAlAs heterostructure patterned in a 5 nm thick GaAs layer and fabricated in

Germany by U.F. Keyser *et al.* [KEY 02]. The oscillation period with respect to the magnetic field strictly follows the expression that we established in the previous section. It is worth noting that in this experiment the conductance modulation is really huge, exceeding 50%. From the atomic force microscope image in Figure 8.2 we can see that the ring calculated from the Fourier component of the conductance spectrum exactly fits inside the semiconductor ring. The side gates IPGA and IPGB are used to adjust the dephasing and maximize the oscillation amplitude (note that to obtain such a curve the temperature T must be quite low, equal to 25 mK, and the *ac* voltage used for measuring the device conductance is equal to 5 μV). Every 62 mT a flux quantum is added through the ring defined by the structure. The transmission magnitude is also controlled by the gate voltage applied onto contact IPGA, which can reduce or enhance the effective entrance constrictions at the drain and source.

8.5. 1D localization ◑

8.5.1. *Interference effects when λ_Φ exceeds the distance between impurities*

Scatterers (e.g. impurities) are randomly located in mesoscopic conductors. Thus, the interference pattern and conductance depend on each particular scatterer configuration. In devices such as those in Figure 8.3, a different configuration of defects and impurities leads to different transmission conductances, even with seemingly identical samples. It is no longer possible to define a unique quantum resistance $R(L)$ as a function of the conductor length. Calculating or predicting a conductance value only has a statistical meaning (we take a large number of seemingly identical samples and the conductance of all samples is averaged). However, a given sample is not necessarily well described by the average behavior, and statistical fluctuations can become very large!

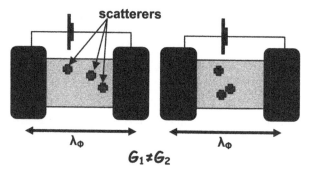

Figure 8.3. *Two seemingly identical mesoscopic conductors with different impurity distributions exhibit different transmission conductances*

The situation is thus quite different from what we expect when measuring macroscopic devices. In the latter case, the departure from the statistical average is much more limited, because in one sample the influence of the many impurities which make the microscopic conductivity exhibit local fluctuations is already averaged, since the conductivity itself is averaged over the conductor length, which is much larger than the typical fluctuation length.

Ohm's law is obeyed if the phase relaxation length is shorter than the average distance between scattering centers, but if λ_φ is larger then the conductor can be assimilated to the series combination of phase-coherent units, each of them with many scatterers and a particular interference pattern.

8.5.2. 1D localization

Here we are going to average the resistance that we (statistically) expect when we increase the device length L of a conductor for which the coherence length is larger than the average distance between the scatterers. In section 5.3 we obtained that putting two scatterers in series in a 1D channel leads to a complex transmission coefficient (or probability amplitude) equal to

$$t = \frac{t_1 t_2}{1 - r_1 r_2} \tag{8.17}$$

and to a transmission probability equal to

$$T = |t|^2 = \frac{T_1 T_2}{1 - 2\sqrt{R_1 R_2}\,\cos\varphi + R_1 R_2}, \tag{8.18}$$

where the angle φ is the phase shift acquired when bouncing back and forth once in between the two scatterers. We also defined the scattering resistance R as $R_0(1-T)/T$ with $R_0 = h/2e^2$ (the overall resistance of the sample is $R + R_C$ with R_C the contact resistance $R_C = R_0$).

Divide a given device into two parts, 1 and 2, and assume that the phases of t_1 and t_2 are completely uncorrelated, so that φ is equally distributed between 0 and 2π. On average this leads to

$$\langle R \rangle = \frac{1}{2\pi} \int_0^{2\pi} \frac{1 - T(\varphi)}{T(\varphi)}\,d\varphi \tag{8.19}$$

and to the relationship

$$\langle R \rangle = \langle R_1 \rangle + \langle R_2 \rangle + \frac{2\langle R_1 \rangle \langle R_2 \rangle}{R_0}. \tag{8.20}$$

If R is small compared to $R_0 = h/2e^2$ we obtain the comforting "ohmic" property $\langle R \rangle = \langle R_1 \rangle + \langle R_2 \rangle$ and everything seems to be "as usual" in a macroscopic conductor. However, now suppose that the device is big enough for $\langle R_1 \rangle$ and $\langle R_2 \rangle$ to be much larger than R_0. Then, from equation (8.20) it clearly appears that the overall average resistance is larger than $\langle R_1 \rangle + \langle R_2 \rangle$! Repeating the same step again and again means that the average resistance increases much faster than with a simple proportionality of the resistance to the device length. Equation (8.20) clearly implies a localization behavior. To go one step further we should examine whether it is possible to deduce or approximate the variation of the average resistance as a function of the channel length. To do this, divide a sample of given length into L (large) and ΔL, small enough for $R(\Delta L)$ to be small with regards to R_0. For the ΔL portion we know that the resistance is additive and thus we demonstrate (Ohm's law from Landauer formula) that

$$T(\Delta L) = \frac{L_0}{\Delta L + L_0}. \tag{8.21}$$

Along with equation (8.20) we can thus write

$$\langle R(L + \Delta L) \rangle = \langle R(L) \rangle + \frac{\Delta L}{L_0} \left(R_0 + 2\langle R(L) \rangle \right), \tag{8.22}$$

so that

$$\frac{d\langle R \rangle}{dL} = \frac{1}{L_0} \left(R_0 + 2\langle R \rangle \right). \tag{8.23}$$

Integrating this simple differential equation leads to

$$\langle R \rangle = \frac{R_0}{2} \left(e^{\frac{2L}{L_0}} - 1 \right). \tag{8.24}$$

We obtained the striking result that in a coherent 1D sample, the resistance increases exponentially with its length L! In other words, the electrons inside a long, imperfect 1D sample do not propagate and are localized. This phenomenon is called 1D localization.

At this point it is worth noting that we voluntarily chose a "good" quantity to be averaged $((1-T)/T)$, to obtain a qualitatively correct formula, but the result would have been quite different when averaging, e.g., $\langle G \rangle$ (for which the result is not amenable to any scaling), or $\langle t^2 \rangle$ (for which we obtain Ohm's law), and saying that $\langle R \rangle$ is, e.g., $1/\langle G \rangle$. This is due to the fact that in general $\langle f(x) \rangle \neq f(\langle x \rangle)$). To be correct the result must be obtained through averaging a variable whose distribution is not singular, and the correct variable is $ln(1+R/R_0)$ [AND 80]. Repeating similar calculations then leads to the correct formula

$$\langle R \rangle = R_0 \left(e^{L/L_0} - 1 \right).$$

(8.25)

In the case of a multi-channel device with M channels it has been demonstrated that the same exponential behavior is obeyed, provided that its overall resistance is larger than R_0. Since for small L $R \approx (1/M)R_0 L/L_0$, the critical length appearing in the exponential is $\lambda_C = ML_0$.

To summarize this section, we can state that quantum wires whose coherence length λ_Φ exceed the localization length λ_C exhibit *strong localization* phenomena (but do not forget that this requires low T!) and very large fluctuations from sample to sample (which are observable with semiconductor devices with small M, but not with metal devices for which M remains extremely high even with very small sections).

8.6. Weak localization ⦿

Another interference effect which lends itself to an easily amenable treatment using the path integral formulation is called weak localization (see, e.g., [BER 83] for a physical interpretation of the effect). In a 2D system, at low T the phase relaxation length may become much larger than the mean free path. In such a case an electron wave can experience many elastic scattering events before losing its phase. Let us adopt the point of view of the path integrals. If an electron is located at a site labeled A, we can calculate the probability amplitude of moving to another site B, or returning back to the initial position A (see Figure 8.4). The probability amplitude for going from A to B is obtained by summing over all the probability amplitudes corresponding to all possible paths that can be followed by the particle. This is given by

$$P(A \to B) = \left| \sum_{all\ paths} A_i \right|^2 = \sum |A_i|^2 + \sum_{i \neq j} A_i A_j^* \tag{8.26}$$

In equation (8.26) we sum complex numbers, so that we have a contribution corresponding to the sum of the transmission probabilities and a contribution corresponding to crossed terms. If the possible paths are numerous and random these crossed terms average to zero. However, the situation is quite different when we study the probability amplitude to go back to A: each possible path can be followed in two directions, and of course the two corresponding probability amplitudes interfere constructively. This constructive interference makes the probability of returning much higher:

$$P(A \to A) = 2 \sum_{\substack{all\ paths \\ from\ A\ to\ A}} |A_i|^2 = 2P_{classical}(A \to A). \tag{8.27}$$

We say that backscattering is enhanced.

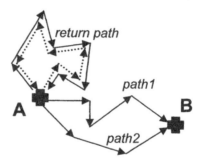

Figure 8.4. *Possible quantum paths followed by a particle originating in point A, to go either to point B or to return to point A. For returning two paths always interfere constructively, because one loop can be followed in two directions*

Putting a carrier somewhere at time *t=0* and calculating the probability distribution resulting from quantum diffusion after a time *t* leads to something like in Figure 8.5.

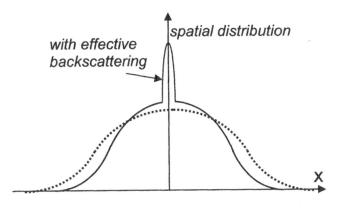

Figure 8.5. *Probability density to find a particle somewhere after a time t if it was at the origin at time t=0. Weak localization enhances the probability close to the origin*

We can heuristically assess the weak localization correction in 1D. Expanding equation (8.25) with small L leads to

$$\langle R \rangle = \frac{R_0 L}{L_C}\left(1 + \frac{L}{2L_C}\right) = R_{classical}\left(1 + \frac{R_{classical}}{2R_0}\right). \tag{8.28}$$

Thus, in 1D the quantum corrections to the resistance and the conductance are of order

$$\Delta R \approx \frac{R_{classical}^2}{2R_0} \tag{8.29}$$

so that the correction to the conductance is of order

$$\Delta G \approx -\frac{e^2}{h}. \tag{8.30}$$

Although we shall not demonstrate it, we give below the correction corresponding to a 2D system:

$$\Delta\sigma_{2D} = -\frac{e^2}{\pi h}\ln\left(\frac{\tau_\Phi}{\tau_m}\right). \tag{8.31}$$

In equation (8.31), τ_Φ is the phase relaxation time and τ_m is the momentum relaxation time. Application of a magnetic field destructs the constructive interference between two return paths traversed in two opposite directions, because the vector potential does not induce the same dephasing for paths with opposite directions. This leads to a negative magnetoresistance correction and is in fact a practical means to determine the phase relaxation time.

8.7. Universal conductance fluctuations ○

In disordered systems operated at low T, the conductance exhibits random fluctuations as a function of magnetic field, chemical potential, or from sample to sample. Experimentally, the RMS value of these fluctuations is "universal", since in all samples it remains of order $\delta G \approx e^2/h$. For this reason they have been called Universal Conductance Fluctuations (UCFs). At a low enough temperature, these fluctuations are independent of the degree of disorder and of the sample dimensions. The only condition to be fulfilled is that the inelastic scattering mean free path be larger than any of the sample dimensions.

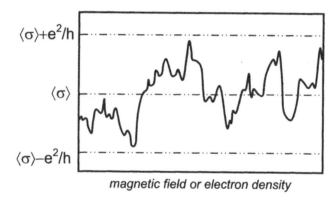

magnetic field or electron density

Figure 8.6. *In disordered mesoscopic systems the conductance exhibits universal fluctuations of order e^2/h around the mean value*

Typically, UCFs exhibit a shape such as that depicted in Figure 8.6. It is rather difficult to rigorously demonstrate this result and here we shall just give a heuristic derivation due to P.A. Lee [LEE87], for which we will once again appeal to Landauer's formula. Label the incoming 1D channels with index a and the outgoing channels with index b. We have for a given sample (or a given magnetic field):

$$G = \frac{2e^2}{h} \sum_{a,b=1}^{N} |t_{a,b}|^2 \qquad (8.32)$$

where $|t_{a,b}|^2$ is the transmission probability which connects incoming channel a to channel b. $|t_{a,b}|^2$ is the sum of all probability amplitudes associated with each Feynman path that connects channel a to channel b:

$$t_{a,b} = \sum_{i=1}^{M} A_{a,b}(i).$$

(8.33)

Now suppose that we can make these amplitudes fluctuate under the effect of a magnetic field or a change in the impurity configuration. We cannot calculate in a straightforward way the moments of the conductance fluctuations from the transmission amplitudes because there is a correlation between the transmission probabilities for different pairs of incident and outgoing channels (transmission through the disordered regions involves a large number of collisions, and thus a given sequence of scattering events is shared by different channels). However, it is more reasonable to assume that the reflection probabilities are uncorrelated, for reflection in the source must be dominated by only a few scattering events. We can thus calculate the variance of G by expressing it as a function of the reflection probabilities through the current conservation relationship:

$$\sum_{a,b=1}^{N} |t_{a,b}|^2 + \sum_{a,b=1}^{N} |r_{a,b}|^2 = N.$$

(8.34)

The conductance expression becomes

$$G = \frac{2e^2}{h}\left(N - \sum_{a,b=1}^{N} |r_{a,b}|^2\right).$$

(8.35)

Now we just remind ourselves that if X is a random variable and $f_X(x)$ is the corresponding probability distribution, the average or expectation value of a function $g(x)$ is given by

$$E(g(x)) = \langle g(x)\rangle = \int g(x) f_X(x) dx.$$

(8.36)

and the variance or mean square deviation of a random variable X is defined as

$$\text{var}(X) = \left\langle X - \langle X\rangle^2\right\rangle = \sigma_X^2$$

(8.37)

where σ_X is the standard deviation. If two random variables X and Y are independent, then it is easy to show that we have the relationship

$$\sigma_{X+Y}^2 = \sigma_X^2 + \sigma_Y^2. \tag{8.38}$$

From equation (8.35) the variance of the conductance fluctuations is

$$\sigma_G^2 = \left(\frac{2e^2}{h}\right)^2 \sum_{a,b=1}^{N} \left\langle |r_{a,b}|^4 - \left\langle |r_{a,b}|^2 \right\rangle^2 \right\rangle. \tag{8.39}$$

The sum above contains N^2 terms. We can arrange the terms corresponding to the reflection amplitudes as

$$\left\langle |r_{a,b}|^2 \right\rangle = \left\langle \left| \sum_i A_{a,b}(i) \right|^2 \right\rangle = \left\langle \sum_{i,j=1}^{M} A_{a,b}(i) A_{a,b}^*(j) \right\rangle$$
$$= \sum_{i,j=1}^{M} \left\langle A_{a,b}(i) A_{a,b}^*(j) \right\rangle = \sum_{i=1}^{M} \left\langle |A_{a,b}(i)|^2 \right\rangle. \tag{8.40}$$

The last equality is obtained by stating that if $i \neq j$ then the phases are random and uncorrelated, so that $\langle A_{a,b}(i) A_{a,b}^*(j) \rangle = 0$. We also have to assess the quantity

$$\left\langle |r_{a,b}|^4 \right\rangle = \left\langle \left| \sum_i A_{a,b}(i) \right|^4 \right\rangle = \left\langle \sum_{i,j,k,l=1}^{M} A_{a,b}(i) A_{a,b}^*(j) A_{a,b}(k) A_{a,b}^*(l) \right\rangle$$
$$= \sum_{i,j,k,l=1}^{M} \left\langle |A_{ab}(i)|^2 \right\rangle \left\langle |A_{ab}(j)|^2 \right\rangle \left\langle \delta_{i,j}\delta_{k,l} + \delta_{i,l}\delta_{j,k} \right\rangle \tag{8.41}$$
$$= 2 \left(\sum_{i=1}^{M} \left\langle |A_{a,b}(i)|^2 \right\rangle \right)^2 = 2 \left\langle |r_{a,b}|^2 \right\rangle^2$$

The trick used to obtain the second line in equation (8.41) is that not only equal i and j and equal k and l lead to modules, but also equal i and l and equal j and k. Thus, we arrive at

$$\sigma_G^2 = \left(\frac{2e^2}{h}\right)^2 N^2 \left\langle |r_{a,b}|^2 \right\rangle^2. \tag{8.42}$$

Along with the relation

$$\langle G \rangle = \frac{2e^2}{h} \frac{\lambda}{NL} \tag{8.43}$$

we finally obtain

$$\delta G = \sigma_G = \frac{2e^2}{h}, \tag{8.44}$$

which is incorrect by a factor of 2, but nevertheless gives us the same order of magnitude as the more exact result $\delta G \approx e^2/h$.

8.8. Bibliography

[AND 80] ANDERSON P.W., THOULESS D.J., ABRAHAMS E., FISHER D.S, "New method for a scaling theory of localization", *Physical Review B*, vol. 22, no. 8, 1980, p. 3519-3526.

[BER 83] BERGMANN G., "Physical interpretation of weak localization: a time-of-flight experiment with conduction electrons", *Physical Review B*, vol. 28, no. 6, 1983, p. 2914-2920.

[COH 77] COHEN-TANNOUDJI C., DIU B., LALOE F., *Quantum Mechanics*, Hermann and Wiley & Sons, 1977.

[DAT 95] DATTA S., *Electronic Transport in Mesoscopic Systems*, Cambridge, Cambridge University Press, 1995.

[FEY 48] FEYNMAN R.P., "Space-time approach to non-relativistic quantum mechanics", *Reviews of Modern Physics*, vol. 20, no. 2, 1948, p. 367-387.

[FEY 65] FEYNMAN R.P., HIBBS A.R., *Quantum Mechanics and Path Integrals*, New York, McGraw-Hill, 1965.

[KEY 02] KEYSER U.F., BORCK S., HAUG R.J., BICHLER M., ABSTREITER G., WEGSCHEIDER W., "Aharanov-Bohm oscillations of a tuneable quantum ring", *Semiconductor Science and Technology*, vol. 17, 2002, p. L22-L24.

[LEE 87] LEE P.A., "Universal conductance fluctuations in disordered metals", *Physica A*, vol. 140A, no. 1-2, 1986, p. 169-174.

[SCH 81], SCHULMAN L.S., *Techniques and Applications of Path Integration*, New York, Wiley, 1981.

[YAC 96] YACOBY A., SCHUSTER R., HEIBLUM M., "Phase rigidity and h/2e oscillations in a single-ring Aharanov-Bohm experiment", *Physical Review B*, vol. 53, no. 15, 1996, p. 9583-9586.

Chapter 9

Graphene and Carbon Nanotubes

9.1. Introduction ●

In most semiconductor nanostructures, quantum-coherent or ballistic effects are observed at low temperature, and people are struggling to fabricate devices which would exhibit such effects at 300 K, with the hope that they might lead to new electron device concepts. In 1991, the fortuitous discovery of carbon nanotubes by S. Iijima [IIJ 91] and a rapid rate of progress in the determination of their astonishing band structure and transport properties gave a new impulse to the field[1]. These nano-objects are composed of carbon atoms forming a hexagonal lattice suitably rolled in the form of a cylinder, with no dangling bond (except at its extremities).

The good quantum confinement and the remarkable values attained by the coherence length in such devices produced a great deal of interest in the research community. This is due to a number of reasons, among which two prevail. Firstly, in contrast with any other semi-conducting material, the carbon nanotubes do not exhibit dangling bonds or defects at their surface and thus are not as sensitive to interface defects as is the case with conventional surface devices. Secondly, in carbon nanotubes the phonon mean free path can exceed one micrometer at room temperature, with acoustic phonon energies larger than 150 meV, so that low-voltage transport may be ballistic even at 300 K [PUR 07].

---

1 In fact, although not a nanotube, the first carbon fiber was developed by Thomas A. Edison in 1892, and vapor-grown carbon tubules with diameters less than 10 nm were already observed by M. Endo in 1975; see [SAI 98] for more historical details and a review.

However, there are still major impediments to a successful development of carbon nanotube electronics, mainly (but not only) because at the present time nobody is capable of assembling many of them in a way enabling us to organize a very complex circuit architecture. Unlike silicon-based integrated circuits, it is not possible to define small elementary units using etching steps and oxidation processing, so that most often nanotubes have to be put at the desired position one-by-one. People have not yet found a way to grow them at pre-defined locations in an acceptable, controllable and *industrial* way, and with reproducible electronic properties, or a way to produce some kind of self-organization which would allow us to produce circuits.

Figure 9.1. *A carbon nanotube is formed of carbon atoms hexagonally packed in a plane which is subsequently rolled into the form a cylinder. There is no dangling bond*

Graphene is a 2D equivalent of carbon nanotubes (see Figure 9.2), and is in fact a very old material, since what is now called graphene is nothing but basic carbon sheets which when stacked are better known as graphite; the same kind of graphite as you use in your pencil (note again that carbon nanotubes themselves are nothing but coaxially rolled 2D graphene sheets). A free-standing 2D lattice cannot be maintained in a plane and must exhibit deformations; however, it has been realized quite recently that graphene can be grown or deposited on a substrate, so that it can be maintained in a single plane and yet keep most of the electronic properties which would be expected from a pure 2D crystal. By singling out one such 2D mono-atomic layer it is possible to form a medium in which the coherence length can also reach quite attractive values, even at room temperature. Provided that we can form a good 2D graphene layer onto a foreign substrate, it becomes feasible to pattern such a layer just with the same principles as with silicon or gallium arsenide technology (with the obvious restriction that right now graphene processing is not very mature, and taking into account that it is not so easy either to make good ohmic contacts to a mono-atomic layer, or to produce just a single graphene layer by, e.g., silicon carbide graphitization or by exfoliating it from bulk graphite materials).

A good coherence length is not the only remarkable property of graphene, and its band structure is also extremely specific: unlike most semi-conducting materials, it possesses a zero bandgap, so that the valence and conduction bands are connected together by a single point in k-space and form a unique band (without intentional doping the Fermi level is positioned just at this particular point, so that we talk about a zero-gap semiconductor rather than a metal). Thus, there is no real minimum in the "conduction band", so that the derivative of the energy E versus wavevector k is not required to vanish at the intersection point to ensure velocity continuity at $k=0$, as it does at the bottom of a conventional conduction band. It turns out that the relationship between energy and wavevector is linear, and we can write $E=\alpha|k|$. Therefore electrons in graphene behave somewhat like photons, as their group velocity is constant and independent of energy. Graphene thus offers the possibility of performing some "photon-like" physics with electrons, but maintaining some essential features of electron dynamic as well as their fermionic properties. This leads to quite remarkable consequences or "anomalies" in a number of phenomena already detailed in the previous sections for conventional semiconductors. The aim of this chapter is both to introduce such remarkable features and to illustrate through a concrete example many of the previously developed physical notions.

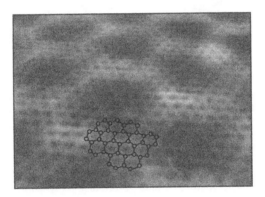

Figure 9.2. *2D graphene layer grown on silicon carbide and observed from the top by scanning tunneling microscopy (STM). The bright areas with a higher current correspond to the carbon atoms and form a regular hexagonal lattice (the larger scale hexagons are due to corrugation by the underlying substrate); by courtesy of P. Mallet and J.Y. Veuillen (Institut Néel, CNRS)*

9.2. Graphene band structure ●

Graphene is made of carbon atoms and crystallizes in the hexagonal form of a honeycomb lattice (see Figure 9.2). Each carbon atom contains six electrons. The first two electrons are not involved in bonding because they are strongly bonded to the first atomic level (*1s* state). The remaining four electrons lie in the next levels, whose orbitals are thus linear combinations of the *2s* or the *2p* kind (see Figure 9.3).

Depending on the location of surrounding atoms, these four electrons will lie in states which to first order are linear combinations of one *2s* orbital and three *2p* orbitals. In a hexagonal 2D lattice one carbon atom forms three strong bonds with neighboring atoms. It turns out that these planar bonds form angles of $2\pi/3$, and this corresponds to an orbital hybridization of the sp^2 type (each hybrid orbital is the combination of one *2s* wave and two p_x and p_y waves with the adequate coefficients). These electrons lie in extremely narrow bands which are fully occupied, and thus cannot contribute to electron transport. As a consequence, there remains one electron per carbon atom not involved in strong bonding, and the only remaining orbital is of the $2p_z$-type. We are thus going to apply the tight-binding method developed in section 2.7 to those p_z wave functions to determine the gross features of the graphene band structure around the Fermi energy.

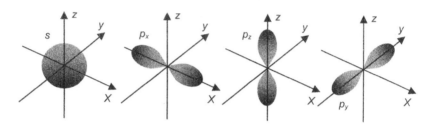

Figure 9.3. *2s and 2p orbitals of a carbon atom. In a graphene sheet aligned along the xy plane only the p_z orbital is not involved in strong bonding and gives rise to delocalized energy bands*

We can choose as a primitive unit cell that of Wigner-Seitz, as illustrated in Figure 9.4. The two basis vectors $\vec{a}_1$ and $\vec{a}_2$ are also shown in Figure 9.4. One unit cell obviously contains two inequivalent carbon atom sites, which we design by A and B. The full lattice is formed by two interpenetrated triangular sublattices corresponding to atoms A and atoms B, respectively. The carbon-carbon distance is equal to $a_0=1.42$ Å, and a rapid trigonometric calculation gives a length $a=a_1=a_2=a_0\times\sqrt{3}$ (see Figure 9.4). Choosing A as the origin, the lattice vectors and the reciprocal lattice vectors (use equation (10.17) of section 10.2) are given by

$$\vec{a}_1 = \begin{pmatrix} a\dfrac{\sqrt{3}}{2} \\ -\dfrac{a}{2} \end{pmatrix} \quad \vec{a}_2 = \begin{pmatrix} a\dfrac{\sqrt{3}}{2} \\ \dfrac{a}{2} \end{pmatrix} \quad \vec{b}_1 = \begin{pmatrix} \dfrac{2\pi}{a\sqrt{3}} \\ -\dfrac{2\pi}{a} \end{pmatrix} \quad \vec{b}_2 = \begin{pmatrix} \dfrac{2\pi}{a\sqrt{3}} \\ \dfrac{2\pi}{a} \end{pmatrix}. \tag{9.1}$$

$\vec{b}_1$ and $\vec{b}_2$ have the same magnitude and their direction is indicated in Figure 9.5. By drawing all vectors that are integer linear combinations of those two reciprocal vectors in a plane we should be easily convinced that we obtain a face-centered 2D hexagonal lattice (see Figure 9.5). Each point of the reciprocal lattice sees exactly the same neighbors, so that we have a one-point-basis reciprocal lattice and the first Brillouin zone is a hexagon.

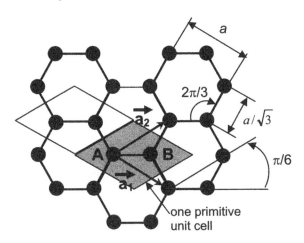

Figure 9.4. *2D hexagonal graphene lattice with two inequivalent atoms per unit cell. Two observers positioned at atomic sites A and B and looking towards the same direction do not see the same landscape*

We are now ready to use these parameters to assess the band structure through the tight-binding method developed in section 2.7. Since the lattice has a two-point basis, for each wavevector $\vec{k}$ we must use a linear combination $\psi_k(\vec{r}) = \alpha_A \psi_k^A(\vec{r}) + \alpha_B \psi_k^B(\vec{r})$ of tight-binding Bloch waves corresponding to the sums over the inequivalent sites A or B, respectively (see section 2.7):

$$\psi_k(\vec{r}) = \alpha_A \sum_{\vec{R}_A} e^{i\vec{k}\vec{R}_A} \phi_{at}^A(\vec{r} - \vec{R}_A) + \alpha_B \sum_{\vec{R}_B} e^{i\vec{k}\vec{R}_B} \phi_{at}^B(\vec{r} - \vec{R}_B) \qquad (9.2)$$

As thoroughly described in section 10.6, minimizing the energy expectation value gives rise to two bands, with a dispersion relationship $E = H_{11} \pm |H_{12}|$, given by equation $(2.86)^2$. In equation (2.86) we have neglected the overlap integrals of the denominator. Thus, we have to calculate the integrals H_{11} and H_{12} defined by equation (2.87). H_{11} can be explicitly written as

---

2 In order to understand this result, it is advisable to read sections 2.7, 10.5 and 10.6, before going on with this section!

$$H_{11} = \left\langle \varphi_k^A \left| H \right| \varphi_k^A \right\rangle = \frac{1}{N} \sum_A \sum_{A'} \int e^{-i\vec{k}\vec{r}_A} \psi_{at}^* \left(\vec{r} - \vec{r}_A \right) H \psi_{at} \left(\vec{r} - \vec{r}_{A'} \right) e^{i\vec{k}\vec{r}_{A'}} d^3\vec{r} \qquad (9.3)$$

also equivalent to

$$H_{11} = \frac{1}{N} \left(N \int \psi_{at}^* \left(\vec{r} \right) H \psi_{at} \left(\vec{r} \right) d^3\vec{r} + N \sum_{n=1}^{6} e^{i\vec{k}\vec{r}_n} \int \psi_{at}^* \left(\vec{r} \right) H \psi_{at} \left(\vec{r} - \vec{r}_n \right) d^3\vec{r} \right) \qquad (9.4)$$

where we separated the terms with $A=A'$ and $A \neq A'$, and kept only the sum over each of the six nearest neighbors for $A \neq A'$. For each lattice site A this corresponds to six vectors $\vec{r}_n$ as defined in Figure 9.6 (left).

We define E_0 and γ_0 as the first and second integrals appearing on the right-hand-side of equation (9.4) (the second integral does not change with the index n of the nearest-neighbor):

$$E_0 = \int \psi_{at}^* \left(\vec{r} \right) H \psi_{at} \left(\vec{r} \right) d^3\vec{r}, \qquad \gamma_0 = -\int \psi_{at}^* \left(\vec{r} \right) H \psi_{at} \left(\vec{r} - \vec{r}_n \right) d^3\vec{r} . \qquad (9.5)$$

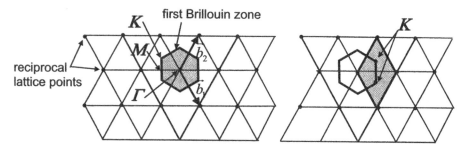

Figure 9.5. *Graphene reciprocal lattice, reciprocal lattice vectors and first Brillouin zone*

These integrals should be numerically calculated from the knowledge of the carbon atomic orbitals and the full potential. The $\vec{r}_n$ s are given by

$$\vec{r}_1 = \begin{pmatrix} 0 \\ -a \end{pmatrix} \quad \vec{r}_2 = \begin{pmatrix} \dfrac{a\sqrt{3}}{2} \\ \dfrac{a}{2} \end{pmatrix} \quad \vec{r}_3 = \begin{pmatrix} \dfrac{a\sqrt{3}}{2} \\ \dfrac{a}{2} \end{pmatrix}, \quad etc. \qquad (9.6)$$

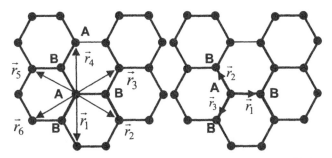

Figure 9.6. *A-type nearest neighbors (left) and B-type nearest-neighbors (right)*
of an A-type carbon atom in graphene, and corresponding vectors

The imaginary parts of the exponentials appearing in equation (9.4) mutually
cancel each other out since each of the six vectors has an exactly opposed
counterpart, and simple trigonometric simplifications of the real part give

$$H_{11} = E_0 - 2\gamma_0 \left(\cos\left(k_y a\right) + 2 \cos\left(k_y a / 2\right) \cos\left(k_x a \sqrt{3} / 2\right) \right). \quad (9.7)$$

H_{12} can be explicitly written as

$$H_{12} = \left\langle \varphi_k^A \middle| H \middle| \varphi_k^B \right\rangle = \frac{1}{N} \sum_A \sum_B \int e^{-i\vec{k}\left(\vec{r}_A - \vec{r}_B\right)} \psi_{at}^*\left(\vec{r} - \vec{r}_A\right) H \psi_{at}\left(\vec{r} - \vec{r}_B\right) d^3r, \quad (9.8)$$

also equivalent to

$$H_{12} = \sum_{n=1}^{3} e^{i\vec{k}\vec{r}_n} \int \psi_{at}^*\left(\vec{r}\right) H \psi_{at}\left(\vec{r} - \vec{r}_n\right) d^3\vec{r} \quad (9.9)$$

if for each atom A we keep only the hopping integrals corresponding to nearest
neighbors of type B, so that we now have to consider the three vectors indicated in
Figure 9.6 (right). This "new" set of $\vec{r}_n$ vectors is given by

$$\vec{r}_1 = \begin{pmatrix} \dfrac{a}{\sqrt{3}} \\ 0 \end{pmatrix} \quad \vec{r}_2 = \begin{pmatrix} -\dfrac{a}{2\sqrt{3}} \\ \dfrac{a}{2} \end{pmatrix} \quad and \quad \vec{r}_3 = \begin{pmatrix} -\dfrac{a}{2\sqrt{3}} \\ -\dfrac{a}{2} \end{pmatrix}. \quad (9.10)$$

Defining β_0 as the integral in the right-hand-side of equation (9.9) (which does not change with n), explicit calculation then leads to

$$|H_{12}|^2 = \beta_0^2 \left(1 + 4\cos^2\left(k_y a / 2\right) + 4\cos\left(k_y a / 2\right)\cos\left(k_x a\sqrt{3} / 2\right) \right) \qquad (9.11)$$

(express the exponential terms in equation (9.9) as $\cos(\vec{k}\vec{r_n})$+isin($\vec{k}\vec{r_n}$), calculate the square modulus by summing the square of the real and imaginary parts and proceed to the appropriate trigonometric simplifications). It can be noted that for an atom at a site A, the nearest neighbors of the same type are farther away than the nearest-neighbors on sites B, so that we can expect that $\beta_0 >> \gamma_0$. Besides, from equation (9.11) it can be checked that at any corner point K of the Brillouin zone, H_{12} vanishes, so that the valence and conduction bands are touching one another. Neglecting γ_0 leads to the simplified dispersion relationship

$$E = E_0 \pm \beta_0 \sqrt{1 + 4\cos^2\left(k_y a / 2\right) + 4\cos\left(k_y a / 2\right)\cos\left(k_x a\sqrt{3} / 2\right)} \qquad (9.12)$$

The value of β_0 is around 3.12 eV. Plotting the two bands as a function of the main crystallographic orientations leads to Figure 9.7 (in which γ_0 has been neglected in front of β_0). Also shown in Figure 9.7 are the energy contours for the conduction band (left). There is a discontinuity in the energy derivative at the corner points K of the Brillouin zone. Since there are two carbon atoms per unit cell, each giving two electrons to the p_z bands, with spin degeneracy we expect the electrons to fill exactly the valence band so that the Fermi level is positioned at the intersection point[3]. In addition, even if we take γ_0 into account, close to a K point (i.e. for small $|k-k_c|$ values where k_c is the wavevector of a point K, for instance with coordinates $(2\pi/a\sqrt{3}, 2\pi/3a)$), the second order terms are negligible with regards to the first order terms and we have

$$E - E_0 = \beta_0 \frac{\sqrt{3}}{2} a \left| \vec{k} - \vec{k_c} \right| \qquad (9.13)$$

Thus, we did not only find that the valence and conduction bands are connected together by one point and that graphene is gapless, but also that the energy wavevector dispersion relationship is linear around this point.

---

3 This point of contact between the valence and conduction bands explains why the color of graphite, which is a stacking of graphene planes, is black, since any energy transition is acceptable and all visible photons can be absorbed.

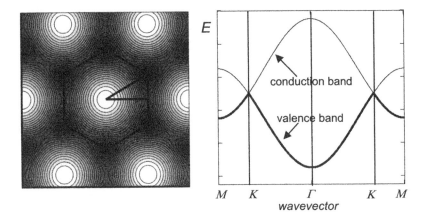

Figure 9.7. *Energy contours of the graphene conduction band (left) and electron valence and conduction bands plotted in some particular crystallographic directions (right), and calculated in a simple tight-binding approximation. The bold line is for the valence band, and the simple line is for the conduction band*

From equation (2.34) this means that the electron velocity no longer depends on energy and is a constant v_c, just as for photons or any massless particle. We can re-write equation (9.13) by taking the origin of energy at E_0 and the origin of momentum at point K:

$$E = v_c \hbar k. \tag{9.14}$$

The point in k-space with $k=0$ is called the Dirac point (due to the analogy with massless relativistic particles described by a Dirac equation). The value for v_C is about 1,000 km/s (the light velocity c divided by 300).

Of course, electrons in graphene are not actually massless. The effective mass describes how the electron wave answers both to an external force and to the forces exerted by the atomic lattice. Its vanishing indicates that whatever the electron kinetic energy is, the lattice acts so as to maintain the velocity at a constant value. In a periodic atomic array the electron wave propagates without attenuation because the scattered waves interfere constructively. We can understand the constant velocity as the fact that although the wavevector increases under the action of the field, the constructive interference changes so as to maintain as constant the group velocity in the direction of propagation. We can develop a somewhat artificial and oversimplified but clear analogy by conceiving a game for which the more we accelerate, the more we are forced to make zig-zags along a straight trajectory, so that the velocity calculated through the straight trajectory does not change. Thus, although there is no acceleration in the force direction, this does not mean that an electron does not gain energy from an electric field (as in our hypothetical game, we can imagine that the constructive interference results from longer scattering paths in directions other than that of propagation as the wavevector value is increased). In addition, in contrast with relativistic, massless particles here the electron velocity is not independent of the reference frame, and there is, in fact, one preferential reference frame which is that of the atom lattice, playing here for the electrons the role which was wrongly ascribed to the aether as a support for electromagnetic wave propagation, before the advent of special relativity.

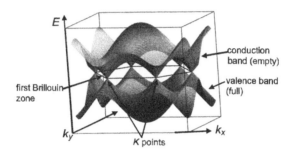

Figure 9.8. *A pictorial view of the graphene band structure as calculated in the tight-binding approximation*

It is worth noting that hexagonality is not enough to obtain a linear dispersion curve, but the inequivalent sites must be occupied by the same type of atoms. For example, boron nitride (BN) can crystallize in the same structure as graphene, but sites A are occupied by boron atoms and sites B by nitrogen atoms. H_{11} and H_{22} are thus different, so that calculating the dispersion relationship using equation (10.105) of section 10.6 without making the simplification $H_{11}=H_{22}$ leads to the apparition of a gap between the valence and conduction bands. Boron nitride thus owns a large bandgap, which explains why it is not conductive and exhibits a whitish aspect.

A pictorial view of the graphene band structure is given in Figure 9.8, in which we can see that at each corner K of the Brillouin zone, the valence band, which is completely filled with electrons, touches the empty conduction band. Each of the six conduction band cones is shared between three cells, so that on average we can count two of these cones per cell in k-space (to obtain only two cones per cell and not six thirds we might choose another cell, such as that depicted in Figure 9.5, right). Thus the conduction band has a two-fold valley degeneracy $g_V=2$, to be added to the two-fold spin degeneracy $g_s=2$. The overall degeneracy is thus equal to four.

There is a mirror symmetry between the electron and hole states. Taking into account the linear dispersion relationship, the density of states in the conduction band is straightforwardly obtained by using a reasoning similar to that used in section 2.10:

$$D(E) = \frac{2}{\pi \hbar^2 v_c^2} E .$$

(9.15)

This density of states vanishes at the Dirac point. A quite interesting feature is that at the intersection point; there are four occupied valence band states in the first Brillouin zone, but also four occupied conduction band states. Thus, at the cone apex electrons and holes are degenerate.

9.3. Integer quantum hall effect in graphene ⑩

A full treatment of this effect would require the use of group theory and irreducible representations to establish the Dirac-like equation which must be obeyed by the electrons and holes. However, even a simple use of the semi-classical approximation can give us the correct Landau level formula. Proceeding as in section 2.18.4.1, we can look for the conditions that the wavevector has to fulfill to allow for the existence of closed orbits. First, by accounting for the constancy of the group velocity v_c we can re-write equation (2.46) as

$$\hbar \frac{d\vec{k}}{dt} = -e\left(\vec{v}_c \wedge \vec{B}\right),$$

(9.16)

so that we obtain the system

$$\frac{dk_x}{dt} = -\frac{eBv_c^2}{E} k_y = -\omega_c k_y ,$$

$$\frac{dk_y}{dt} = \omega_c k_x$$

(9.17)

which, solved for k_x and k_y in the very same way as in section 2.18.3, leads to

$$k_x = k \cos\left(\omega_c t + \phi\right)$$
$$k_y = k \sin\left(\omega_c t + \phi\right),$$

(9.18)

Thus, the wavevector components rotate with a rotation speed equal to

$$\omega_c = \frac{eBv_c^2}{E},$$
(9.19)

which is not independent of energy as in the parabolic dispersion case but inversely proportional to it (it is thus highly pathological when k tends to zero, which is precisely the point which should be treated in a more rigorous fashion). If we add the phase shift due to the wavevector and that due to the magnetic field as in section 2.18.4.1, the orbits in real space must verify the condition

$$2\pi r_c k - \frac{eB}{\hbar}\pi r_c^2 = 2n\pi .$$
(9.20)

where the cyclotron radius r_c is equal to v_c/ω_c. From equations (9.20) and (9.14) it is straightforward to find that the energy levels are given by

$$E = \pm v_c \sqrt{2eB\hbar} \times \sqrt{n},$$
(9.21)

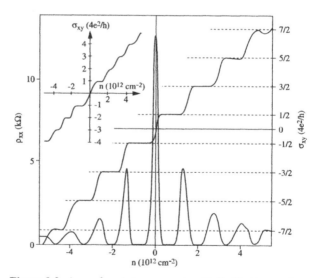

Figure 9.9. *Anomalous integer quantum Hall effect in graphene*

in marked contrast with the conventional case. The plus or minus sign depends on whether the valence or conduction band is considered, but here with the restriction that the index $n=0$ should not be authorized as, in the semi-classical scheme, it is forbidden by the uncertainty principle (the electron cannot be precisely localized at a space point with a zero wavevector). As a matter of fact, not only an equation similar to equation (9.21) is also obtained by a more rigorous approach, but then even the index $n=0$ is allowed. Solving the problem in a more rigorous fashion requires us to apply to the envelope wave functions a system of equations similar to equation (2.113), but in which the possibility of degeneracy has been taken into account, since the conduction and valence bands touch at the K points[4].

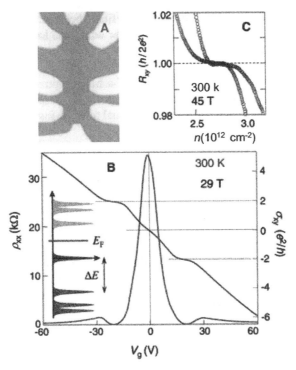

Figure 9.10. *Room temperature quantum Hall effect in graphene; from Novoselov et al., Science, vol. 315, p. 1379 (2007), reprinted with permission from AAAS [NOV 07]*

---

4 Treating such a degeneracy results in a set of coupled differential equations with relatively simple expressions, formally analogous to the relativistic massless Dirac equation. However, the derivation of this equation requires much ingenuity in the general case, and we shall not reproduce the result without the demonstration in order to maintain the self-contained character of the book. Thus, we refer any reader with an interest in going beyond this simple heuristic derivation to more advanced textbooks or articles; see, e.g., [DIV 84] and [ZHE 02].

Not only this anomalous quantum Hall effect was observed (see Figure 9.9, and the steps in σ_{XY} equal to $4e^2/h$ instead of the conventional $2e^2/h$ value, due to the valley degeneracy), but it can be evidenced even at room temperature (Figure 9.10), projecting a research field confined in the low temperature world during decades onto more human conditions (but the magnetic field value).

The possibility of observing such a ballistic effect at 300 K is due to the combined action of a number of factors. Firstly, it is possible to populate the lowest Landau bands even under very high magnetic fields because the electron densities can be quite high, so that we can apply fields for which the Landau level separation exceeds the thermal energy. Secondly, the mobility remains high enough to fulfil the condition $\mu B > 1$ even at 300K. A striking point is the existence of the zero energy Landau level, which leads to the formation of a half-quantized conductance value for the first plateau, in marked contrast with the parabolic dispersion case (as can be seen in Figure 9.9, where the first plateau is equal to $2e^2/h$ instead of $4e^2/h$ as in the next steps).

9.4. Carbon nanotube band structure ●

The band structure of single wall carbon nanotubes can be simply deduced from that of graphene. A carbon nanotube is nothing but a cylinder formed by the thought operation consisting of rolling up a strip cut from a 2D graphene sheet. This must be achieved by ensuring that each carbon atom initially lying at the edge of the graphene strip is effectively bonded to three other carbon atoms after the roll-up operation. This can be performed in different ways, two of them being pictorially illustrated in Figure 9.11 (in practice such perfect or almost perfect nanotubes can be formed by a number of techniques, such as arc discharge, laser ablation or crystalline growth by chemical vapor deposition).

A nanotube formed as in Figure 9.11 (left) is called an armchair nanotube, and a nanotube formed as in Figure 9.11 (right) is called a zig-zag nanotube. Each of these nanotubes can be indexed by the number of lattice vectors $\vec{a}_1$ and $\vec{a}_2$ which must be combined to recover $\vec{L}$, a vector parallel to the axis to be rolled and with a length equal to the width of the strip (thus L is the circumference of the final cylinder, see Figure 9.11). An armchair-type nanotube has equal indexes (n,n). A zigzag-type nanotube is such that vector $\vec{L}$ is parallel to one of the basis vectors, and thus has indexes in the form $(n,0)$. These two nanotube families are not chiral, because if we look at them in a mirror we obtain the same nanotube. With non-zero indexes and $m \neq n$ the nanotube exhibits a chirality, and its mirror image is different from the original. It is said to be chiral and exhibits a spiral symmetry.

It is worth noting that a nanotube is nothing but a realization of the up-to-now hypothetical BVK boundary conditions presented in section 2.10. To ensure the continuity of the wave function and its derivative, the wavevector projection along the circumference must obviously fulfill the condition

$$\vec{k}.\vec{L} = k_x L_x + k_y L_y = 2p\pi ,$$

(9.22)

where from equation (9.1) the (L_x, L_y) coordinates are ($(m+n)a\sqrt{3}/2, (-m+n)a/2$), so that L can also be expressed as

$$L = a\sqrt{m^2 + mn + n^2} ,$$

(9.23)

and p is an integer. Note that if we define the transverse wavevector $k_\perp$ as the projection of the wavevector k onto the circumference, the condition above can be re-written

$$k_\perp L = 2p\pi .$$

(9.24)

so that for a given nanotube, to one value of p corresponds only one value of $k_\perp$.

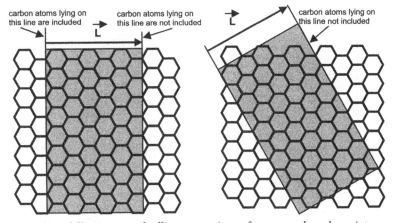

Figure 9.11. *Two different ways of rolling up a strip cut from a graphene layer into a carbon nanotube. Left figure: an "armchair-type" (3,3) nanotube and right figure: a "zig-zag type" (0,5) nanotube*

The two Cartesian coordinates of the wavevector are not independent, and the energy, given by equation (9.12) can thus be expressed as a function of only one of them and the index p. Each p value gives rise to two 1D subbands (one from the valence band and one from the conduction band), which are simply obtained by slicing the energy bands drawn in Figure 9.8 along the straight lines defined by equation (9.22) when p is spanned over all possible integer values which correspond to lines crossing the first Brillouin zone. If the line defined by equation (9.22) does not intersect one of the K points, there is a gap between the highest energy slice extracted from the graphene valence band and the lowest energy slice cut from the conduction band. Thus, the nanotube is semi-conductive (see Figure 9.12). If the line intersects one of the K points, then there is no gap and the nanotube is said to be metallic. For instance, if the point K of coordinates $(2\pi/a\sqrt{3}, 2\pi/3a)$ is considered, equation (9.22) becomes

$$m + 2n = 3p .$$

(9.25)

This can be advantageously replaced by the condition

$$m - n = 3l .$$

(9.26)

by withdrawing $3n$ on each side, and noting that if p is arbitrary, p-n spans over the same values and can be replaced by any arbitrary integer l. Repeating the same procedure for all six K points leads to the general condition

$$|m - n| = 3l ,$$

(9.27)

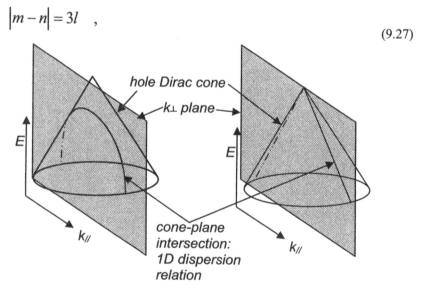

Figure 9.12. *The intersection between the hole (or electron) Dirac cone and the plane corresponding to a quantized $k_\perp$ value determines the nanotube 1D subband dispersion relation. If the plane does not intersect the Dirac point, there is a forbidden energy gap; dispersion is parabolic and determines an effective mass (left). If the plane intersects the Dirac point the nanotube is metallic and the electrons are massless*

where l is any integer. If equation (9.27) is fulfilled, the nanotube is metallic, and if not it is semi-conductive (it is worth noting that an armchair nanotube is always metallic, but zig-zag nanotubes or nanotubes with an arbitrary chirality can be either metallic or semi-conductive). To obtain the dispersion curves as a function of the wavevector projection onto the nanotube axis $k_{//}$, which is perpendicular to the circumference and corresponds to the wave propagation axis, first, as detailed in section 10.10 we can note that after some cumbersome calculations we can express k_x and k_y as a function of $k_{//}$:

$$k_x = \frac{1}{L}\left(- L_y k_{//} + 2 p \pi \frac{L_x}{L} \right). \tag{9.28}$$

$$k_y = \frac{1}{L}\left(L_x k_{//} + 2 p \pi \frac{L_y}{L} \right)$$

Then, injecting the formulae above into equation (9.12) gives us the energy as a function of the wavevector $k_{//}$ in each subband of index p. Typical 1D dispersion curves for a semi-conductive armchair nanotube has been plotted in Figure 9.13 along with the straight lines corresponding to the variation of $k_{//}$ for each p (and thus $k_\perp$) value in the Brillouin zone. Figure 9.14 shows the 1D dispersion curves for two different zig-zag nanotubes, one metallic and one semi-conductive. Figure 9.15 shows the 1D dispersion curve of a nanotube with arbitrary chirality.

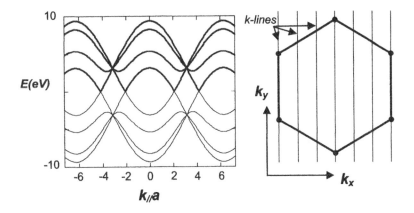

Figure 9.13. *1D subbands of a (3,3) armchair nanotube (right) and corresponding lines spanned by the wavevector in reciprocal space*

In the idealized description of the nanotube band structure which was developed above, a number of aspects have been neglected, which should nevertheless be considered to give quantitatively accurate dispersion curves. First, the curvature along the circumference becomes increasingly important as the nanotube diameter is reduced, and this curvature has two important effects. On the one hand, it may induce a substantial hybridization between p and s orbitals, and the simple model involving only p-type orbitals is no longer appropriate. On the other hand, the hopping integrals between nearest neighbors are no longer necessarily equivalent. Thus, for small diameters we expect important deviations from a simple tight-binding model. For example, if armchair nanotubes are always metallic, other supposedly metallic nanotubes may exhibit a small bandgap as a consequence of the curvature. There are thus a number of exceptions to our ideal results and in any nanotube with a small diameter the band structure should indeed be determined by more accurate methods than a simple tight-binding calculation.

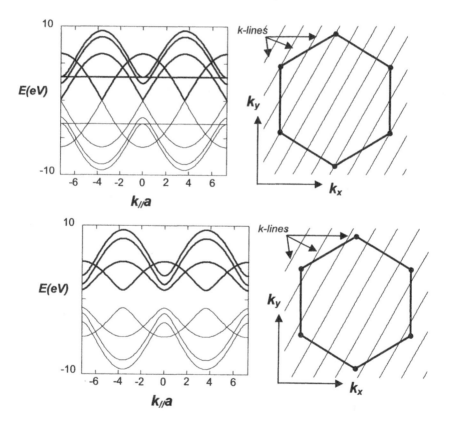

Figure 9.14. *1D subbands of a (6,0) metallic, zigzag nanotube (top left) and of a (5,0) semi-conducting zigzag nanotube (bottom left). The lines spanned by the wavevector in reciprocal space are shown in the right figures*

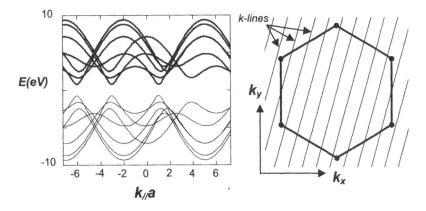

Figure 9.15. *1D subbands of a (6,2) nanotube with arbitrary chirality (left) and lines spanned by the wavevector in k-space(right)*

In addition, here we just described single-wall carbon nanotubes, but in the real world we also find multi-wall carbon nanotubes. In such nested nanotubes the distance between each neighboring tube is quite comparable to the distance between graphene planes in graphite (around *0.31 nm*), and a realistic modeling of such devices implies to account for the interactions taking place between each tube. We can also find curiosities such as junctions between nanotubes with different chiralities or between three nanotubes.

However, the nanotube picture given by the simple model above actually reflects much of the reality. Scanning tunneling microscopy on armchair nanotubes allows us to directly visualize the wave function square modulus of the Fermi electrons, and Fourier transforming the real space images gives points exactly located at the predicted K points. A careful analysis of the images obtained at different gate voltage values even enables a reconstruction of the dispersion relation, which turns out to be linear for metallic nanotubes, as predicted by the tight-binding model. These really nice experiments were reported in [LEM 01].

9.5. Carbon nanotube bandgap ●

The bandgap of a semi-conducting carbon nanotube is obviously given by finding in the Brillouin zone the points that lie closest to a K point on a 1D k-line as defined by equation (9.22), and then by multiplying the corresponding energy by a factor of *2* (see Figure 9.16). There is a general analytical formula which approximates the nanotube bandgap as a function of the nanotube radius, and here we shall first demonstrate it in the simpler case of semi-conducting zigzag nanotubes. For such nanotubes there is an angle of $-\pi/6$ between the k_x axis and the transverse wavevector, so that on a line parallel to the transverse wavevector the coordinates of the K points in the first Brillouin zone are equally spaced and equal to $-4\pi/3a$, $-2\pi/3a$, $2\pi/3a$ and $4\pi/3a$ (see Figure 9.16).

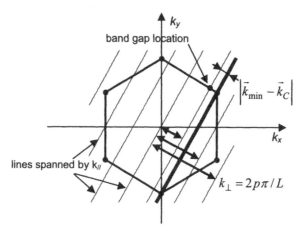

Figure 9.16. *1D wavevector lines in k-space for a (4,0) zigzag nanotube and location of the point closest to a Dirac point K (for m=4 there are 4 such points and only one is indicated)*

Thus we have to solve the problem of finding the integer value of p for which $k_\perp = 2p\pi/L = 2p\pi/ma$ lies closest to one of these K_c coordinates. For example, let us consider the point $2\pi/3a$. The distance to be minimized is given by

$$\Delta k = \left| k_\perp - \frac{2\pi}{3a} \right| = \left| p\frac{2\pi}{ma} - \frac{2\pi}{3a} \right| = \frac{2\pi}{ma}\left| p - \frac{m}{3} \right|, \qquad (9.29)$$

and we thus have to find the p value which lies closest to $m/3$. If the nanotube is semi-conducting, $m/3$ is not an integer, but either $(m-1)/3$ or $(m+1)/3$ is. Thus, we have $p=(m+1)/3$ or $p=(m-1)/3$, depending on which of these two values represents an integer. In both cases the wavevector difference Δk is equal to $2\pi/3ma$. If the corresponding wavevector lies in the region where the energy varies linearly with the wavevector, from equation (9.13) we obtain the bandgap as

$$E_G = \frac{\beta_0 a}{\sqrt{3}R} \qquad (9.30)$$

where $R=L/2\pi$ is the nanotube radius. Choosing any other K point for the calculation leads to the same result.

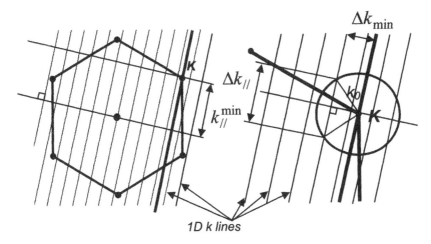

Figure 9.17. *Left: 1D wavevector lines in k-space for a (7,3) nanotube and right: enlarged view of the same figure around a K point, showing in addition a constant energy circle of radius k_0 around a Dirac point and its two intersections with a k-line*

We now demonstrate the validity of the formula with any *(m,n)* combination not leading to a metallic state. First use equation (9.28) to replace k_x and k_y by their expression as a function of $k_{//}$ to calculate the distance $\left|\vec{k}-\vec{k_c}\right|$ (choose for instance the same point *K* as before). Differentiate with respect to $k_{//}$ to find the $k_{//}$ value which renders this distance minimum. The calculation is cumbersome but basic and the corresponding value is

$$k_{//}^{\min} = \frac{2m\pi}{\sqrt{3}L} = \frac{2\pi}{a\sqrt{3}} \frac{m}{\sqrt{m^2+mn+n^2}}, \qquad (9.31)$$

independent of *p* (this property is quite obvious when considering Figure 9.17, and we can check that making *m=n* gives us the k_y coordinate of *K* as in Figure 9.13, independent of *m*). Insert this value into the expression of the distance $\left|\vec{k}-\vec{k_c}\right|$ as a function of $k_{//}$. After some tedious but simple simplifications, we obtain the minimum distance between a *k*-line of index *p* and the point *K* as

$$\Delta k_{\min} = \left|\vec{k}-\vec{k_c}\right|_{\min} = \frac{2\pi}{3L}\sqrt{\left|9p^2-6(m+2n)p+m^2+4mn+4n^2\right|}. \qquad (9.32)$$

The minimum in the expression above is obtained for $p_{opt}=(m+2n)/3$, which gives us once again the condition equation (9.26) for which the nanotube is metallic as we require *p* to be an integer. If p_{opt} is not an integer, the nanotube is semi-conductive and the closest *p* value is either *(m+2n-1)/3* or *(m+2n+1)/3*. Inserting either of both values in equation (9.32) gives a distance

$$\Delta k \,^{\min}_{\min} = \frac{2\pi}{3L} = \frac{1}{3R}.$$ (9.33)

Insert this value into the energy expression equation (9.13) and multiply by a factor of 2 to obtain the bandgap and to complete the proof. The bandgap can also be conveniently expressed as a function of m and n, which shows that the bandgap is in fact only indirectly dependent on a (through the coupling β_0), and is just a function of β_0, m and n:

$$E_G = \frac{\beta_0 a}{\sqrt{3}R} = \frac{2\pi\beta_0}{\sqrt{3\left(m^2 + mn + n^2\right)}}$$ (9.34)

A plot of the bandgap versus nanotube radius is shown in Figure 9.18. Of course, for a "small" radius we expect some departures from this ideal result, due to the increased influence of the curvature, but for a larger radius, the bandgap should follow this curve more closely.

Eventually, it is worth noting that only the armchair nanotubes remain fully metallic down to small radius values. In other metallic-like nanotubes the curvature slightly modifies the band structure, so that they exhibit a small bandgap (typically a few tens of meV for a radius in the nm range), inversely proportional to the *square* of the radius. More accurate calculations than our simple tight-binding model also predict that small zigzag nanotubes can be semi-metallic instead of being semi-conductive as predicted in the simplest approach. Thus, as the nanotube diameter is reduced down to a very few nanometers it is clearly preferable to rely upon a more advanced formalism

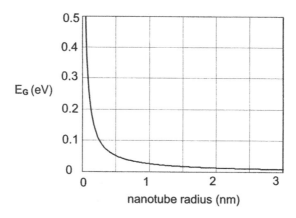

Figure 9.18. *Forbidden bandgap versus nanotube radius, as calculated with the approximate tight-binding model*

9.6. Carbon nanotube density of states and effective mass ●

From the calculations already carried out in section 9.5 we can easily estimate the density of states in each 1D subband. As depicted by Figure 9.17, consider an index p such that its k-line crosses the circle of radius k_0 in k-space which corresponds to a constant energy E. This circle is of course centered around a point K, which is the location of the graphene conduction band minimum. In the vicinity of point K, the interval $\Delta k_{//}$ over which any $k_{//}$ corresponds to an energy lower than E is given by

$$\left(\frac{\Delta k_{//}}{2}\right)^2 + \Delta k_{min}^2 = k_0^2 \ . \tag{9.35}$$

where Δk_{min} is given by equation (9.32). As in section 2.10, we consider that each $k_{//}$ value occupies a space $2\pi/L_0$ on a k-line, where L_0 is now the nanotube length. Thus, the number of admissible $k_{//}$ values is given by

$$N(k_0) = 2\times 2\times\frac{\Delta k_{//}}{\dfrac{2\pi}{L_0}}\times\frac{1}{L_0} = \frac{4}{\pi}\sqrt{k_0^2 - \Delta k_{min}^2} \ . \tag{9.36}$$

where both valley and spin degeneracies have been accounted for. The derivative is equal to

$$\frac{dN}{dk_0} = \frac{4k_0}{\pi\sqrt{k_0^2 - \Delta k_{min}^2}} \ . \tag{9.37}$$

and the density of states is obtained as usual by calculating $D(E)=(\partial N/\partial k)\times(\partial k/\partial E)$, using the linear dispersion relationship equation (9.13) along with equation (9.37). This straightforwardly leads to the following formula:

$$\frac{dN}{dE} = \frac{4}{\pi\,\beta_0\,a\,\sqrt{3}}\times\frac{E-E_C}{\sqrt{(E-E_C)^2 - (E_m-E_C)^2}} \ , \tag{9.38}$$

where $E_m - E_C$ is given by

$$E_m - E_C = \frac{\sqrt{3}}{2}\beta_0 a\Delta k_{min} \tag{9.39}$$

and represents the difference between the bottom energy of the 1D subband minimum and the minimum of the graphene conduction band E_C (see Figure 9.19). Of course this expression is valid for $E>E_m$, and below it it is equal to zero. Due to the mirror symmetry of the graphene band structure, the 1D density of states of a k-line in the valence band is evidently given by the same expression, but reverting the energy sign. The overall density of states for an energy E is obtained by summing the densities of states of all 1D subbands with $E_m<E$.

From equation (9.38) the *1D* density of states is not equal to the conventional expression derived for a *1D* subband with a parabolic dispersion relation. However, it is clearly as pathological as the conventional *1D* density, and when the energy becomes close to the subband bottom (i.e. for $E-E_m$ small) the reader can check from equation (9.38) that the dependence tends to that of a conventional *1D* density of states. Thus, a semi-conducting carbon nanotube exhibits the Van Hove singularities typical of a *1D* system. An example of the density of states variation as a function of energy is given in Figure 9.19 for a carbon nanotube with *(m,n)=(22,12)*. It is worth noting that these peaks in the density of states have in fact been observed by measuring the tunneling current between a nanotube and the probe of a scanning tunneling microscope.

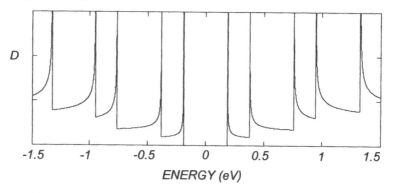

Figure 9.19. *Density of states of a carbon nanotube with (m,n)=(22,12), as calculated from equation (9.38) (arbitrary units for D)*

The effective mass describing the dispersion curve of any semi-conducting 1D subband can be easily derived by noting that close to a subband bottom the energy can be expressed as

$$E = \frac{a\sqrt{3}}{2} \beta_0 \sqrt{\left(k_{//} - k_{//}^{min}\right)^2 + \Delta k_{min}^2} \qquad (9.40)$$

by using equation (9.1) and considering Figure 9.17. Factorizing Δk_{min}^2 out of the square root and developing the square root to first order leads to a parabolic relation between energy and wavevector $k_{//}-k_{//}^{min}$, and from the proportionality coefficient and equation (2.43) the effective mass can easily be deduced as

$$m_{eff} = \frac{2\hbar^2}{a\beta_0\sqrt{3}}\Delta k_{min}, \tag{9.41}$$

which in the case of the lowest electron and hole subbands of a semi-conducting nanotube reduces to

$$m_{eff}^{1^{st} band} = \frac{2\hbar^2}{3a\beta_0\sqrt{3}}\frac{1}{R} \tag{9.42}$$

(inject equation (9.33) into equation (9.41)). This result is quite reasonable, since as the radius tends towards infinity the curvature tends towards zero and the circumference tends towards infinity, so that we recover the graphene band structure and the effective mass must vanish, as indicated by equation (9.42). For a radius $R=10$ nm, the effective mass already drops down to 0.0198 m_0. Of course, the reserves that we already expressed about the usefulness of the analytical model to calculate the bandgap for too small a radius also hold for the mass.

9.7. Electron transport in and quantum dots from carbon nanotubes ●

Since we are able to produce carbon nanotubes with quite a small radius, we expect to observe features characteristic of 1D electron transport. In the ballistic regime and owing to the fact that the valley degeneracy is equal to 2, we expect from equation (3.3) the low voltage conductance to be quantized as

$$G = \frac{4e^2}{h}N \tag{9.43}$$

where N is the number of 1D open modes, and the nanotube is metallic or semi-conductive (gate voltage control must be achieved in the latter case)[5]. This has in fact already been observed, but contradictory experimental data have led the

---

5 The demonstration that a linear or a parabolic dispersion lead to the same conductance quantization has already been offered as an exercise to the reader (see Chapter 3, exercise 3.14.2).

community to the conclusion that two other important factors were also contributing to the determination of the overall conductance value. On the one hand, not all the modes can be connected in a satisfactory way to the contacts, depending on the type of nanotube considered and on the nature of the contact materials (e.g., titanium, considered to be the best metal for this purpose, or graphite). For instance, armchair nanotubes are expected to be coupled through only one mode, due to a wavevector mismatch at the interface. On the other hand, the angle between the nanotube and the contact may also modify the number of modes which can be satisfactorily coupled to the contacts. The situation is thus a bit more complicated than when considering III-V or silicon devices, for which there is generally no interface or discontinuity between the constriction and the 2D or the 3D gas used to form the wider contacts.

If the carbon nanotubes were operated only a low temperature, their intrinsic band structure properties would make them quite interesting objects of study. However, there is much more than this in them: carbon nanotubes can be shown to remain ballistic and quantum-coherent even at room temperature, offering to the physicist and perhaps even to the device engineer the possibility of playing with effects that were up to now most often confined to the world of liquid helium temperature. In general for moderately doped materials at 300 K the scattering is dominated by the lattice vibrations, whose quantified modes are called phonons. In a graphene plane those vibrations can take the form of the usual transverse and longitudinal phonons, with in and out-of-plane modes. Roughly summarizing, the phonon modes in nanotubes are calculated by the same kind of zone-folding procedure as the one we followed to establish the electron band structure, but this results in somewhat unusual modes. To the common acoustic and optical phonon modes we must add a "radial breathing mode", which corresponds to contractions and elongations with respect to the tube axis, and twisting modes, which correspond to rotations [SAI 98]. At 300 K nanotube transport is limited both by defects and phonons, as for many other materials, but it turns out that the calculated and measured mean free paths are most often longer than one micrometer, provided that the voltage is kept below the optical phonon energy. A first reason is that the defects are in general not part of the nanotube lattice, which remains perfectly crystalline, but consist of adsorbed species or more remote impurities and defects (in contrast, for instance, with the Si-SiO_2 system where the interface incorporates many electrically active structural defects, even if such a system was good enough to initiate and enable the prodigious development of digital electronics). A second reason is a low coupling between the electrons and the acoustic phonons. At low longitudinal electric field, the intervalley scattering processes are due to transverse, longitudinal and "radial breathing mode" acoustic phonons, and it turns out that the coupling with these modes is weak enough for the mean free path to exceed one micron (to be compared with that of silicon, smaller than 10 nm). The more effective optical and twisting phonon modes typically occur at energies larger than 150 meV, so that they do not impede electron transport at small bias. Intervalley scattering occurs at higher voltages, but even in this case defects are still not prevailing, because they are not located on the lattice but farther away and most generally their potential will remain long-range with respect to the wavevector difference between the two valleys (this wavevector difference is of order a^{-1}). Thus, momentum conservation cannot be ensured in a collision and defects are ineffective to scatter the electrons, because their power spectrum does not own components with the required wavevector value.

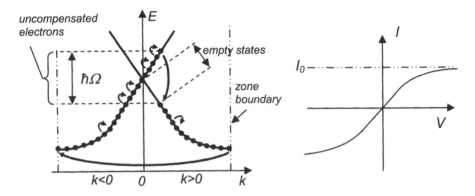

Figure 9.20. *Phonon-limited transport mechanism in a metallic carbon nanotube; due to the action of the electric field the wavevector of all electrons increases linearly with time until they can lose the energy of one phonon. This leads to experimental current-voltage curves as in the right figure*

Due to the reasons mentioned above, metallic nanotubes can exhibit a huge current drive capability, and are actually possible candidates to increase the current density of vertical interconnect accesses (vias) in integrated circuits. Let us examine a simplified model in order to understand the phenomenon at play [YAO 00]. When the conducting electrons of a metallic nanotube are submitted to an electric field, they are given an additional drift momentum, but due to the two bands the situation is not adequately depicted as in Figure 2.27, but rather as in Figure 9.20. First note that in the conduction band for positive k the electron velocity is constant and positive, but in the valence band its value changes sign, and for *positive k* the velocity is *negative* for an energy below the Dirac point (use equation (2.34)). Then, in a semi-classical approximation and without collisions (use equation (2.45)), due to the action of the electric field ε the electron momentum of all electrons first increases proportionally to $e\varepsilon/\hbar t$, where t is the time. This means that the electrons lying on the $E(k)$ curve part with a positive slope see their energy increasing, whereas the electrons lying on the curve with a negative slope see their energy decreasing until they pass into the curve with a positive slope at the opposite zone boundary or Bragg point (see Figure 9.20). This occurs as long as the electrons cannot experience lossy collisions, i.e. as long as the energy gain remains below the threshold for a phonon-mediated transition. If we assume that the phonon coupling is so strong as to immediately backscatter a conduction band electron into the valence band once its energy exceeds the threshold, we obtain the graph in Figure 9.20, and can expect that the current increases with the bias until it reaches a saturation level due to the phonon-limiting transport mechanism. The product of the density of states by the velocity remaining constant, Landauer formula is applicable and we thus find that the saturation current saturates at a value of order $I_0=(4e/h)\hbar\Omega$, where $\hbar\Omega$ is the phonon energy. Thus, the current saturates around $I_0\cong0.64e\Omega$ also equal to *25 μA* for a phonon energy of *160 meV*. The corresponding current density is tremendous, exceeding *1 GA/cm²*. This is the order of magnitude observed by Zhen Yao and co-workers, which was also supported by a more involved Boltzmann's transport formalism [YAO 00], and later confirmed by other groups (see, e.g., [PUR 07] and references therein). There was almost no temperature dependence from liquid helium up to 200 K.

Another point of interest is coherence: consider a metallic carbon nanotube. The Fermi velocity is constant, equal to that of graphene and we can estimate the coherence by calculating the time spent by an electron to travel through a *1 μm* long carbon nanotube, approximated as $\Delta t=L/v_F$. By inserting this value into the uncertainty relation we can estimate a typical energy spreading $\Delta E=\hbar/t=\hbar v_F/L$, with an equivalent temperature $T=\Delta E/k_B\approx10$ K. This means that at not so low a temperature, for instance that of liquid helium, even a 1 μm long nanotube can be fully quantum-coherent [SCH 06]. As a consequence, the energy barriers which can appear at the hetero-contact between the nanotube and the connecting materials result in the formation of a coherent quantum dot (Figure 9.21). The dot states are standing waves inside the nanotube. In an ideal metallic sample of length L, the energy levels can be assessed by assuming that the standing wave must vanish at the extremities, so that the wavevector is given by $k_n=n\pi/L$ with n integer. Since the energy is linear with k (equation (9.14)), the quantized levels are equally spaced and given by

$$E_n = n\frac{hv_F}{2L} \quad ,$$

(9.44)

as schematically represented in Figure 9.21.

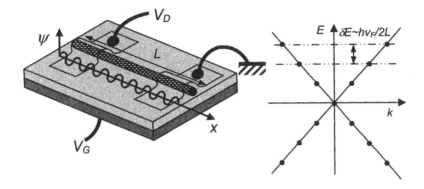

Figure 9.21. *Quantum dot made from a metallic carbon nanotube with standing waves (left) and ideal energy levels (right)*

A nice example which illustrates this aspect is given in Figure 9.22, which is reproduced from [BAB 04] (this figure is also discussed in a review article [SCH 06]). There is in fact a lot of physics in this single curve, some of which have already been discussed in this book. The gate voltage does not only change the nanotube electron density, but controls the tunneling coupling Γ to the contacts, so that a number of different regimes are apparent.

At high gate voltage, the nanotube is weakly coupled to the contact and behaves as a quantum dot in the Coulomb blockade regime (discussed in Chapter 6), with equally spaced conductance peaks. As the voltage decreases, the coupling between the nanotube and the contacts increases, the resonances become wider and progressively wipe out the Coulomb blockade. The resonances first exhibit a phenomenon known as the Kondo effect, which is not discussed in this book because it is a many-body effect with high order contributions to the tunneling process. Then, when the coupling is still enhanced the charge effects become negligible, and the resonant tunneling peaks get close to the maximum theoretical value of $4e^2/h$, ascribed to the combination of spin and valley degeneracies. In these latter peaks we can also clearly distinguish Fano resonances (discussed at length in Chapter 5), so that there is really a lot of fundamental effects in this single conductance-voltage curve, measured from a single-wall carbon nanotube!

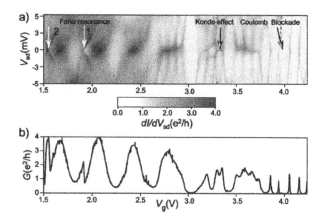

Figure 9.22. *Conductance plot from a single wall carbon nanotube, showing different electron transport regimes; reproduced with permission from Babic B. et al., Phys. Rev. B 70, p. 195408 (2004)[BAB 04], copyright (2004) by the American Physical Society*

From this single example we can imagine the wealth of interesting physical effects which can be studied by manipulating such small objects. There is already an incredibly large amount of scientific literature published on the subject: in 2007, by using a commonly used scientific research engine on the web, more than 12,500 published references could be identified, with numbers in excess of 2,000 per year in more recent years. This is already much more than a single physicist can read in his whole life (and definitely more than can be included in an introductory book). Here we shall just quote a number of interesting physical effects or potential applications, not restricting ourselves to mesoscopic physics: in contrast with most other nanostructures, carbon nanotubes can be used in conjunction with superconducting device parts, so as to form very sensitive superconducting quantum interference devices or to study exotic physics such as the Kondo effect or Andreev reflections. Semi-conducting nanotubes can be doped by the insertion of substitutional impurities such as boron or nitrogen, or by adsorbed potassium. Therefore, they can be used to fabricate junctions or field-effect transistors of both kinds (*n*- or *p*-type). They can also be used to produce field emitters for display, or be inserted into polymers to fabricate reinforced composites. They can be used as X-ray emitters or as scanning probe tips for atomic force microscopy and nanoscale metrology. In integrated circuits, their remarkable conductivity makes them attractive to fill the small vias which connect the different metallization planes, in the form of bundles, etc. There is no doubt that although not yet implied in large scale electronic production, they will find some useful applications. Note, however, that to become serious competitors of something as efficient as CMOS Si technology, nanotubes will have to overcome the laboratory craft stage, a step remaining far from achievement.

9.8. Exercises

9.8.1. *Exercise*

This problem does not involve 2D graphene or 1D carbon nanotubes, but its aim is to assess the band structure of the "next step" configuration, i.e. carbon "lines". Such organic polymers are in fact increasingly used in a number of electronic applications (e.g. for light emission). The calculation is simpler than for graphene, and will allow us to check that we have correctly understood the principles involved in the determination of the graphene and carbon nanotube band structures. It will also enable us to understand the principle at the heart of the conduction and light-emission properties of the conjugated polymers.

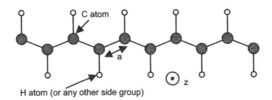

Figure 9.23. *Hypothetical polymer chain with equally spaced carbon atoms*

Consider an ideal carbon chain as shown in Figure 9.23, where the atoms are identically spaced, and where the $2s$ and the $2p$ orbitals in the molecule plane are involved in strong bonding, so that only the $2p_z$ orbital is involved in the formation of extended states (for the sake of simplicity you can consider that all atoms are on a line).

1) Calculate the band structure of the carbon chain in the tight-binding approximation.

2) What is the nature of this system?

3) What is the shape of the wave function at the bottom and the top of the conduction band?

Figure 9.24. *Conjugated polymer*

In fact, it can be demonstrated that equally spaced atoms do not lead to the minimum energy state when the energy of all atoms and electrons is accounted for. The lowest energy state favors an arrangement where the atoms form pairs[6], as shown in equation (9.23), and with hydrogen side atoms this polymer is called polyacetylene. Define a' as the chain period. The first interatomic distance is $\eta a'$ and the second is $(1-\eta)a'$, with $\eta<1$.

4) Calculate the band structure of the carbon chain, and shows that the pair configuration leads to a splitting of the previously calculated band.

5) What is the nature of this "conjugated" polymer?

6) Suggest practical and theoretical reasons why it is not possible to use a single conjugated chain as a ballistic conductor.

9.8.2. *Exercise*

1) Consider a (5,5) carbon nanotube. Which nanotube family does it belong to? Is it suitable for emitting radiation?

2) Consider a (40,19) carbon nanotube. Is it metallic or semi-conductive?

3) Consider a (13,9) nanotube. Is it metallic or semi-conductive?

4) For the same nanotube as in 3), calculate the bandgap and the levels of the first three subbands.

5) Calculate the density of carriers in the first subband of a semi-conductive nanotube as a function of the Fermi energy.

6) From this question consider a (20,20) carbon nanotube of length L embedded into SiO_2 and controlled by an underlying planar, degenerate silicon gate at a distance $d=10$ nm. Assume that the nanotube control capacitance can be approached by the expression of the capacitance between a metallic cylinder of radius R and a metal plane at a distance d from the cylinder axis

$$C/L = \frac{2\pi\varepsilon_{OX}}{\ln\left(\dfrac{d+\sqrt{d^2-R^2}}{R}\right)} .$$

What is the density of states in the first subband?

---

6 This phenomenon is known as a "Peierl transition".

7) Consider a large L. Express the Fermi level value as a function of the gate voltage. Numerical application: find E_F for V_G-V_T=5 V, where V_T is the gate voltage corresponding to a Fermi level at the K point (*i.e.* the flat band condition).

8) Consider a length L=500 nm. Compare the quantized level splitting and Coulomb blockade. Which effect prevails?

9.8.3. *Exercise*

Consider a metallic nanotube. Assume that the electrons lose all the energy acquired from the electric field once it exceeds the optical phonon energy $E=\hbar\Omega$

1) Explain qualitatively how the ballistic character of the electrons is progressively lost as the nanotube length or the electric field is increased.

2) What is the length l_a over which an electron accelerates in the field before emitting an optical phonon?

3) Propose an approximate formula for expressing the sample resistance (hint: consider the formula established in another chapter to derive Ohm's law from Landauer's formula)?

4) From the previous answer show that the current saturates with the applied voltage. Plot the shape of the I-V curve and describe it. Numerical application: what is the expected saturation current with an optical phonon energy around 0.2 eV?

5) As a matter of fact, after an electron attains a sufficient energy for emitting one optical phonon the transition is not instantaneous but random and requires further electron drift over an average length l_Ω before losing the energy gained in the field. Can you find conditions for which this delayed transition affects the previous result? Plot current-voltage and resistance-voltage curves for various lengths and comment.

This exercise is based on results published by J.Y. Park *et al.* and further details can be found in [PAR 04].

9.9. Bibliography

[ANA 00], ANANTRAM M.P., DATTA. S., XUE Y., "Coupling of carbon nanotubes to metallic contacts", *Physical Review B*, vol. 61, no. 20, 2000, p. 14219-14224.

[ANA 06] ANANTRAM M.P., LEONARD F., "Physics of carbon nanotube electronic devices", *Report on Progress in Physics*, vol. 69, 2006, p. 507-561.

[BAB 04] BABIC B., SCHONENBERGER C., "Observation of Fano resonances in single-wall carbon nanotubes", *Physical Review B*, vol. 70, no. 195408, 2004, p. 1-7.

[DIV 84] DiVINCENZO D.P., MELE E.J., "Self-consistent effective-mass theory for intralayer screening in graphite intercalation compounds", *Physical Review B*, vol. 29, no. 4, 1984, p. 1685-1694.

[IIJ 91] IIJIMA S., "Helical microtubules of graphitic carbon", *Nature*, vol. 354, no. 6348, 1991, p. 56-58.

[LEM 01] LEMAY S.G., JANSSEN J.W., VAN DEN HOUT M., MOOIJ M., BRONIKOWSKI M.J., WILLIS P.A., SMALLEY R.E., KOUWENHOVEN L.P., DEKKER C., "Two-dimensional imaging of electronic wave functions in carbon nanotubes", *Nature*, vol. 412, 2001, p. 617-620.

[MAC 56] McCLURE J.W., "Diamagnetism of graphite", *Physical Review*, vol. 104, no. 3, 1956, p. 666-671.

[MIN 98] MINTMIRE J.W., WHITE C.T., "Universal density of states for carbon nanotubes", *Physical Review Letters*, vol. 81, no. 12, 1998, p. 2506-2509.

[NOV 05] NOVOSELOV K.S., McCANN E., MOROZOV S.V., FAL'KO V.I., KATNELSON M.I., ZEITLER U., JIANG D., SCHEDIN F., GEIM A.K., "Unconventional quantum hall effect and Berry's phase of 2π in bilayer graphene", *Nature Physics*, vol. 2, no. 3, 2005, p. 177-180.

[NOV 07] NOVOSELOV K.S., JIANG Z., ZHANG Y., MOROZOV S.V., STORMER H.L., ZEITTER U., MAAN J.C., BOEBINGER G.S., KIM P., GEIM A.K., "Room-temperature quantum Hall effect in graphene", *Science*, vol. 315, no. 5817, 2007, p. 1379.

[PAR 04] PARK J.Y., ROSENBLATT S., YAISH Y., SAZONOVA V., USTUNEL H., BRAIG S., ARIAS T.A., BROUWER P.W., MCEUEN P.L., "Electron-phonon scattereing in metallic single-walled carbon nanotubes", *Nanoletters*, vol. 4, no. 3, p. 517-520, 2004.

[PUR 07] PUREWAL M.S., HONG B.H., RAVI A., CHANDRA B., HONE J., KIM P., "Scaling of resistance and electron mean free path of single-wall carbon nanotubes", *Physical Review Letters*, vol. 98, no. 186808, 2007, p. 1-4.

[SAI 98] SAITO R., DRESSELHAUS G., DRESSELHAUS M.S., *Physical Properties of Carbon Nanotubes*, London, Imperial College Press, 1998.

[SCH 06] SCHONENBERGER C., "Charge and spin transport in carbon nanotubes", *Semiconductor Science and Technology*, vol. 21, 2006, p. S1-S9.

[SLO 58] SLONCZEWSKI J.C., WEISS P.R., "Band structure of graphite", *Physical Review*, vol. 109, no. 2, 1958, p. 272-279.

[YAO 00] YAO Z., KANE C.L., DEKKER C., "High-field electrical transport in single-wall carbon nanotubes", *Physical Review Letters*, vol. 84, no. 13, 2000, p. 2941-2944.

[ZHE 02] ZHENG Y., ANDO T., "Hall conductivity of a two-dimensional graphite system", *Physical Review B*, vol. 65, no. 245420, 2002, p. 1-11.

Chapter 10

Appendices

10.1. The uncertainty principle ○

Consider two observables described by two non-commuting Hermitian operators A and B with commutator $AB-BA=iC$, where C is also a Hermitian operator. Define the quantities $\delta A=A-\langle A\rangle$ and $\delta B=B-\langle B\rangle$ where the terms in brackets represent the average values of A and B. It is trivial to show that (i) δA and δB are also Hermitian (see the definition by equation (2.3)) and (ii) they obey the same commutation relation as A and B.

Now let us examine the positive quantity

$$f(\lambda) = \int \left|(\lambda\delta A - i\delta B)\psi\right|^2 d^3r \geq 0 , \tag{10.1}$$

where λ is an arbitrary, real parameter. Why define this? Just wait for a few lines … we can write

$$f(\lambda) = \int (\lambda\delta A - i\delta B)*\psi *(\lambda\delta A - i\delta B)\psi d^3r \tag{10.2}$$

and by virtue of the hermiticity of δA and δB (see equation (2.3)) the formula above can easily be turned into

$$f(\lambda) = \int \psi *(\lambda\delta A + i\delta B)(\lambda\delta A - i\delta B)\psi d^3r . \tag{10.3}$$

Thus, we have

$$f(\lambda) = \int \psi *\left(\lambda^2 \delta A^2 + i\lambda(\delta B\,\delta A - \delta A\,\delta B) + \delta B^2\right)\psi d^3r \tag{10.4}$$

and from the commutation relation between δA and δB we arrive at

$$f(\lambda) = \int \psi * \left(\lambda^2 \delta A^2 + \lambda C + \delta B^2 \right) \psi d^3 r .$$ (10.5)

If we take the notation $(\Delta A)^2 = \langle (\delta A)^2 \rangle$ and $(\Delta B)^2 = \langle (\delta B)^2 \rangle$ for the squared uncertainties in the measurements of A and B, equation (10.5) can be re-written as

$$\lambda^2 (\Delta A)^2 + \lambda \langle C \rangle + (\Delta B)^2 \geq 0 .$$ (10.6)

Since this inequality is valid for any value of λ we can deduce that the polynomial in λ defined by the left hand side above has no real roots, and thus the corresponding discriminant must be negative. This gives us

$$\langle C \rangle^2 - 4(\Delta A)^2 (\Delta B)^2 \leq 0 ,$$ (10.7)

which reduces to

$$\Delta A \Delta B \geq \langle C \rangle / 2 .$$ (10.8)

Equation (10.8) is the uncertainty relationship resulting from measurements of the observables described by operators A and B. These quantities could be found by carrying out many measurements with identically prepared systems, measuring alternately the observables A or B. It is worth noting that we obtained a factor of 2 difference with equation (2.11), and in textbooks we can find either one relation or the other. In fact both are valid but do not correspond to the same situation. The more restrictive equation (2.11) corresponds to a simultaneous measurement of both A and B. Although we shall not demonstrate it here in such a case the uncertainty is further increased.

10.2. Crystalline lattice; some definitions and theorems ◉

This section covers the basic notions used to describe a periodic lattice which are required to derive or understand some of the results expounded in other parts of the book (e.g. to establish graphene band structure). Here we shall not derive the laws which govern the physical arrangement of the atoms in a perfectly ordered lattice, but instead explain in a simple way how this arrangement must be described and accounted for in order to derive the free electron properties of a given material. The purpose of this section is therefore not to introduce the science of crystallography but instead to maintain the self-contained character of the book.

First of all, a crystalline lattice can be generated by the periodic repetition of a unit cell (Figure 10.1). Such a cell is not uniquely defined, but is characterized by three fundamental translation vectors $\vec{a}_1$, $\vec{a}_2$ and $\vec{a}_3$ in three dimensions, two such vectors in two dimensions, etc. The very existence of these basic vectors lies in the translational symmetry of a crystal. The crystal structure is unaffected by any translation through any integer linear combination of these base vectors

$$\vec{R} = n_1\vec{a}_1 + n_2\vec{a}_2 + n_3\vec{a}_3 \qquad (10.9)$$

$\vec{R}$ is a lattice vector. A lattice with only one atom per unit cell (or lattice site) is called a Bravais lattice. If there are several atoms per unit cell the lattice is said to have a basis. When the lattice owns a symmetry point, it may be convenient to construct a special unit cell, the center of which is precisely this symmetry point. This unit cell is called the Wigner-Seitz cell. It is constructed by plotting the lines joining the symmetry point to the nearest equivalent lattice sites, and then by drawing the bisector planes of those lines (Figure 10.1). The volume enclosed by these planes is a unit cell and gives all points closer to the symmetry point considered than to any other equivalent symmetry point of the lattice.

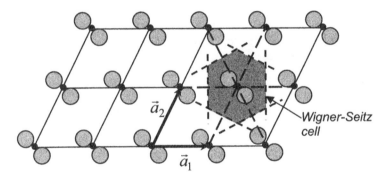

Figure 10.1. *2D periodic lattice with a two-atom basis; the two basic vectors and the Wigner-Seitz cell construction are shown*

Any possible lattice is conserved after symmetry operations which are specific to this lattice (invariance under reflection through a given plane, or under rotation around a specific axis, etc.). These operations form a symmetry group and the number of such groups is in fact finite. Although this is not the topic of this book, we should know that a proper account of the group properties of a given lattice can tremendously help in calculating, for instance, the band structure of given materials, because it restricts the form of the wave functions and determines some of their properties. The reason the tight-binding approximation gives so close a result to the reality in the case of graphene (section 9.2) is that it obeys the symmetry rules of the lattice, even if we do not pay attention to them when we carry out the calculation. The classification of all possible crystal lattices was an initial purpose of crystallography.

The symmetry properties of the crystal lattice in real space have their counterpart in reciprocal space, where a corresponding reciprocal lattice can be defined. To do this, first remember that in one dimension any periodic function $F(x)$ with period L can be defined by its Fourier series:

$$F(x) = \sum_n a_n \exp\left(\frac{2i\pi nx}{L}\right) = \sum_g a_g \exp(gx) \quad , \tag{10.10}$$

where the first summation is over all integer values and the quantity $g=n\times2\pi/n$ is a reciprocal lattice scalar. In 1D the full set of g values defines the reciprocal lattice. As is well known, the corresponding a_g coefficients are given by

$$a_g = \frac{1}{L} \int_{one\ period} F(x) \exp(-igx) dx \quad . \tag{10.11}$$

Note that we have $exp(igL)=exp(2i\pi n)=1$ for any value of g. In fact exactly the same Fourier formalism can be developed in 3D. Any function with the periodicity of the lattice can be expanded as

$$F(\vec{r}) = \sum_{\vec{g}} a_{\vec{g}} \exp(\vec{g}.\vec{r}) \quad , \tag{10.12}$$

where vectors $\vec{g}$ are all possible reciprocal lattice vectors. Note that they have the same dimension as wavevectors. To form a Fourier series and make the function F periodic with the periodicity of the lattice, as in 1D the vectors $\vec{g}$ must verify the relation

$$\exp(i\vec{g}.\vec{R}) = 1 \quad , \tag{10.13}$$

where $\vec{R}$ is any lattice vector, as defined by equation (10.9). The reciprocal lattice vectors can be determined as follows: consider the three vectors

$$\vec{b}_1 = 2\pi \frac{\vec{a}_2 \wedge \vec{a}_3}{\vec{a}_1.(\vec{a}_2 \wedge \vec{a}_3)} \qquad \vec{b}_2 = 2\pi \frac{\vec{a}_3 \wedge \vec{a}_1}{\vec{a}_2.(\vec{a}_3 \wedge \vec{a}_1)} \qquad \vec{b}_3 = 2\pi \frac{\vec{a}_1 \wedge \vec{a}_2}{\vec{a}_3.(\vec{a}_1 \wedge \vec{a}_2)} \quad . \tag{10.14}$$

Form the integer linear combination

$$\vec{g} = n_1\vec{b}_1 + n_2\vec{b}_2 + n_3\vec{b}_3 \quad . \tag{10.15}$$

As an exercise for the reader, vector $\vec{g}$ verifies equation (10.13). The three basic vectors of the reciprocal lattice are therefore given by equation (10.14). Equation (10.12) is generalized as

$$a_{\vec{g}} = \frac{1}{V_{cell}} \int_{one\ cell} F(\vec{r})\exp(-i\vec{g}\vec{r})d\vec{r} \quad , \tag{10.16}$$

where V_{cell} is the cell volume. In two dimensions, it is also easy to demonstrate that equation (10.13) is verified by using a linear combination of two base vectors generating the 2D reciprocal lattice as in equation (10.15), now calculating these two base vectors as

$$\vec{b}_1 = 2\pi\frac{\vec{a}_2 \wedge (\vec{a}_1 \wedge \vec{a}_2)}{(\vec{a}_1 \wedge \vec{a}_2)^2} \qquad \vec{b}_2 = 2\pi\frac{\vec{a}_1 \wedge (\vec{a}_2 \wedge \vec{a}_1)}{(\vec{a}_1 \wedge \vec{a}_2)^2} \quad , \tag{10.17}$$

where a third dimension is momentarily taken into account in order to calculate the vectorial products which appear as calculation intermediates. The reciprocal lattice vectors verify a number of interesting properties that the reader can find in any solid-state physics textbook, but since we do not need them we refer the interested reader to those books (among them is for example the fact that the reciprocal lattice of the reciprocal lattice is the lattice itself). To derive the band structure of graphene we only have to use the basic vectors of the reciprocal lattice.

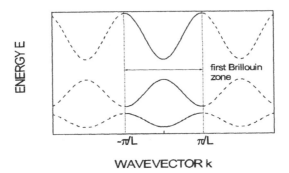

Figure 10.2. *Typical 1D dispersion relationship; wavevectors separated by a reciprocal lattice vector describe the same state; thus the first Brillouin zone contains all necessary information*

The Wigner-Seitz cell of the reciprocal lattice is called the first Brillouin zone.
The last important result which we shall demonstrate in this section is that it suffices
to consider wavevectors belonging to the first Brillouin zone to fully describe the
energy-wavevector dispersion relation as described in section 2.2. Consider a
crystalline lattice and the associated Bloch wave functions $\psi_{\vec{k},n}(\vec{r})$. More
specifically, given a lattice vector $\vec{R}$, we know that these functions verify

$$\psi_{\vec{k},n}(\vec{r} + \vec{R}) = e^{i\vec{k}\vec{R}} \psi_{\vec{k},n}(\vec{r}) \quad . \tag{10.18}$$

Now consider the wavevector

$$\vec{k}' = \vec{k} - \vec{g} \tag{10.19}$$

where $\vec{g}$ is any reciprocal lattice vector. From equation (10.13) we can re-write
equation (10.18) as

$$\psi_{\vec{k},n}(\vec{r} + \vec{R}) = e^{i(\vec{k}'+\vec{g})\vec{R}} \psi_{\vec{k},n}(\vec{r}) = e^{i\vec{k}'\vec{R}} \psi_{\vec{k},n}(\vec{r}) \quad . \tag{10.20}$$

The result above means that the wavevector $\vec{k}'$ satisfies the Bloch theorem for
the *same* wave as that corresponding to $\vec{k}$. In other words, the wave function
associated with the wavevector $\vec{k}$ is also associated with any wavevector obtained
from $\vec{k}$ by a translation through a vector of the reciprocal lattice. This implies that
a given eigenstate is not associated with a single wavevector, but with an infinite
number of such vectors. For instance, a consequence of equation (10.20) is that in
1D the associated energies are a periodic function of wavevector (see also Figure
10.2), and $\vec{k}$ is defined modulo $2\pi/L$. Thus, a unit cell of the reciprocal lattice
contains a whole set of $\vec{k}$ values which allows us to recover the full dispersion
relationship and set of eigenfunctions. It is convenient (and usual) to consider only
the first Brillouin zone, since any value of $\vec{k}$ can be reduced to another wavevector
value comprised inside it.

10.3. The harmonic oscillator ○

The harmonic oscillator is a 1D quantum well exhibiting a potential of the form

$$V(x) = \frac{1}{2} K x^2 \quad . \tag{10.21}$$

We define a pulsation $\omega_0 = (K/m)^{1/2}$ as in the classical case. The Hamiltonian of the system is

$$H = \frac{p^2}{2m} + \frac{Kx^2}{2}.$$ (10.22)

To simplify the notations we define two new reduced operators P and Q as

$$P = \frac{p}{\sqrt{\hbar m \omega_0}} \quad \text{and} \quad Q = \sqrt{\frac{K}{\hbar \omega_0}} x.$$ (10.23)

Then the Hamiltonian can be written as a function of P and Q:

$$H = \frac{\hbar \omega_0}{2} \left(P^2 + Q^2 \right).$$ (10.24)

From the canonical commutation relation

$$\left[x, p \right] = i\hbar$$ (10.25)

We immediately obtain

$$\left[Q, P \right] = i.$$ (10.26)

From P and Q we define two (non-Hermitian) operators a and $a^\dagger$ by

$$a = \frac{Q + iP}{\sqrt{2}} \quad \text{and} \quad a^\dagger = \frac{Q - iP}{\sqrt{2}}.$$ (10.27)

From the commutation relation equation (10.26) and the definition of a and $a^\dagger$ we can deduce that

$$2aa^\dagger = P^2 + Q^2 + 1$$ (10.28)

and

$$2a^\dagger a = P^2 + Q^2 - 1,$$ (10.29)

which imply in turn that

$$\left[a, a^\dagger\right] = aa^\dagger - a^\dagger a = 1.$$

(10.30)

If we define the operator $N = a^\dagger a$, it is not difficult to express H as a function of N *(see equations (10.24) and (10.29))*:

$$H = \hbar\omega_0\left(N + \frac{1}{2}\right).$$

(10.31)

From equation (10.31) it is obvious that if a state vector $|\psi\rangle$ is an eigenstate of the operator N with an associated eigenvalue n, it is also an eigenstate of the Hamiltonian with an associated energy level $E=(n+1/2)\hbar\omega_0$.

Multiplying the left sides of equation (10.30) by $a^\dagger$ leads to

$$Na^\dagger - a^\dagger N = a^\dagger$$

(10.32)

and multiplying the right sides of equation (10.30) by a gives

$$aN - Na = a.$$

(10.33)

Since $|\psi\rangle$ is an eigenfunction of N with eigenvalue n, we can write

$$N|\psi\rangle = n|\psi\rangle.$$

(10.34)

Multiplying the formula above by a on the left sides and taking equation (10.33) into account we obtain

$$N\left(a|\psi\rangle\right) = (n-1)\left(a|\psi\rangle\right).$$

(10.35)

From equation (10.35) either $a|\psi\rangle=0$ or $a|\psi\rangle$ is an eigenstate of the operator N with eigenvalue $(n-1)$. From equations (10.32) and (10.34) we could obtain in the same manner that $a^\dagger|\psi\rangle$ is an eigenstate of the operator N with eigenvalue $(n+1)$ if it is not equal to zero.

If we consider the ground state $|\psi_0\rangle$ of the system, i.e. the one which exhibits the lowest quantized energy, we necessarily have

$$a|\psi_0\rangle = 0.$$

(10.36)

Otherwise, from the considerations above, n_0-1 would also be an eigenvalue, and since we consider the ground state this is impossible because this eigenvalue is lower than n_0. Now if we multiply the left sides of equation (10.36) by $a^\dagger$ we obtain

$$N|\psi_0\rangle = 0 , \tag{10.37}$$

which indicates that $n_0=0$. The ground energy of the harmonic oscillator is equal to

$$E_0 = \frac{1}{2}\hbar\omega_0 . \tag{10.38}$$

Thus, by acting repeatedly on $|\psi_0\rangle$ with the operator $a^\dagger$ we obtain eigenstates of the Hamiltonian with eigenvalues

$$E_n = \left(n + \frac{1}{2}\right)\hbar\omega_0 . \tag{10.39}$$

with n integer. We have thus found the harmonic oscillator levels (a result which ought to be known). The eigenstates exhibit equally spaced energy levels, all levels being separated from the next ones by a quantum $\hbar\omega_0$, where $\omega_0/2\pi$ is the natural frequency of the classical harmonic oscillator. Now we must determine the eigenfunctions. From $p=-i\hbar \nabla$ and equation (10.23) we can write

$$P = -i\frac{\partial}{\partial Q} . \tag{10.40}$$

Thus, from equation (10.27) we can also write

$$a = \frac{1}{\sqrt{2}}\left(Q + \frac{\partial}{\partial Q}\right) \tag{10.41}$$

and

$$a^\dagger = \frac{1}{\sqrt{2}}\left(Q - \frac{\partial}{\partial Q}\right) . \tag{10.42}$$

However, from equations (10.36) and (10.41) we have

$$\left(Q + \frac{\partial}{\partial Q} \right) \psi_0 = 0, \tag{10.43}$$

which can be easily integrated to give

$$\psi_0(Q) = C e^{-Q^2/2} \tag{10.44}$$

Taking into account the definition of Q (equation (10.23)), the normalization of the wave function is obtained by integrating its square modulus with respect to x. This determines the constant in front of the exponential and eventually we obtain

$$\psi_0(Q) = \left(\frac{m\omega_0}{\hbar\pi} \right)^{\frac{1}{4}} e^{-Q^2/2}. \tag{10.45}$$

If Ψ_n is the eigenfunction associated with the energy level E_n, we have demonstrated that $a^\dagger \Psi_n$ is also an eigenfunction of the level E_{n+1} (though not necessarily normalized). Acting on an eigenstate with this operator results in adding an energy quantum to the system, and for this reason the operator $a^\dagger$ is called the creation operator (reciprocally, the operator a is called the annihilation operator, because its action "destroys" exactly one energy quantum). We thus have

$$a^\dagger \left| \psi_n \right\rangle = \alpha_n \left| \psi_{n+1} \right\rangle \tag{10.46}$$

where α_n is an unknown coefficient. This coefficient is determined by stating that Ψ_{n+1} must be normalized, and along with equation (10.32) we obtain

$$\left| \alpha_n^2 \right| = \int dx \left(a^\dagger \psi_n \right)^* a^\dagger \psi_n = \int dx \, \psi_n^* \left(a a^\dagger \psi_n \right) = \int dx \, \psi_n^* \left[\left(1 + a^\dagger a \right) \psi_n \right] = n + 1 \tag{10.47}$$

The above relationship does specify the phase factor value, and to simplify the discussion we can restrict α_n to real, positive values:

$$a^\dagger \left| \psi_n \right\rangle = \sqrt{n+1} \left| \psi_{n+1} \right\rangle. \tag{10.48}$$

From equation (10.48) we obtain

$$|\psi_n\rangle = \frac{\left(a^\dagger\right)^n}{\sqrt{n!}} |\psi_{n+1}\rangle \tag{10.49}$$

and taking into account equations (10.41), (10.42) and (10.45) we obtain

$$\psi_n(Q) = \left(\frac{m\omega_0}{\hbar\pi}\right)^{1/4} \frac{1}{\left(n!2^n\right)^{1/2}} \left(Q - \frac{d}{dQ}\right)^n e^{-Q^2/2}. \tag{10.50}$$

Noting that

$$\left(Q - \frac{d}{dQ}\right) f(Q) = -e^{Q^2/2} \frac{d}{dQ} e^{-Q^2/2} f(Q) \tag{10.51}$$

and thus that

$$\left(Q - \frac{d}{dQ}\right)^n = (-1)^n e^{Q^2/2} \left(\frac{d}{dQ}\right)^n e^{-Q^2/2} \tag{10.52}$$

leads to the wave function expression

$$\psi_n(Q) = \left(\frac{m\omega_0}{\hbar\pi}\right)^{1/4} \frac{(-1)^n}{\left(n!2^n\right)^{1/2}} e^{Q^2/2} \frac{d^n}{dQ^n} e^{-Q^2}. \tag{10.53}$$

Introducing the orthogonal Hermitian polynomials as

$$H_n(Q) = \left(\frac{m\omega_0}{\hbar\pi}\right)^{1/4} \frac{1}{\left(n!2^n\right)^{1/2}} (-1)^n e^{Q^2} \frac{d^n}{dQ^n} e^{-Q^2}, \tag{10.54}$$

we eventually obtain

$$\psi_n(Q) = e^{-Q^2/2} H_n(Q). \tag{10.55}$$

10.4. Stationary perturbation theory ⦾

10.4.1. *Non-degenerate perturbation theory*

Consider a stationary system with Hamiltonian H_0, energy eigenvalues E_n^0 and eigenstates $|\psi_n^0\rangle$. Suppose that we are able to calculate all those quantities. Is it possible to assess in a simple way the modification brought to this system by a small perturbation W? Assume that the new Hamiltonian is given by $H=H_0+W=H_0+\alpha V$, where α is a very small, dimensionless quantity and the eigenvalues of V have the same order of magnitude as the E_Ns. Also assume that the system is not degenerate (i.e. all eigenvalues correspond to only one eigenvector). The Schrödinger equation reads

$$\left(H_0 + \alpha V\right)|\psi_n\rangle = E_n|\psi_n\rangle. \tag{10.56}$$

Assume that both the eigenvalues and eigenfunctions can be expanded with respect to the small parameter α. We can write for a state $|\psi_n\rangle$

$$E_n = \varepsilon_0 + \alpha\varepsilon_1 + \alpha^2\varepsilon_2 + ... + \alpha^k\varepsilon_k + ... \tag{10.57}$$

and

$$|\psi_n\rangle = |u_0\rangle + \alpha|u_1\rangle + \alpha^2|u_2\rangle + ... + \alpha^k|u_k\rangle + ... \tag{10.58}$$

The problem is to find the $|u_n\rangle$s and the ε_ns. Replace E_n and $|\psi_n\rangle$ by their expressions (equations (10.57) and (10.58), respectively) in equation (10.56):

$$\left(H_0 + \alpha V\right)\left(|u_0\rangle + \alpha|u_1\rangle + \alpha^2|u_2\rangle + ...\right) =$$
$$\left(\varepsilon_0 + \alpha\varepsilon_1 + \alpha^2\varepsilon_2 + ...\right)\left(|u_0\rangle + \alpha|u_1\rangle + \alpha^2|u_2\rangle + ...\right) \tag{10.59}$$

Identify the terms which have the same power in the expansion. The first three orders (0, 1 and 2) give

$$H_0|u_0\rangle = \varepsilon_0|u_0\rangle, \tag{10.60}$$

$$\left(H_0 - \varepsilon_0\right)|u_1\rangle + \left(V - \varepsilon_1\right)|u_0\rangle = 0 \tag{10.61}$$

and

$$\left(H_0 - \varepsilon_0\right)\left|u_2\right\rangle + \left(V - \varepsilon_1\right)\left|u_1\right\rangle - \varepsilon_2\left|u_0\right\rangle = 0. \tag{10.62}$$

Equation (10.60) implies that $|u_0\rangle$ is an eigenstate of H_0. In addition, making α vanish in equation (10.57) corresponds to the unperturbed situation and indicates that $E_n^0 = \varepsilon_0$. Thus, we can write that $|u_0\rangle$ is any of the $|\psi_n^0\rangle$ (note that our reasoning is valid because the system is non-degenerate and thus a given ε_0 corresponds to only one eigenstate in equation (10.60)). Projecting equation (10.61) onto eigenstate $|u_0\rangle = |\psi_n^0\rangle$ and taking into account that $\left\langle u_0 \middle| H_0 - \varepsilon_0 \middle| u_1\right\rangle = \left(\varepsilon_1 - \varepsilon_0\right)\left\langle u_1 \middle| u_0\right\rangle = 0$, we obtain

$$\varepsilon_1 = \left\langle \psi_n^0 \middle| V \middle| \psi_n^0\right\rangle, \tag{10.63}$$

so that to first order, perturbation theory gives us

$$E_n = E_n^0 + \left\langle \psi_n^0 \middle| W \middle| \psi_n^0\right\rangle. \tag{10.64}$$

To find the eigenstate $|\psi_n\rangle$ to first order, we note that $|u_1\rangle$ and $|u_0\rangle = |\psi_n^0\rangle$ are orthogonal, so that we can expand $|u_1\rangle$ on the basis formed by the $|\psi_p^0\rangle$'s with $p \neq n$, writing

$$\left|u_1\right\rangle = \sum \left\langle \psi_p^0 \middle| u_1\right\rangle \left|\psi_p^0\right\rangle. \tag{10.65}$$

To calculate $\langle \psi_p^0 | u_1 \rangle$ we project equation (10.61) onto each eigenstate $|\psi_p^0\rangle$. This leads to

$$\left\langle \psi_p^0 \middle| u_1\right\rangle = \frac{\left\langle \psi_p^0 \middle| V \middle| \psi_n^0\right\rangle}{E_n^0 - E_p^0} \tag{10.66}$$

and from equation (10.58) we obtain

$$\left|\psi_n\right\rangle = \left|\psi_n^0\right\rangle + \sum_{p \neq n} \frac{\left\langle \psi_p^0 \middle| W \middle| \psi_n^0\right\rangle}{E_n^0 - E_p^0}\left|\psi_p^0\right\rangle. \tag{10.67}$$

If the first order perturbation is equal to zero it may be necessary to expand the eigenvalues up to the second order. To do this, we must project equation (10.62) onto $|\psi_n\rangle$. The second order perturbation result is

$$E_n = E_n^0 + \langle \psi_n^0 | W | \psi_n^0 \rangle + \sum_{p \neq n} \frac{\left| \langle \psi_p^0 | W | \psi_n^0 \rangle \right|^2}{E_n^0 - E_p^0} \qquad (10.68)$$

From equation (10.67) or (10.68) we can deduce that the modification brought to the system by the perturbation W is enhanced if the initial energy level spacings are small, since energy difference terms appear at the denominators. To give but one example, this explains why wide band gap materials most often exhibit a lower dielectric constant. We know that application of an electric field to a semiconductor leads to an energy increase which is proportional to the dielectric constant. Thus, the dielectric constant is nothing but a measure of the extent to which an electric field perturbs the medium. As a consequence, if the unperturbed levels are widely separated we do expect from a formula such as equation (10.68) that the dielectric constant be small.

10.4.2. Degenerate perturbation theory

Now we assume that one level E_n^0 is initially degenerate, with a set of K eigenstates labeled as $|\psi_{n,1}^0\rangle$, $|\psi_{n,2}^0\rangle$,..., $\rangle$,$|\psi_{n,K}^0\rangle$[1]. If we look back at equation (10.60), which is the zero[th] order equation, the eigenstate $|u_0\rangle$ which corresponds to the initially degenerate state cannot be determined from this expression, as any $|\psi_{n,k}^0\rangle$ or linear combination of those degenerate states is suitable. Therefore, to be lifted, the degeneracy must be examined using the first order equation (10.61). Project equation (10.61) onto one of the initially degenerate states:

$$\begin{aligned} 0 &= \langle \psi_{n,k}^0 | H_0 - \varepsilon_0 | u_1 \rangle + \langle \psi_{n,k}^0 | V - \varepsilon_1 | u_0 \rangle \\ &= \langle u_1 | H_0 - \varepsilon_0 | \psi_{n,k}^0 \rangle^* + \langle \psi_{n,k}^0 | V | u_0 \rangle - \langle \psi_{n,k}^0 | \varepsilon_1 | u_0 \rangle \quad . \end{aligned} \qquad (10.69)$$

---

[1] n indexes the energy level, the upper 0 stands for the unperturbed states and the right numbers $1,2,...,K$ index the K degenerate states with energy E_n^0.

The first term of the second line is obtained from the hermiticity of H_0, and obviously reduces to zero, since by virtue of equation (10.60) $\varepsilon_0=E_n$. Thus, we obtain the equation

$$\left\langle \psi_{n,k}^0 \left| V \right| u_0 \right\rangle = \left\langle \psi_{n,k}^0 \left| \varepsilon_1 \right| u_0 \right\rangle \quad . \tag{10.70}$$

Examine the equation above. We cannot calculate the left-hand side term since we do not know $|u_0\rangle$, which is the unknown. Thus, we must do something more. A good point would be to succeed in calculating the decomposition of $|u_0\rangle$ as a function of the $|\psi_{n,k}^0\rangle$'s, which are known. This can be achieved from equation (10.70) by inserting the unity operator written as in equation (2.6) between V and $|u_0\rangle$:

$$\sum_p \left\langle \psi_{n,k}^0 \left| V \right| \psi_p^0 \right\rangle \left\langle \psi_p^0 \left| u_0 \right\rangle = \varepsilon_1 \left\langle \psi_{n,k}^0 \left| u_0 \right\rangle \right. \right. , \tag{10.71}$$

where the sum is carried over all the eigenstates of H_0. Since $|u_0\rangle$ is a linear combination (to be determined) of the degenerate states $|\psi_{n,k}^0\rangle$, any $|\psi_p^0\rangle$ which is not one of those degenerate eigenstates is orthogonal to $|u_0\rangle$, and the only terms which survive in the sum are those which correspond to any of the K degenerate eigenstates. Multiplying both sides by α, we obtain the secular system of equations

$$\sum_{k'} \left\langle \psi_{n,k}^0 \left| W \right| \varphi_{n,k'}^0 \right\rangle \left\langle \varphi_{n,k'}^0 \left| u_0 \right\rangle = \Delta E \left\langle \psi_{n,k}^0 \left| u_0 \right\rangle \right. \right. , \tag{10.72}$$

where index k' runs over all initially degenerate states, and $\Delta E = E_n - E_n^0$ is the energy variation induced by the perturbation. Since k can also be varied from 1 to K, the number of equations is equal to K. We can obtain the decomposition of $|u_0\rangle$ over the subspace of the initially degenerate eigenstates just by solving this system. For instance, if the system is doubly degenerate, we obtain the secular equation

$$\begin{pmatrix} W_{11} & W_{12} \\ W_{21} & W_{22} \end{pmatrix} \begin{pmatrix} \left\langle \psi_1^0 \left| u_0 \right\rangle \right. \\ \left\langle \psi_2^0 \left| u_0 \right\rangle \right. \end{pmatrix} = \Delta E \begin{pmatrix} \left\langle \psi_1^0 \left| u_0 \right\rangle \right. \\ \left\langle \psi_2^0 \left| u_0 \right\rangle \right. \end{pmatrix} \quad , \tag{10.73}$$

in which W_{ij} is defined as

$$W_{ij} = \left\langle \psi_i^0 \left| W \right| \varphi_j^0 \right\rangle \quad . \tag{10.74}$$

If by chance we can neglect W_{11} and W_{22} and $W_{12}=W_{21}$ (a fact which may occur in practice, as we shall see in a few lines, and happens still more often in exercises), the system reduces to

$$\begin{pmatrix} 0 & W_{12} \\ W_{12} & 0 \end{pmatrix}\begin{pmatrix} c_1 \\ c_2 \end{pmatrix} = \Delta E \begin{pmatrix} c_1 \\ c_2 \end{pmatrix} \quad , \tag{10.75}$$

where we replaced the scalar products to be determined by the simpler notations c_1 and c_2. This system is easily solved and leads to the two energies

$$E^+ = E - W_{12} \qquad E^- = E + W_{12} \tag{10.76}$$

and eigenstates

$$|-\rangle = \frac{1}{\sqrt{2}}\left(\left|\psi_1^0\right\rangle + \left|\psi_2^0\right\rangle\right) \qquad |+\rangle = \frac{1}{\sqrt{2}}\left(\left|\psi_1^0\right\rangle - \left|\psi_2^0\right\rangle\right). \tag{10.77}$$

In practice, it often turns out that we do not have to work with a single initial Hamiltonian and one perturbation, but instead we have to deal with a situation in which we consider two identical systems initially located very far apart, i.e. without interaction, and then we put them closer to one another (for example two identical atoms). Let us affect the index l for the "left" system and r to the "right system". When the two systems are brought together, the potential corresponding to one of the systems acts as a perturbation on the other. We can write the full Hamiltonian either as $H=H_l+W_l$ or as $H=H_r+W_r$, where W_l is the potential added by the *right* system and seen by the left system as a perturbation. We can now write equations similar to equations (10.60), (10.61) and (10.62), which can be written in two different forms, depending on whether we write H as H_l+W_l or as H_r+W_r. In particular, the first order equation can be written either as

$$\left(H_r - \varepsilon_0\right)\left|u_1\right\rangle + \left(V_r - \varepsilon_1\right)\left|u_0\right\rangle = 0 \tag{10.78}$$

or as

$$\left(H_l - \varepsilon_0\right)\left|u_1\right\rangle + \left(V_l - \varepsilon_1\right)\left|u_0\right\rangle = 0 \quad . \tag{10.79}$$

Following exactly the same reasoning as in the degenerate case already discussed above, and multiplying the first order equation either by $\langle\psi_r|$ (in which case we use equation (10.78)) or by $\langle \psi_l |$ (in which case we use equation (10.79)), we obtain the system

$$\begin{pmatrix} W_r & W_{rl} \\ W_{lr} & W_l \end{pmatrix} \begin{pmatrix} c_1 \\ c_2 \end{pmatrix} = \Delta E \begin{pmatrix} c_1 \\ c_2 \end{pmatrix} \quad , \tag{10.80}$$

where the matrix elements are defined as

$$\begin{aligned} W_r &= \langle \psi_r | V_r | \psi_r \rangle & W_l &= \langle \psi_l | V_l | \psi_l \rangle \\ W_{rl} &= \langle \psi_r | V_r | \psi_l \rangle & W_{lr} &= \langle \psi_l | V_l | \psi_r \rangle \end{aligned} \quad . \tag{10.81}$$

Note that in general, an element such as W_r is expected to remain quite small, since the right wave function is located far apart from the perturbation V_r, and it is generally reasonable to neglect W_r and W_l. If we consider two identical systems, W_{rl} and W_{lr} are also identical, and therefore we find, as in the former case (see equations (10.76) and (10.77)), the energies

$$E^+ = E - W_{rl} \qquad E^- = E + W_{rl} \tag{10.82}$$

and eigenstates

$$|-\rangle = \frac{1}{\sqrt{2}} \big(|\psi_d\rangle + |\psi_g\rangle \big) \qquad |+\rangle = \frac{1}{\sqrt{2}} \big(|\psi_d\rangle - |\psi_g\rangle \big). \tag{10.83}$$

The initially degenerate energy has been split into two different levels, separated by an amount equal to twice the interaction matrix element. We say that degeneracy has been lifted by the interaction, as schematically illustrated by Figure 10.3.

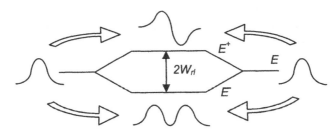

Figure 10.3. *Hybridization of initially degenerate states occurring when two far apart identical systems are led to interact by putting them closer to one another*

It is worth noting that this is exactly the physical process at the origin of the chemical bond: just consider two initially separated Hydrogen atoms. As they move closer to one another, since one energy state can be populated by two electrons with spins up and down, the overall energy can be reduced down to $2E^-$ instead of $2E$, and the atoms will become bonded (note that to explain why there is a position with a minimum energy requires a little more work; consult any QM textbook to go further). This also corresponds to the case of two identical quantum wells lying close to one another, which is discussed in section 2.5.2.

10.5. Method of Lagrange multipliers O

Assume that we want to find the extrema of a function $F(x_1, x_2, ..., x_n)$. If the n x_n variables were independent, we would just have to solve the n independent equations obtained by partial differentiation of F with respect to each variable $\partial F / \partial x_i = 0$. The situation is different if we must fulfil an additional constraint relating the x_ns together[2], such as a relation of the form $G(x_1, x_2, ..., x_n) = 0$. The simplest way to solve for the non-independent x_ns is to use the method of Lagrange multipliers, which we demonstrate below. Suppose that we can express the variable x_n as a function of the n other variables. The extremum is found by solving the $(n-1)$ remaining equations with now $(n-1)$ independent variables

$$\frac{\partial F}{\partial x_i} + \frac{\partial F}{\partial x_n} \frac{\partial x_n}{\partial x_i} = 0 \; , \tag{10.84}$$

i ranging from 1 to $n-1$. To assess $\partial x_n / \partial x_i$, we differentiate the constraint G:

$$dG = \sum \frac{\partial G}{\partial x_i} dx_i = 0 \tag{10.85}$$

where the sum runs from 1 to n and the differential is equal to zero because the constraint does not change during the minimization process. If we keep constant all variables but x_n and x_i, we obtain

$$\frac{\partial x_n}{\partial x_i} = -\left(\frac{\partial G}{\partial x_n}\right)^{-1} \frac{\partial G}{\partial x_i} \; . \tag{10.86}$$

---

2 The most famous physical example is the determination of thermodynamic equilibrium, which corresponds to entropy maximization while respecting the other constraints imposed upon the system (e.g. particle number conservation, etc.).

If we substitute equation (10.86) for $\partial x_n/\partial x_i$ in equation (10.84) we obtain

$$\frac{\partial F}{\partial x_i} - \frac{\partial F}{\partial x_n}\left(\frac{\partial G}{\partial x_n}\right)^{-1}\frac{\partial G}{\partial x_i} = 0 . \tag{10.87}$$

Thus, if we define the Lagrange multiplier λ as

$$\lambda = \left(\frac{\partial G}{\partial x_n}\right)^{-1}\frac{\partial F}{\partial x_n} . \tag{10.88}$$

the *(n-1)* equations become

$$\frac{\partial F}{\partial x_i} - \lambda\frac{\partial G}{\partial x_i} = 0 . \tag{10.89}$$

However, we can note that from the definition of λ (equation (10.88)) we also have

$$\frac{\partial F}{\partial x_n} - \lambda\frac{\partial G}{\partial x_n} = 0 . \tag{10.90}$$

so that we have n equations with a similar form, and we are not forced to treat the variable x_n on a particular footing. We can thus define a Lagrangian

$$L = F - \lambda G \tag{10.91}$$

and partially differentiate with respect to each x_n while maintaining λ constant, so as to recover the set formed by equation (10.89) and equation (10.90) in the unified form

$$\frac{\partial L}{\partial x_i} = 0 . \tag{10.92}$$

where i now ranges from 1 to n. From this set of equations we can find the extremum as a function of the Lagrange multiplier. Then λ is found by demanding that its value verifies the constraint $G(x_1(\lambda),x_2(\lambda),...,x_n(\lambda))=0$. The method is of course generalizable to the case of more than one constraint, just by forming as many Lagrange multipliers as required by the number of constraints.

10.6. Variational principle ◉

Consider a Hamiltonian operator H with its eigenstate basis $|\psi_n\rangle$ and corresponding energies E_ns. Let $|\psi\rangle$ be a trial wave function. It can be developed on the eigenstate basis as

$$|\psi\rangle = \sum_n a_n |\psi_n\rangle. \tag{10.93}$$

Taking into account the orthogonality of the eigenstates and that $H|\psi_n\rangle = E_n|\psi_n\rangle$, it is easy to show that

$$\langle \psi |H|\psi\rangle = \sum_n a_n a_n^* E_n \geq E_0 \sum_n a_n a_n^* = \langle \psi |\psi\rangle E_0. \tag{10.94}$$

so that

$$\langle H\rangle = \frac{\langle \psi |H|\psi\rangle}{\langle \psi |\psi\rangle} \geq E_0 \tag{10.95}$$

which in fact constitutes the variational theorem. In practice this means that if we have a good intuition of what the ground state wave function should look like, we can improve its determination by minimizing the quantity $\langle E\rangle = \langle \psi|H|\psi\rangle / \langle \psi|\psi\rangle$ with respect to the (initially) arbitrary parameters which determine its shape, or by minimizing the quantity $\langle E\rangle = \langle \psi|H|\psi\rangle$ with the constraint $\langle \psi|\psi\rangle = 1$.

In practice, if the variational function $|\psi\rangle$ depends on some parameters λ_is, we have to find the precise set of parameter values which makes $\langle E\rangle$ a minimum. This method can be very precise for determining the energy if the trial wave function is well chosen, but unfortunately it is quite difficult to assess the error made. The variational theorem can be extended to the Ritz theorem, which establishes that the average value of the Hamiltonian is stationary at the neighborhood of any discrete eigenvalue. Thus, the variational method can be generalized, and any local extremum of the trial wave function (min or max) represents an approached value of a quantized energy level.

An interesting development of the variational method can be used when the variational function is a linear combination of atomic orbitals. Here we restrict the discussion to the case of two functions but it is easily generalizable to a larger number. Assume that we approach two atoms initially very far apart. As long as they stay at not too close a distance we can assume that the atomic levels are only weakly perturbed. If we consider one of those energy levels, which initially is doubly degenerate (one per atom), a good guess for the variational wave function associated with the perturbed level is to use a linear combination of the two atomic orbitals of the separated atoms. More generally, we can use a linear combination of two functions:

$$|\psi\rangle = c_1|\varphi_1\rangle + c_2|\varphi_2\rangle ,\qquad(10.96)$$

Let H be the full Hamiltonian of the system (e.g. the sum of the atomic potential of our two atoms). Assuming that the c_is are real, we can define the overlap integral

$$S_{12} = \langle\varphi_1|\varphi_2\rangle = S_{21}^* .\qquad(10.97)$$

From equation (10.96) the norm of the wave function is easily turned into

$$\langle\psi|\psi\rangle = c_1^2 + c_2^2 + c_1 c_2 \left(S_{12} + S_{21}\right) = 1 .\qquad(10.98)$$

Defining the transfer integral

$$H_{ij} = \langle\varphi_i|H|\varphi_j\rangle = H_{ji}^* ,\qquad(10.99)$$

where the last right-hand term is obtained from the hermiticity of H, it is straightforward to arrive at the energy expectation value

$$\langle H\rangle = \langle\psi|H|\psi\rangle = c_1^2 H_{11} + c_2^2 H_{22} + c_1 c_2 \left(H_{12} + H_{21}\right) .\qquad(10.100)$$

To minimize this expression we cannot write the usual condition $\partial\langle H\rangle/\partial c_1 = 0$ and $\partial\langle H\rangle/\partial c_2 = 0$, for the two coefficients c_1 and c_2 are not independent but linked by the normalization condition $\langle\psi|\psi\rangle = 1$ (equation (10.98)). We thus have to minimize an expression which depends on two variables, these two variables being related to one another by one constraint. This problem can be solved by using the method of Lagrange multipliers summarized in section 10.5. We replace the energy expectation value by the Lagrangian, itself formed by the linear combination $\langle H\rangle - \lambda\langle\psi|\psi\rangle$, where λ is a Lagrange multiplier, and we differentiate with respect to c_1 and c_2, respectively. This leads to

$$\frac{\partial L}{\partial c_1} = 2c_1\left(H_{11} - \lambda\right) + c_2\left(H_{12} + H_{21} - \lambda\left(S_{12} + S_{21}\right)\right) = 0$$

$$\frac{\partial L}{\partial c_2} = 2c_2\left(H_{22} - \lambda\right) + c_1\left(H_{12} + H_{21} - \lambda\left(S_{12} + S_{21}\right)\right) = 0 .\qquad(10.101)$$

Multiply the first equation above by a factor $c_1/2$, the second equation by $c_2/2$ and take the sum of the two expressions. Factorize λ. Then if we compare the resulting equation with equations (10.98) and (10.100) it becomes obvious that the Lagrange multiplier is equal to the energy expectation value, $\lambda = \langle H \rangle = E$. Replacing λ by E in equation (10.101) gives us two new equations:

$$(H_{11} - E)c_1 + \frac{1}{2}(H_{12} + H_{21} - E(S_{12} + S_{21}))c_2 = 0$$

$$\frac{1}{2}(H_{12} + H_{21} - E(S_{12} + S_{21}))c_1 + (H_{22} - E)c_2 = 0 \quad . \tag{10.102}$$

To obtain solutions different from zero the determinant of this system must be equal to zero, so that we obtain the secular equation

$$\begin{vmatrix} H_{11} - E & \frac{1}{2}(H_{12} + H_{21} - E(S_{12} + S_{21})) \\ \frac{1}{2}(H_{12} + H_{21} - E(S_{12} + S_{21})) & H_{22} - E \end{vmatrix} = 0 \quad , \tag{10.103}$$

which gives us two energy values. If the overlap and transfer integrals are real we have $H_{12} = H_{21}$ and $S_{12} = S_{21} = S$, so that equation (10.103) simplifies to

$$\begin{vmatrix} H_{11} - E & H_{12} - ES \\ H_{12} - ES & H_{22} - E \end{vmatrix} = 0 \quad . \tag{10.104}$$

In general the overlap integrals are quite small compared to the transfer integrals, as when one wave amplitude is large, i.e. close to the atom to which it is attached, the other wave is exponentially small. Thus, if some of the values are complex but if we can neglect the overlap integrals, from equation (10.103) the energy is given by

$$E = \frac{1}{2}\left(H_{11} + H_{22} \pm \sqrt{(H_{11} - H_{22})^2 + 4(\mathrm{Re}(H_{12}))^2} \right) \tag{10.105}$$

where we solved the secular equation by neglecting S_{12} and S_{21}, and accounted for the fact that $(H_{12} + H_{21})/2 = (H_{12} + H_{12}{}^*)/2 = \mathrm{Re}(H_{12})$. If H_{11} is equal to H_{22} this further reduces to

$$E = H_{11} \pm \sqrt{\mathrm{Re}(H_{12})^2} \quad . \tag{10.106}$$

We can replace the latter quantity by

$$E = H_{11} \pm |H_{12}| \tag{10.107}$$

if the wave functions are given by tight-binding expressions such as in equation (2.85). To demonstrate this we can straightforwardly express H_{12} as

$$H_{12} = \sum_A \sum_B \exp\left(i\vec{k}.\left(\vec{R}_B - \vec{R}_A\right)\right) \int \phi_{at}\left(\vec{r} - \vec{R}_A\right) H \phi_{at}\left(\vec{r} - \vec{R}_B\right), \tag{10.108}$$

where use has been made of equation (2.85), and accounting for the fact that the atomic wave functions are real and thus equal to their complex conjugate. Then we write the square modulus as

$$\begin{aligned}
H_{12}H_{12}^* = &\sum_{A,A',B,B'} \exp\left(i\vec{k}\left(\vec{R}_B - \vec{R}_A - \vec{R}_{B'} + \vec{R}_{A'}\right)\right) \times \\
&\times \int \phi_{at}\left(\vec{r} - \vec{R}_A\right) H \phi_{at}\left(\vec{r} - \vec{R}_B\right) \int \phi_{at}\left(\vec{r} - \vec{R}_{A'}\right) H \phi_{at}\left(\vec{r} - \vec{R}_{B'}\right)
\end{aligned} \tag{10.109}$$

We can expand the exponential and drop the imaginary contribution, because we calculate a square modulus and know that its imaginary part is equal to zero. This leads to

$$\begin{aligned}
H_{12}H_{12}^* = &\sum_{A,A',B,B'} \int \phi_{at}\left(\vec{r} - \vec{R}_A\right) H \phi_{at}\left(\vec{r} - \vec{R}_B\right) \int \phi_{at}\left(\vec{r} - \vec{R}_{A'}\right) H \phi_{at}\left(\vec{r} - \vec{R}_{B'}\right) \times \\
&\times \left(\cos\left(\vec{k}\left(\vec{R}_B - \vec{R}_A\right)\right)\cos\left(\vec{k}\left(\vec{R}_{B'} - \vec{R}_{A'}\right)\right) + \sin\left(\vec{k}\left(\vec{R}_B - \vec{R}_A\right)\right)\sin\left(\vec{k}\left(\vec{R}_{A'} - \vec{R}_{B'}\right)\right)\right)
\end{aligned} \tag{10.110}$$

where the terms with sine functions mutually cancel each other out. Take the origin at a lattice site; for each (A,B) couple it is obvious that there is also another (A'',B'') couple for which $\vec{R}_B - \vec{R}_A = -\left(\vec{R}_{B''} - \vec{R}_{A''}\right)$, so that the sine values corresponding to these two particular (A,B) couples mutually cancel each other out (note that the transfer integrals do not change under this operation). We thus arrive at

$$\begin{aligned}
H_{12}H_{12}^* = &\sum_{A,A',B,B'} \int \phi_{at}\left(\vec{r} - \vec{R}_A\right) H \phi_{at}\left(\vec{r} - \vec{R}_B\right) \int \phi_{at}\left(\vec{r} - \vec{R}_{A'}\right) H \phi_{at}\left(\vec{r} - \vec{R}_{B'}\right) \times \\
&\times \cos\left(\vec{k}\left(\vec{R}_B - \vec{R}_A\right)\right)\cos\left(\vec{k}\left(\vec{R}_{B'} - \vec{R}_{A'}\right)\right).
\end{aligned} \tag{10.111}$$

However, using equation (10.108) we can also straightforwardly calculate

$$\begin{aligned}
\left(\frac{H_{12} + H_{12}^*}{2}\right)^2 = &\sum_{A,A',B,B'} \int \phi_{at}\left(\vec{r} - \vec{R}_A\right) H \phi_{at}\left(\vec{r} - \vec{R}_B\right) \int \phi_{at}\left(\vec{r} - \vec{R}_{A'}\right) H \phi_{at}\left(\vec{r} - \vec{R}_{B'}\right) \times \\
&\times \cos\left(\vec{k}\left(\vec{R}_B - \vec{R}_A\right)\right)\cos\left(\vec{k}\left(\vec{R}_{B'} - \vec{R}_{A'}\right)\right)
\end{aligned} \tag{10.112}$$

and we see that equations (10.111) and (10.112) are the same. Thus, we have

$$|H_{12}|^2 = (\text{Re}(H_{12}))^2 .$$

(10.113)

This is just what we needed to prove equation (10.107).

10.7. Wiener-Khintchine theorem ◉

The Wiener-Khintchine theorem states that the power spectral density $\Phi_x(f)$ of a random and stationary process $x(t)$ is the Fourier transform of its autocorrelation function

$$\Phi_x(f) = \int_{-\infty}^{+\infty} R_x(\tau) \exp(-2i\pi f\tau) d\tau .$$

(10.114)

This can be demonstrated as follows. The power spectral density of a random signal realization $x_i(t)$ on the interval T can be expressed as

$$\Phi_i^{obs}(f) = \frac{1}{T} X_i^{obs}(f) \left(X_i^{obs}(f) \right)*$$

(10.115)

$$= \frac{1}{T} \iint_{-\infty}^{+\infty} x_i^{obs}(t) x_i^{obs}(t') \Pi\left(\frac{t}{T}\right) \Pi\left(\frac{t'}{T}\right) \exp(-2i\pi f(t-t')) dt dt'$$

where the function $\Pi(t/T)$ is equal to 1 when $-T/2 < t < T/2$ and zero elsewhere. As the signal is stationary, only the difference between t and t' is significant. Thus, making the variable change $\tau = t - t'$, averaging over all realizations and inverting the summation over time and over all realizations, we obtain

$$\left\langle \Phi_i^{obs}(t,T) \right\rangle = \frac{1}{T} \iint_{-\infty}^{+\infty} \left\langle x_i^{obs}(t) x_i^{obs}(t+\tau) \right\rangle \Pi\left(\frac{t}{T}\right) \Pi\left(\frac{t+\tau}{T}\right) \exp(-2i\pi f\tau) dt d\tau$$

$$= \frac{1}{T} \iint_{-\infty}^{+\infty} R_x(\tau) \Pi\left(\frac{t}{T}\right) \Pi\left(\frac{t+\tau}{T}\right) \exp(-2i\pi f\tau) dt d\tau .$$

(10.116)

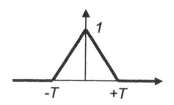

Figure 10.4. *Triangular function $\Lambda(t/T)$*

Integrating over t leads to

$$\left\langle \Phi_i^{obs}(t,T) \right\rangle = \int_{-\infty}^{+\infty} R_x(\tau)\Lambda\left(\frac{t}{\tau}\right)\exp(-2i\pi f\tau)d\tau \quad , \tag{10.117}$$

where the triangular function $\Lambda(t/T)$ is defined in Figure 10.4. Making T tend towards infinity leads to equation (10.114) and completes the proof.

10.8. Binomial probability law O

The binomial law can be defined as follows. It is the statistical distribution which characterizes the number of successes which are obtained by repeating N times an experiment in which the successful event probability is equal to a given number p.

Let us calculate this law. Suppose that N experiments have been achieved. The probability of obtaining k successes for the first k experiments and then $(N-k)$ failures is equal to $p^k(1-p)^{N-k}=p^k q^{N-k}$ (with $q=1-p$). Any other combination leading to k successes has the same probability, and the number of such combinations is equal to

$$C_N^k = \frac{N!}{k!(N-k)!} \quad . \tag{10.118}$$

Thus, the probability of obtaining k successes is equal to

$$P(N,k) = C_N^k p^k q^{N-k} \tag{10.119}$$

The mean number of successes can be easily calculated by noting that it is the product of p by the derivative of Newton's binomial:

$$\langle n \rangle = \sum_{k=0}^{k=N} k C_N^k p^k q^{N-k} = p\frac{\partial}{\partial p}\sum_{k=0}^{k=N} C_N^k p^k q^{N-k} \tag{10.120}$$

$$= p\frac{\partial}{\partial p}(p+q)^N = Np.$$

Calculating the variance using a similar approach gives

$$\sigma_n^2 = Np(1-p). \tag{10.121}$$

10.9. Random Poisson process ⊙

A Poisson process is a random process for which a succession of independent events can occur at any time with the same probability, the probability of obtaining more than two such events on an infinitesimal time interval dt being equal to zero, or assumed to be negligible compared to the probability of obtaining just one or just zero. We define λ as the average number of events per unit time, and $P(N,t)$ the probability of obtaining exactly N events over a time t. Since $P(N,dt)=0$ if $N>1$, we have $P(1,dt)=\lambda dt$ and $P(0,dt)=1-\lambda dt$, and we can also write

$$
\begin{aligned}
P(N,t+dt) &= P(N,t)\times P(0,dt) + P(N-1,t)\times P(1,dt) \\
&= P(N,t)(1-\lambda)dt + \lambda P(N-1,t) \quad,
\end{aligned}
\tag{10.122}
$$

from which we can obtain the differential equation

$$
\frac{dP(N,t)}{dt} = \lambda\bigl(P(N-1,t) - P(N,t)\bigr) \quad.
\tag{10.123}
$$

In the same fashion, we have

$$
P(0,t+dt) = P(0,t)\times P(0,dt)
\tag{10.124}
$$

which leads to the differential equation

$$
\frac{dP(0,t)}{dt} = -\lambda P(0,t) \quad.
\tag{10.125}
$$

Taking $P(0,t=0)=1$, the solution of the latter equation is

$$
P(0,t) = \exp(-\lambda t) \quad.
\tag{10.126}
$$

$P(0,t)=\exp(-\lambda t)$. Re-writing equation (10.123) for N=1 then gives us

$$
\frac{dP(1,t)}{dt} = -\lambda\bigl(e^{-\lambda t} - P(1,t)\bigr) \quad,
\tag{10.127}
$$

the solution of which is

$$
P(1,t) = \lambda t \exp(-\lambda t) \quad.
\tag{10.128}
$$

Incrementing N leads to the general solution

$$P(N,t) = \frac{(\lambda t)^N}{N!}\exp(-\lambda t) = \frac{\langle N \rangle}{N!}\exp(-\langle N \rangle). \qquad (10.129)$$

From the latter equation it is possible to calculate the standard deviation, which is in fact equal to the mean[3], $\sigma_N^2 = \langle N^2 \rangle - \langle N \rangle^2 = \langle N \rangle$. This is an important property of a Poisson process. If we define the random variable τ as the time interval separating two successive events, the probability corresponding to $d\tau$ is given by

$$p(\tau)d\tau = P(0,\tau) \times P(1, d\tau) = e^{-\lambda \tau} \lambda d\tau \quad . \qquad (10.130)$$

The distribution of such time intervals is thus equal to

$$p(\tau) = \lambda e^{-\lambda \tau} \qquad (10.131)$$

This is an exponential function which is also called Laplace distribution.

10.10. Transformation of the Cartesian wavevector coordinates into transverse and parallel components ○

We wish to express the wavevector Cartesian coordinates as a function of the wavevector component parallel to the axis of a carbon nanotube with circumference vector $\vec{L}$. Consider a straight line in 2D k-space, defined by the equation

$$k_y = \alpha k_x + \beta \; . \qquad (10.132)$$

The distance between the origin and a point M with coordinates (k_x, k_y) lying on this line is given by

$$|\vec{k}| = \sqrt{(\alpha k_x + \beta)^2 + k_x^2} \; . \qquad (10.133)$$

---

3 This is an exercise for the reader; the trick is to differentiate the expansion of an exponential e^x twice and to note that the variance can be obtained by bringing the standard deviation into a form analogous to this expansion.

Minimizing with respect to k_x gives the coordinates of the point M_{min} closest to the origin and defined in Figure 10.5:

$$M_{min} = \left(\frac{-\alpha\beta}{1+\alpha^2}, \frac{\beta}{1+\alpha^2} \right).$$

(10.134)

Calculating the distance $\left| MM_{min} \right|$ leads to

$$\left| MM_{min} \right|^2 = k_{//}^2 = \left(1+\alpha^2\right)\left(k_x + \frac{\alpha\beta}{1+\beta^2} \right)^2.$$

(10.135)

from which we can deduce that in the case of Figure 10.5, with $\alpha>0$ and $\beta>0$, we can write

$$k_x = \frac{1}{\sqrt{1+\alpha^2}} k_{//} - \frac{\alpha\beta}{1+\alpha^2}.$$

(10.136)

From equation (9.22) we also have

$$\alpha = -\frac{L_x}{L_y}, \quad \beta = \frac{2p\pi}{L_y},$$

(10.137)

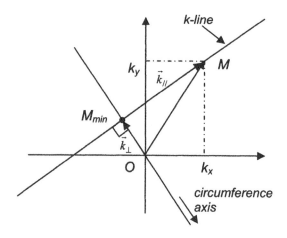

Figure 10.5. *Wavevector with its Cartesian, parallel and transverse coordinates; here we have $\alpha>0$, $\beta>0$, $L_x>0$ and $L_y>0$*

which in conjunction with equation (10.136) gives us

$$k_x = \frac{-L_y}{L} k_{//} + 2p\pi \frac{L_x}{L^2} , \qquad (10.138)$$

taking into account the fact that in the considered case (see Figure 10.5) we have $L_y<0$. Then, replacing k_x by the above expression in equation (10.132) we obtain

$$k_y = \frac{L_x}{L} k_{//} + 2p\pi \frac{L_y}{L^2} . \qquad (10.139)$$

Considering any other sign combination for α and β would lead to the same result. Equations (10.138) and (10.139) are identical to equation (9.28).

10.11. Useful physical constants ○

Planck constant:	h=6.6261_$\times10^{-34}$ J.s
Reduced Planck constant:	$\hbar$=$h/2\pi$
Electron mass in vacuum:	m=9.10953$\times10^{-31}$kg
Vacuum permittivity:	ε_0=8.854188$\times10^{-12}$F/m
Boltzmann constant:	k_B=1.38066$\times10^{-23}$J/K
Electron charge:	e=1.602189$\times10^{-19}$Cb
Light speed:	c=2.99792458$\times10^8$ms^{-1}

Solutions to Exercises

Exercise 2.19.1

1)
$$\Delta x = \frac{L}{2}\sqrt{\frac{1}{3} - \frac{2}{\pi^2 n^2}} \quad .$$

2) By expressing the sine function appearing in the wave function as a sum of exponential terms, the Fourier transform can be readily calculated and expressed under the proposed form. The probability density in momentum space $\varphi_n(p)\ \varphi_n^*(p)$ is an even function of p. It exhibits one central peak for $n=1$ but as n increases it tends to a bimodal distribution, with two diffraction-like peaks occurring at $p=\pm n\pi\hbar/L$. These two values are the same as those which would be obtained from the classical motion in the well with the same energy.

3) Since the probability density is even we have $\langle P \rangle = 0$. It is faster to calculate $\langle P^2 \rangle$ by integrating by parts in real space the quantity

$$\langle P^2 \rangle = \int \varphi_n(x)\left(-i\hbar\frac{\partial}{\partial x}\right)^2 \varphi_n(x)dx \quad .$$

We find

$$\langle P^2 \rangle = \left(\frac{n\pi\hbar}{L}\right)^2 \quad .$$

and therefore
$$\Delta p = \sqrt{\langle P^2 \rangle - \langle P \rangle^2} = \frac{n\pi\hbar}{L} \quad .$$

The uncertainty product is given by

$$\Delta p \Delta x = \frac{\hbar}{2} \sqrt{\frac{n^2 \pi^2}{3} - 2} \quad .$$

For $n=1$ $\Delta p \Delta x = 0.568$, just above the Heisenberg limit $\hbar /2$. As n is enlarged the uncertainty becomes almost proportional to n, following the variation of Δp; this is to be expected since the wave function in real space tends to look like a plane wave.

Exercise 2.19.2

1) $L=8.0117$ $nm \cong 8$ nm.

2) The numerical values is $E_2=191.2$ meV.

3) For an infinite square well we have E1=87.3 meV and E2=349.1 meV, which grossly overestimate the finite well values.

4) From the thermal de Broglie length definition we find
$T = \dfrac{h^2}{2mk_B L^2} \cong 4050K$ much above any practical measurement temperature range.

5) See equations (2.54) and (2.55).

6) Integrating the square of the wave function and requiring normalization gives

$$2 \int_{L/2}^{+\infty} D^2 \exp(-2\beta x) dx + \int_{-L/2}^{L/2} B^2 \cos^2(kx) = 1 \quad .$$

The first integral gives a contribution equal to

$$\frac{D^2}{\beta} \exp(-\beta L) = \frac{1}{\beta} \frac{k^2}{k^2 + \beta^2} B^2 ,$$

where the latter expression is obtained by noting that using equations (2.57), and (2.55) and the relation $\cos^2 a=1/(1+\tan^2 a)$, we can write

$$D^2 = \exp(\beta L) \frac{k^2}{k^2 + \beta^2} B^2 \quad .$$

The second integral is readily calculated as

$$\int_{-L/2}^{L/2} B^2 \cos^2(kx)dx = \int_{-L/2}^{L/2} \frac{B^2}{2}\left(1 + \cos(2kx)\right)dx = \frac{B^2}{2}\left(L + \frac{\sin(kL)}{L}\right).$$

Using equation (2.57) and the trigonometric identity $\sin(2a)=2\tan(a)/(1+\tan^2(a))$, this term can still be reduced to

$$\frac{B^2}{2}\left(L + \frac{2\beta}{k^2 + \beta^2}\right) .$$

Imposing that the sum of the two integral must be equal to one then leads to the required formula for B, and D is easily deduced from the relation between B and D.

7) $B_1=1.366\times10^4$, $D_1=9.828\times10^4$, $B_2=1.309\times10^4$ and $D_2=8.474\times10^4$.

8) $E_1=150$ meV and $E_2=291.2$ meV. The degeneracy of E_2 is equal to 3.

Exercise 2.19.3

1) From the Si ellipsoid shape, it is obvious that for the same energy the larger mass corresponds to the "elongated" ellipsoid axis. Thus, the two valleys along the axis k_x perpendicular to the interface exhibit a larger mass m_L in the k_x direction, and they will be populated first. The corresponding valley degeneracy is 2. For these valleys the transverse mass m_T corresponds to free motion in the interface plane and must be used to calculate the 2D density of states.

2) The Schrödinger equation reads

$$-\frac{\hbar^2}{2m_L}\frac{\partial^2\varphi}{\partial x^2} + (e\mathcal{E}x - E)\varphi = 0 .$$

We assume that the variation of the charge as a function of x occurs on a distance small enough for the field to be considered as a constant. A more rigorous treatment should calculate the electric field as a function of the charge distribution, and thus as a function of the position.

3) Make the variable change

$$u = \left(\frac{2m_L e\mathcal{E}}{\hbar^2}\right)^{\frac{1}{3}}\left(\frac{E}{e\mathcal{E}} - x\right)$$

to find the prescribed equation.

4) In reciprocal space it is easy to prove from the definition of the reciprocal Fourier transform

$$\varphi(u) = \frac{1}{\sqrt{2\pi}} \int_{-\infty}^{+\infty} \varphi(\omega) e^{+iu\omega} d\omega$$

and by integrating by parts that the inverse Fourier transform of the derivative $\varphi'(\omega)$ is $-iu\varphi(u)$. It is also easily proved that the Fourier transform of the derivative $\varphi'(u)$ is equal to $i\omega\varphi(\omega)$. With the help of these two relations we can transform the differential equation

$$\varphi''(u) + u\varphi(u) = 0$$

by finding its Fourier transform:

$$-\omega^2 \varphi(\omega) + i\varphi'(\omega) = 0.$$

The latter is easily integrated as

$$\varphi(\omega) = K e^{-i\omega^3/3}.$$

The inverse Fourier transform is

$$\varphi(u) = \frac{K}{\sqrt{2\pi}} \int_{-\infty}^{+\infty} e^{i(\omega u - \omega^3/3)} d\omega.$$

Making the variable change $p=-\omega$ and taking into account the definition of the Airy function we arrive at

$$\varphi(u) = C Ai(-u)$$

where C is a normalization constant to be determined.

This can be re-written as

$$\varphi_n(x) = C \times Ai\left(-\left(\frac{2m_L e\mathcal{E}}{\hbar^2}\right)^{\frac{1}{3}}\left(\frac{E_n}{e\mathcal{E}} - x\right)\right)$$

to make the electric field dependence explicit.

5) The energy levels can be calculated by imposing the condition $\varphi_n(0)=0$ (infinite energy barrier at $x=0$). From the approximate formula giving the zeroes of $Ai(u)$ we arrive at

$$E_n \cong \left(\frac{\hbar^2}{2m_L}\right)^{\frac{1}{3}} \left(\frac{3\pi}{2}e\mathcal{E}\left(n-\frac{1}{4}\right)\right)^{\frac{2}{3}}$$

where the first index $n=1$ is for the *ground state*. The wave function of index n is obtained by shifting the Airy function so as to put the nth zero exactly at the interface, keeping only the part with $x>0$ and multiplying by an appropriate normalization factor (which we can calculate).

6) When x tends towards infinity (i.e. deeper inside the Si substrate) the field is equal to zero and from Gauss' theorem

$$\oiint \vec{\mathcal{E}}\, d\vec{s} = \frac{Q}{\varepsilon_s}$$

the electric field at the interface can is given by

$$\mathcal{E}_S = \frac{e n_S}{\varepsilon_S}$$

where n_S is the density of carriers per surface unit. The average field seen by the electrons can be taken as the average between the surface value and the value when the wave function becomes negligible (the latter being close to zero in our case):

$$\mathcal{E} \cong \frac{e n_S}{2\varepsilon_S}$$

(in the case of the Airy function it turns out that the average field created and seen by the electrons is *exactly* equal to this value, and in any case, by applying Gauss' theorem between the interface and the charge centroid it is immediately seen that the expression above gives nothing but the field seen at the charge centroid). In the ground state the charge centroid can be deduced from the numerical integral value given in the text and is equal to

$$x_b = (\alpha - a_1)\left(\frac{\hbar^2}{2m_L e\mathcal{E}}\right)^{\frac{1}{3}}$$

The "inversion" or "accumulation" layer capacitance is

$$C_S = \frac{\varepsilon_S}{x_b} = \frac{\varepsilon_S}{\alpha - a_1}\left(\frac{2m_L e\mathcal{E}}{\hbar}\right)^{\frac{1}{3}}$$

7) With $n_S=1\times10^{12}cm^{-2}$ the first energy levels are $E_1=30.4$ meV, $E_2=53.5$ meV and $E_3=72.3$ meV. Suppose that only the first subband is occupied. Taking the first level as the origin, the Fermi level is given by $E_F=n_S/D_{2D}$ with $D_{2D}=2$ $m_T/\pi\hbar^2$, and equal to 6.3 meV (the factor of 2 in D_{2D} accounts for the valley degeneracy). This value is smaller than E_2-E_1. Thus, only one 2D subband is occupied. The charge centroid is at 2.68 nm from the interface, comparable to the oxide thickness. The inversion capacitance is equal to $C_S=3.92\times10^{-6}$ F/cm^2 and the oxide capacitance $C_{ox}=\varepsilon_{ox}/t_{ox}$ is equal to 1.72×10^{-6} F/cm^2. The overall capacitance is the series association of those two capacitances, and is smaller than C_{ox} by a ~30.5% amount. Thus in today's CMOS technology, for which typical dimensions are of the same order as the data in this exercise, the quantum correction to the gate capacitance is substantial.

Exercise 2.19.4

1) The Hall voltage is negative. Since the current goes from right to left both holes and electrons would be deflected by the Lorentz force towards contact 2. The Hall field opposes to this deflection and since it is directed from 3 to 2, this can only correspond to electrons.

2) $n_s=-IB/eV_H\cong3\times10^{11}cm^{-2}$.

3) The first and second quantized levels in the quantum well are equal to 14 and 56 meV in the hard wall approximation. If only the first subband is occupied this would lead to $E_F=\pi n_s\hbar^2/m=10.7$ meV. Since E_F is smaller than the level spacing the assumption that only one subband is occupied is verified.

4) $B_C=n_s h/2e=6.2T$.

5) No, it is not. It is possible to define a density of states per wavevector value and unit length, with a flat potential the energy is constant in one Landau band (this would change with an added potential).

6) It is easy to check that in the classical case and for a parabolic potential of the form $V=m\omega_0^2/2=Kx^2/2$, the condition for which the Lorentz force for a longitudinal velocity v is perfectly counterbalanced by the spring force is given by

$evB=Ky= m\omega_0^2 y$. Thus the line with abscissa $y=v\omega_C/\omega_0^2$ is a classical orbit for electrons with a velocity v. In the quantum case and Landau gauge insert equation (2.181) in equation (2.196) and use the velocity expression equation (2.194) to find that the wave function is centered at exactly the same abscissa:

$$y = \frac{\omega_C^2}{\omega_C^2 + \omega_0^2} y_k = \frac{\omega_C}{\omega_C^2 + \omega_0^2} \frac{\hbar k}{m} = \frac{\omega_C}{\omega_0^2} \left(\frac{\omega_0^2}{\omega_C^2 + \omega_0^2} \frac{\hbar k}{m} \right) = \frac{\omega_C}{\omega_0^2} v \quad .$$

Thus, the wave center propagates along a classical orbit and in fact follows an equipotential line. When the potential becomes infinitely smooth, from the left expression above we see that the orbits tend to the quantum expression obtained with the infinite 2D plane and Landau gauge. We thus expect the extended quantum states to follow equipotential lines at locations and with an inter-spacing similar to the results obtained in the Landau gauge with a flat potential.

7) From the analysis in question 6, with a potential landscape we expect the electrons to follow the classical equipotential lines which also obey closed-orbit quantization conditions, so that some electrons can circle around the defect. Therefore, the potential peak induces localized states, as depicted by the figure below (this point is important for explaining the integer quantum Hall effect described in section 3.13).

8) The electrons spread on an interval $\Delta y=W\times(B_C/B)=W/2$.

Exercise 3.14.1

1) With $V_3=0$ the conductance matrix is

$$\begin{pmatrix} I_1 \\ I_2 \end{pmatrix} = \begin{pmatrix} G_{12} + G_{13} & -G_{12} \\ -G_{21} & G_{21} + G_{23} \end{pmatrix} \begin{pmatrix} V_1 \\ V_2 \end{pmatrix}.$$

From Onsager reciprocity relations with $B=0$ we obtain $\overline{T}_{21} = \overline{T}_{12} = T_{SC}$, $\overline{T}_{23} = \overline{T}_{32} = T_{WC}$ and $\overline{T}_{31} = \overline{T}_{13} = T_D$. This in turn gives the conductance matrix:

$$\begin{pmatrix} I_1 \\ I_2 \end{pmatrix} = \begin{pmatrix} G_{SC} + G_D & -G_{SC} \\ -G_{SC} & G_{SC} + G_{WC} \end{pmatrix} \begin{pmatrix} V_1 \\ V_2 \end{pmatrix}.$$

2) We expect $G_D >> G_{SC} >> G_{WC}$.

3) $I_2 = 0$ implies $-G_{SC}V_1 + (G_{SC} + G_{WC})V_2 = 0$ and

$$V_2 = \frac{G_{SC}}{G_{SC} + G_{WC}} V_1 \cong V_1.$$

A macroscopic conductor with the same geometry would give $V_2 \cong V_1/2$.

4) From the conductance matrix and $I_2 = 0$ we obtain

$$R_{21} = \frac{V_2}{I_1} = \frac{G_{SC}}{G_{SC}G_{WC} + G_{SC}G_D + G_D G_{WC}} \cong \frac{1}{G_D}.$$

5) $\overline{T}_{21} + \overline{T}_{31} + R_1 = 1 \quad \Rightarrow \quad T_{SC} + T_D + R_1 = 1 \quad \Rightarrow \quad R_1$ is small.
$\overline{T}_{13} + \overline{T}_{23} + R_3 = 1 \quad \Rightarrow \quad T_D + T_{WC} + R_3 = 1 \quad \Rightarrow \quad R_3$ is larger.

From the two relations above we obtain $R_3 - R_1 = T_{SC} - T_{WC} \cong T_{SC}$ so that $R_3 \cong R_1 + T_{SC}$.

An electron wave propagating from contact 3 to the crossing is enlarged at the opening of contact 2, and a large fraction of it is reflected by wall no. 2 and goes back into contact 3 (see the figure below). This reflected part corresponds to the same flow as the transmitted part of a wave originating in contact 1 and transmitted into contact 2, so that the condition $\overline{T}_{13} = \overline{T}_{31}$ is fulfilled (and thus $R_3 \cong R_1 + T_{SC}$).

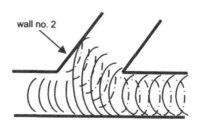

wall no. 2

Exercise 3.14.2

1) $v_G = \dfrac{1}{\hbar} \dfrac{\partial E}{\partial k} = \dfrac{an}{\hbar} k^{n-1}.$

2) $N(k) = \dfrac{2k}{\pi} \quad \Rightarrow \quad D(E) = \dfrac{\partial N}{\partial k} \times \dfrac{\partial k}{\partial E} = \dfrac{2a^{-1/n}}{\pi n} E^{\frac{1-n}{n}}.$

3) $I = \dfrac{e}{2} \displaystyle\int_{\mu_1}^{\mu_2} D(E) v_G(E) dE = \dfrac{2e}{h}(\mu_1 - \mu_2)$, same formula as with a parabolic dispersion relationship.

4) The same formula is obtained.

5) With $E=f(k)$ the product

$$v_G \times D(E) = \dfrac{1}{\hbar} \dfrac{\partial E}{\partial k} \dfrac{\partial N}{\partial k} \dfrac{\partial k}{\partial E} = \dfrac{1}{\hbar} \dfrac{\partial N}{\partial k} = \dfrac{2}{\pi \hbar}$$

is constant and $I=(2e/h)(\mu_1-\mu_2)$. The current formula does not depend on the dispersion relation and is universal.

Exercise 3.14.3

1) See Chapter 2, section 2.12.

2) By using the reciprocity relations we can show that $\overline{T}_{21} = \overline{T}_{13} = T_L$, as detailed by the schema below, and from a quite similar reasoning we obtain $\overline{T}_{31} = \overline{T}_{12} = T_R$. We also define $T_C = \overline{T}_{23}$ and $T'_C = \overline{T}_{23}$.

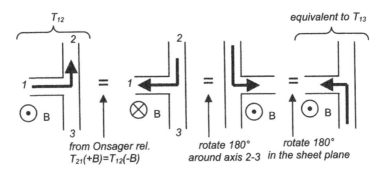

3) From the Büttiker formula we obtain

$$\begin{pmatrix} I_1 \\ I_2 \end{pmatrix} = \begin{pmatrix} G_R + G_L & -G_R \\ -G_L & G_L + G_C \end{pmatrix} \begin{pmatrix} V_1 \\ V_2 \end{pmatrix} \ .$$

4) From the conductance matrix relations we obtain

$$V_2 = \frac{G_L}{G_L + G_C} V_1 \quad and \quad G = G_R + G_L - \frac{G_R G_L}{G_L + G_C} \ .$$

5) If $B=0$ then $G_R=G_L$, so that

$$G = 2G_L \left(1 - \frac{G_L}{2(G_L + G_C)} \right) \ .$$

6) For the ground wave function, whose maximum is in the middle of the constriction, an electron coming from contact 1 will be prevalently reflected by the wall rather than turning left or right, so that we expect T_L to be small. In contrast, an electron coming from contact 2 will pass with a high probability into contact 3, rather than into contact 1, so that we expect T_C to be close to unity. Thus we expect G_L to be small against G_C.

7) The first order expansion is

$$G \cong 2G_L \left(1 - \frac{G_L}{2G_C} \right)$$

and to zero[th] order $G \cong 2G_L$.

8) Everything occurs as if the situation can be described as follows: electrons can pass coherently from 1 to 3 but also from 1 to 2 and then (incoherently) from 2 to 3 (an excess electron in the floating contact 2 must necessarily go back to one of the other "connected" contacts). Probabilities can be summed up since they represent incompatible processes. Other processes are of higher order, with a smaller probability. The probability of the former process is T_L and that of the latter process is $T_L \times T_C/(T_C+T_L)$, with $T_C/(T_C+T_L)$ close to unity. Thus the transmission probability from 1 to 3 is close to $2T_L$, as exactly calculated in the previous question.

9) $I_1=0$ implies $V_1=G_RV_2/(G_R+G_L)$, which implies in turn

$$G = G_L + G_C - \frac{G_LG_R}{G_R + G_L} \quad .$$

with $B=0$ $G_L=G_R$ and $G=G_C+G_L/2$.

10) An electron coming from 2 can go straightforwardly to 3 (probability T_C), or pass through 1 and then to 3. The probability of going to 1 is T_L, and thus the electron, which cannot stay in contact 1, has a ½ probability of ending either in contact 2 or contact 3. Thus, the corresponding probability is $T_L/2$, and the overall probability is $T_C+T_L/2$, so that $G=G_C+G_L/2$, as calculated in the previous question.

11) In the two configurations above, which we can label A and B, we measure $G_A\cong2G_L$, which gives $G_L\cong G_A/2$, and $G_B=G_C+G_L/2$, which gives $G_C=G_B-G_A/4$.

12) Keeping the spin degeneracy, one 1D channel is occupied with $g_s=2$, and if the degeneracy is lifted by the field but the two channels are still occupied there are now two channels with $g_s=1$. In both cases $G_L=2e^2/h$, $G_R=0$ and $G_C=0$, so that the conductance matrix is

$$G = \frac{2e^2}{h}\begin{pmatrix} 1 & 0 \\ -1 & 1 \end{pmatrix} \quad .$$

13) From 12/ we have $V_1=V_2$ and $G=2e^2/h$.

Exercise 3.14.4

1) See the related part in Chapter 3.

2) See the related part in Chapter 3.

3) From the structure symmetries we obtain $T_{14}=T_{21}=T_{32}=T_{43}=T_R$, $T_{41}=T_{12}=T_{23}=T_{34}=T_L$ and $T_{13}=T_{31}=T_{42}=T_{24}=T_F$.

4) From the Büttiker formula and making $V_3=0$ we have

$$\begin{pmatrix} I_1 \\ I_2 \\ I_4 \end{pmatrix} = \begin{pmatrix} G_0 & -G_L & -G_R \\ -G_R & G_0 & -G_F \\ -G_L & -G_F & G_0 \end{pmatrix}\begin{pmatrix} V_1 \\ V_2 \\ V_4 \end{pmatrix} \quad .$$

5) Contacts 2 and 4 are floating so that $I_2=I_4=0$, and we have

$$0 = -G_R V_1 + G_0 V_2 - G_F V_4$$
$$0 = -G_L V_1 - G_F V_2 + G_0 V_4 \quad ,$$

from which we obtain

$$V_4 = \frac{G_R G_F + G_0 G_L}{G_0^2 - G_F^2} V_1, \qquad V_2 = \frac{G_R G_0 + G_L G_F}{G_0^2 - G_F^2} V_1$$

(multiply the top equation by G_F, the bottom one by G_0 and sum to get V_4, multiply the top one by G_0 and the bottom one by G_F and sum to obtain V_2). Then insert these two expressions in the top equation. We obtain

$$G = G_0 - \frac{G_L (G_R G_0 + G_L G_F) + G_R (G_R G_F + G_0 G_L)}{G_0^2 - G_F^2} \quad .$$

Develop everything in the fraction numerator and re-order the terms so as to factorize (G_R+G_L) and simplify. After a few algebraic manipulations we obtain

$$G = \frac{G_0}{2} + \frac{G_F}{2} + \frac{(G_R - G_L)^2}{2(G_R + G_L + 2G_F)}$$

which is the required result.

6) If $B=0$ then $T_R=T_L$ and $G=(e^2/h)(M+T_F)$ (we have $T_R+T_L+T_F=M$ because we assume that the entrance constrictions are perfect).

7) Electrons can pass directly from 1 to 3, or transit through contacts 2 and 4 and then can lose quantum coherence before passing to contact 3 or going back to contact 1. Thus their transmission probability is affected by contacts 2 and 4, which are invasive if transport is ballistic. With macroscopic conductors they would play no role in the determination of the transport properties.

8) If T_F is small and has only one populated channel in the entrance constriction, from 6) we have $G \cong G_0/2 = e^2/h$, as if we put two contact resistances in series: the electrons are forced to transit and loose coherence through contacts 2 and 4, and the overall resistance is the sum of two contact resistances.

9) If T_F prevails then from 6) we have $G=2e^2/h$; contacts 2 and 4 play no role and since transport is ballistic there is only one contact resistance quantum. We do not add two resistances in series.

10) The wave functions are deflected to the left with respect to the propagation direction. Thus T_L increases and T_R decreases.

11) The device is in the quantum Hall regime with M edge channels and a spin degeneracy $g_s=1$. $T_R=T_F=0$ and $T_L=M$. Thus from 5) G reduces to $G=G_L=Me^2/h$. V_4 is equal to V_1 and $V_2=0$.

Exercise 3.14.5

1) If the devices are truly in the ballistic regime, there is no difference between the Si and GaAs samples. However, the answer can be made a little bit more contrasted by noting that in general, the ballistic transport regime will be attained at lower temperatures (or smaller dimensions) in the case of silicon, due to inferior transport properties with respect to that in GaAs.

2) From symmetry considerations we obtain $\overline{T}_{31}=\overline{T}_{23}=\overline{T}_{12}=T_L$ and $\overline{T}_{21}=\overline{T}_{32}=\overline{T}_{13}=T_R$.

3) From Büttiker formula we have

$$\begin{pmatrix} I_1 \\ I_2 \\ I_3 \end{pmatrix} = \begin{pmatrix} G_L+G_R & -G_L & -G_R \\ -G_R & G_R+G_L & G_L \\ -G_L & -G_R & G_R+G_L \end{pmatrix}\begin{pmatrix} V_1 \\ V_2 \\ V_3 \end{pmatrix}.$$

4) With $V_2=0$ the matrix relation reduces to

$$\begin{pmatrix} I_1 \\ I_3 \end{pmatrix} = \begin{pmatrix} G_L+G_R & -G_R \\ -G_L & G_L+G_R \end{pmatrix}\begin{pmatrix} V_1 \\ V_3 \end{pmatrix}$$

and with $I_3=0$ we readily obtain

$$V_3 = \frac{G_L}{G_L+G_R}V_1$$

and then

$$G = G_L+G_R - \frac{G_L G_R}{G_L+G_R}.$$

5) $T_L=M$ and $T_R=0$, so that $G=G_L=M{\times}2e^2/h$.

6) $B=0$ and the device symmetry imply that $T_R=T_L$. From question 4) we then obtain $G=3G_L/2$. This is as if the overall result was the sum of two parallel contributions: a direct, quantum-coherent contribution from electrons going straightforwardly from 1 to 2, with an associated conductance $(2e^2/h)T_L$, and a contribution corresponding to electron transit through contact 3. In the latter case, contact 3 makes the electrons lose quantum coherence and the contribution is the series association of two conductances equal to G_L, thus equal to $G_L/2$. The overall conductance is $G_L+G_L/2=3G_L/2$, as predicted by the formula.

7) By symmetry considerations we obtain

$$\overline{T}_{41} = \overline{T}_{23} = T_l \quad \overline{T}_{34} = \overline{T}_{12} = T_L \quad \overline{T}_{32} = \overline{T}_{14} = T_r \quad \overline{T}_{43} = \overline{T}_{21} = T_R \quad \overline{T}_{13} = \overline{T}_{31} = T_C \quad \overline{T}_{42} = \overline{T}_{24} = T'_C$$

and from the Büttiker formula the conductance matrix is

$$G = \begin{pmatrix} G_L + G_C + G_r & -G_L & -G_C & -G_r \\ -G_R & G_R + G_l + G'_C & -G_l & -G'_C \\ -G_C & -G_r & G_C + G_r + G_L & -G_L \\ -G_l & -G'_C & -G_R & G_l + G'_C + G_R \end{pmatrix}.$$

8) In the quantum Hall regime, with M occupied edge channels per constriction and a lifted spin degeneracy the conductance matrix relation becomes

$$\begin{pmatrix} I_1 \\ I_2 \\ I_4 \end{pmatrix} = M \frac{e^2}{h} \begin{pmatrix} 1 & 0 & 0 \\ 0 & -1 & 0 \\ -1 & 0 & 1 \end{pmatrix} \begin{pmatrix} V_1 \\ V_2 \\ V_4 \end{pmatrix}.$$

From the relation above and since $I_2=0$ we have $V_2=0$. $I_4=0$ leads to $V_4=V_1=V$, and the conductance is $G=Me^2/h$.

Exercise 5.11.1

1) Close to the resonance the current exhibits a peak and is given by $I = \frac{2e}{h} T(E_F)(\mu_1 - \mu_2)$, and the transmission $T(E_F)$ is equal to

$$T(E_F) = \frac{\Gamma_1 \Gamma_2}{\left(\frac{\Gamma_1 + \Gamma_2}{2}\right)^2 + (E_F - E_R)^2}.$$

2) To first order when the perturbing potential is scanned inside the well only the resonance energy E_R is affected, and first order perturbation theory (see section 10.4.1) gives

$$E_R^{perturbed} = E_R^p = E_R + \langle \psi |V| \psi \rangle = E_R + \int \psi^*(x)V(x)\psi(x)dx \cdot$$

3) $G = \dfrac{2e^2}{h} T(E_F) \quad \Rightarrow \quad \Delta R = \dfrac{h}{2e^2} \dfrac{1}{\Gamma_1 \Gamma_2} \left(-2(E_F - E_R)\langle \psi |V| \psi \rangle + \langle \psi |V| \psi \rangle^2 \right)$

so that slightly off the resonance we have to first order

$$\Delta R \cong -\frac{h}{e^2} \frac{(E_F - E_R)}{\Gamma_1 \Gamma_2} \langle \psi |V| \psi \rangle$$

and exactly at the resonance we have

$$\Delta R = \frac{h}{2e^2} \frac{1}{\Gamma_1 \Gamma_2} \langle \psi |V| \psi \rangle^2 \cdot$$

4) If the tip is scanned above the dot, ΔR is only proportional to the product $\psi^*\psi$ provided that the Fermi level is not exactly at the resonance. If V was a constant this would directly give the local density of states or the square modulus of the wave function as a function of tip position.

5) Unfortunately, even if the electric potential applied to the tip is constant, the perturbing potential also varies with the position because in most cases the surface potential difference between the tip and the sample is not a constant.

6) This method is suitable even when considering buried electron gases, which actually represents the case of a vast majority of semiconductor nanostructures, and is known as scanning gate microscopy (SGM). With scanning tunneling microscopy (STM) we measure the tunneling current between the tip and the surface, so that the probed electrons must be located exactly at the surface. With SGM it must be noted that the resolution would be broader due to a larger distance between the tip and sample.

Exercise 5.11.2

1) From equations (5.14) and (5.15) the dephasing φ is given by the argument of $r_1 r_2 exp(2ika)$, and with identical barriers this reduces to $2ka+2arg(r_1)$. Using equation (5.9) and the real quantity $\beta_1 = ik_1$ it is easily found that only the denominator is complex:

$$r_1 = \frac{(k^2 + \beta_1^2)(1 - \exp(-2\beta_1 a))}{(k + i\beta_1^2)^2 - (k - i\beta_1^2)^2 \exp(-2\beta_1 a)}$$

From the formula above the denominator D can be put in the form

$$D = \exp(-\beta_1 a)\cosh(\beta_1 a)\left((k^2 - \beta_1^2)\tanh(\beta_1 a) + 2ik\beta_1\right)$$

from which the dephasing φ is readily obtained:

$$\varphi = 2ka - 2A\tan\left(\frac{2k\beta_1}{(k^2 - \beta_1^2)\tanh(\beta_1 a)}\right) \ .$$

The required resonance condition can be immediately deduced from the latter expression.

2) From the resonance plot below we find E_1=46.8 meV and E_2=192.9 meV, to be compared with E_1=50.1 meV and E_2=191.6 meV in the case of the corresponding finite quantum well. The values are quite close to one another.

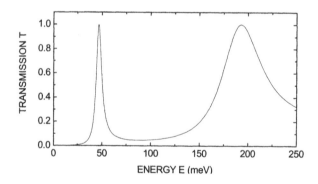

Exercise 5.11.3

1) Fermi's golden rule can be written as

$$\Omega = \frac{2\pi}{\hbar} D_{1D}(E)L\left|\langle\varphi_f|\delta V|\varphi_i\rangle\right|^2$$

where the density of states at wavevector k (or wave function φ_f) in the wide, left well is given by the 1D density of states $D_{1D}(E)$ per unit energy per unit length (equation (2.123)) multiplied by the length L. The coupling term δV is the potential which must be added to the right, finite well potential to obtain the potential of the two wells. It is thus equal to $-V$ for $-L/2 < x < +L/2$, and to zero elsewhere. Thus, the matrix element is calculated over the latter interval, using the wave function $B\cos(kx)$ for φ_f and $D\exp(-\beta(L/2+a-x))$ for φ_i, where the normalizing coefficients B and D are given in the text of exercise 2.19.2. After a somewhat cumbersome calculation, and using equation (2.57) to turn terms such as $\cos(kL/2)$ into $k^2/(k^2+\beta^2)$, we obtain the intermediate term

$$
\left\langle \varphi_f \left| \delta V \right| \varphi_i \right\rangle = VBD \int_{-L/2}^{+L/2} \cos(kx) e^{\beta(x-a-L/2)} dx =
$$

$$
\frac{VBDe^{-a}}{\beta^2 + k^2} \left(\beta\cos(kL/2) + k\sin(kL/2) - \left(\beta\cos(kL/2) - k\sin(kL/2) \right) e^{-\beta L} \right)
$$

$$
\cong \frac{2VBD\beta k^2}{(\beta^2 + k^2)^4} \quad ,
$$

from which we obtain the escape rate (use exercise 2.19.2 for expressing B and D)

$$
\Omega = \frac{64mV^2 k^5 \beta^2 e^{\beta(w-2a)}}{\hbar^3 (k^2 + \beta^2)^5 (w+2/\beta)} \frac{L}{L+2/\beta} \quad .
$$

Passing to the limit $L \to +\infty$ gives the escape rate:

$$
\Omega = \frac{64mV^2 k^5 \beta^2 e^{\beta(w-2a)}}{\hbar^3 (k^2 + \beta^2)^5 (w+2/\beta)} \quad .
$$

2) Use equation (5.14) with intermediate barrier coefficients given by equations (5.8) and (5.9) to obtain a plot as already shown in the text of the exercise.

3) For half-transmission a Lorentzian transmission as in equation (5.23) with two identical barriers gives a peak width $\Delta E = 2\Gamma$.

4) For an $8\ nm$ wide quantum well, Fermi's golden rule gives a ground state level $E = 51.1\ meV$ and a resonance width $\Delta E = 5.8\ meV$. From the plot we obtain $\Delta E \cong 6.7\ meV$. Thus Fermi's golden rule gives a quite reasonable estimation.

Exercise 5.11.4

1) With S-matrices of the form

$$
S_1 = S_2 = \begin{pmatrix} 0 & 1 \\ 1 & 0 \end{pmatrix}
$$

in the lower and upper arms, from equation (4.24) we have $s_{ud}=s_{du}=s_u=s_d=-1$ and $\alpha_u=\alpha_d=0$. Substituting in equation (4.23) gives the undetermined form $0/0$. Thus it is not suitable to calculate the overall transmission.

2) With forward and backward amplitudes a_2 and b_2 identical in the two arms, from the Y-junction S-matrices we can write

$$
\begin{pmatrix} b_1 \\ a_2 \\ a_2 \end{pmatrix} = \begin{bmatrix} c & t & t \\ t & -\frac{1}{2}(1+c) & \frac{1}{2}(1-c) \\ t & \frac{1}{2}(1-c) & -\frac{1}{2}(1+c) \end{bmatrix} \begin{pmatrix} a_1 \\ b_2 \\ b_2 \end{pmatrix} \quad and \quad \begin{pmatrix} b_2 \\ b_2 \\ a_3 \end{pmatrix} = \begin{bmatrix} -\frac{1}{2}(1+c) & \frac{1}{2}(1-c) & t \\ \frac{1}{2}(1-c) & -\frac{1}{2}(1+c) & t \\ t & t & c \end{bmatrix} \begin{pmatrix} a_2 \\ a_2 \\ b_3 \end{pmatrix}.
$$

This leads to $a_3=a_1$, equivalent to $t_T=1$ and therefore $T_T=1$. The transmission is equal to unity, independent of the input transmission t!

3) We find $a_2=a_1/t$. Thus, when the input transmission t tends towards zero the wave amplitude inside the ring becomes infinitely higher than the input amplitude, just as in the 1D resonant tunneling case. Even for a vanishing transmission of the single barrier at resonance the increasingly large wave amplitude inside the well formed by two identical barriers compensates the exponential vanishing of wave penetration in the barriers, so as to maintain a unit transmission. This is clearly the signature of resonant tunneling. Thus, these Y-junctions are the ones which describe resonant tunneling when there is no dephasing in the two symmetric arms.

Exercise 5.11.5

1)

$$
t_T = \frac{i(\Gamma_1 + \Gamma_2)}{2E\left(1 - t^2 + \sqrt{1 - 2t^2}\right) + it^2(\Gamma_1 + \Gamma_2)}.
$$

2) The quantum ring exhibits a single resonance at the same value as in each separate dot. The energy spreading Γ and the lifetime $\tau=\hbar/\Gamma$ are functions of the input transmission:

$$
\Gamma = \frac{(\Gamma_1 + \Gamma_2)t^2}{2\left(1 - t^2 + \sqrt{1 - 2t^2}\right)}.
$$

3) As the input transmission t tends towards zero the lifetime can be approximated as

$$\tau = \frac{2\hbar}{(\Gamma_1 + \Gamma_2)t^2} \quad .$$

At the resonance, if the transmission is exceedingly small, the electrons spend more and more time inside the ring. In other words, the smaller the communication between the ring and the leads, the closer the wave inside the ring is to a stationary state (and the sharpest is the resonance). Once again, this is the usual signature of resonant tunneling.

4) The resonance width of the ring is the average of the resonance widths in each dot.

Exercise 5.11.6

1) We find

$$S = \begin{pmatrix} 0 & \exp(ikL) \\ \exp(ikL) & 0 \end{pmatrix} \quad .$$

2) Applying the general formula equation (4.23) to S-matrices as that in the previous question gives

$$t_T = \frac{2t^2}{\exp(-ikL) + (1 - 2t^2)\exp(ikL)} \quad .$$

3) The resonance condition is $kL = n\pi$, with n integer.

4) The plot below is for t=0.1, 0.166, 0.233, 0.299, 0.365, 0.432, 0.498, 0.564, 0.631 and 0.697.

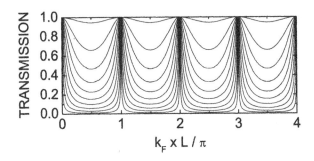

Exercise 7.7.1

1) See the book.

2) From the set of relations

$$Q_1 = C_1V_1$$
$$Q_2 = C_2V_2 \quad \text{we obtain}$$
$$Q_3 = C_3V_3$$
$$Q_2 - Q_3 = N_2e$$
$$Q_1 - Q_2 = N_1e$$
$$V = V_1 + V_2 + V_3$$

$$V_1 = \frac{C_2C_3V + C_3N_1e + C_2(N_1 + N_2)e}{C_1C_2 + C_2C_3 + C_1C_3}$$

$$V_2 = \frac{C_1C_3V - N_1C_3e + N_2C_1e}{C_1C_2 + C_2C_3 + C_1C_3}$$

$$V_3 = \frac{C_1C_2V - N_1C_2e - N_2(C_1 + C_2)e}{C_1C_2 + C_2C_3 + C_1C_3}.$$

3)

$$V_1 = \frac{1}{3}\left(V - \frac{2N_1e}{C} - \frac{N_2e}{C}\right)$$

$$V_2 = \frac{1}{3}\left(V - \frac{N_1e}{C} + \frac{N_2e}{C}\right)$$

$$V_3 = \frac{1}{3}\left(V - \frac{N_1e}{C} - \frac{2N_2e}{C}\right)$$

4)

$$E_C = \frac{1}{2}\left(N_1eV_1 + N_2e(V_1 + V_2) + Q_3V\right) = \frac{1}{6}CV^2 + \frac{e^2}{3C}\left((N_1 + N_2)^2 - N_1N_2\right)$$

5) $N_1 \rightarrow N_1 + 1$, the electron is injected from the cathode.

$$\delta Q_3 = C\delta V_3 = -e/3 \text{ so that } \Delta E_G = -\delta Q_3V = eV/3.$$

6) $\Delta E_C = E_C(N_1 + 1) - E_C(N_1) = \frac{e^2}{3C}(1 + N_2 + 2N_1).$

7) $\Delta E_{tot} = \frac{e}{3}\left(V + \frac{e}{C}(1 + N_2 + 2N_1)\right).$

8) $V > \frac{-e}{C}(1 + 2N_1 + N_2).$

9) One electron is transferred from cell 1 to cell 2:

$$N_1 \rightarrow N_1 - 1 \quad N_2 \rightarrow N_2 + 1$$
$$\delta Q_3 = C\delta V_3 = -e/3 \text{ so that } \Delta E_G = -\delta Q_3 V = eV/3.$$

10) $\Delta E_C = \dfrac{e^2}{3C}(1 - N_1 + N_2).$

11) $V > -\dfrac{e}{C}(1 - N_1 + N_2).$

12) Initially $N_1 = N_2 = 0$. Now assume that we have moved N_1 electrons into dot 1 and N_2 into dot 2. The condition for having $N_1 + 1$ electrons in dot 1 is

$$V > \frac{-e}{C}(1 + 2N_1 + N_2).$$

If this condition is fulfilled then the condition for transferring this charge into dot 2 is given by

$$V > -\frac{e}{C}(1 - (N_1 + 1) + N_2) = \frac{e}{C}(N_1 - N_2).$$

The difference between both expressions is thus equal to

$$\Delta V = \frac{e}{C}(3N_1 + 1) < 0$$

and is always negative. This means that any charge injected into dot 1 is transferred into dot 2. The graph $N_2(V)$ is the same as in the book, but shifted by a constant amount; N_1 is always equal to zero.

Exercise 7.7.2

1) V_D must verify the inequality

$$V_D > \frac{e}{2(C_1 + C_G)}$$

with $C_1 = 1.293 \times 10^{-17} F$ and $C_G = 1.724 \times 10^{-18} F$. This leads to $V_D > 5.47 \ mV$.

2) The gate voltage interval is given by the difference $\Delta V_G = V_{G2} - V_{G1}$ as defined in the figure below.

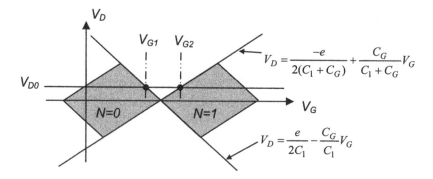

The two lines defining the adjacent Coulomb blockade domains (here with $N=0$ and $N=1$) are described by the equations written in the figure above, from which we obtain

$$V_{G2} - V_{G1} = \left(1 + \frac{2C_1}{C_G}\right) V_{D0}$$

Numerical application gives $\Delta V_G = 8\ mV$.

3) With dimensions reduced down to 5 nm we must assess if the interval between the various quantized levels inside the box are still negligible compared to the Coulomb blockade energies. These intervals are of order $\hbar^2 \pi^2 / 2m_L L^2 = 15.34\ meV$ in the hard wall approximation. Taking $C_G = 1.724 \times 10^{-19} F$, when V_D is close to zero the gate voltage spacing between two conductance peaks is around $e/C_G = 0.928$ V. Thus, we should observe a slight departure from a perfect periodicity.

Exercise 7.7.3

1) $V_{G1} = \left(N - \frac{1}{2}\right) \frac{e}{C_G}$ and $V_{G2} = \left(N + \frac{1}{2}\right) \frac{e}{C_G}$.

2) $\Delta E(N \to N+1) = \frac{e}{C_1 + C_2 + C_G}\left(\left(N + \frac{1}{2}\right)e - C_G V_G\right)$.

To check our result we can verify that for $V_G = V_{G2}$, $\Delta E = 0$ since either of the two situations with N or $N+1$ electrons is equally favorable. The corresponding graph is represented below.

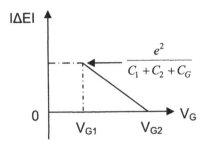

3) $\Delta E(N-1 \to N) = \dfrac{e}{C_1 + C_2 + C_G}\left(\left(N - \dfrac{1}{2}\right)e - C_G V_G\right).$

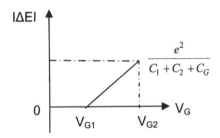

4) ΔE_{max} is given by the intersection of the two preceding curves:

$$V_{G\,max} = N\frac{e}{C_G} \quad and \quad \Delta E_{max} = \frac{e^2}{2(C_1 + C_2 + C_G)}.$$

5) $\Delta E_{max} \geq k_B T$ implies

$$T \leq \frac{e^2}{2k_B(C_1 + C_2 + C_G)}.$$

(Note that we partly check the validity of our formula since for $V_G=0$ we must recover the course expression involving only two capacitors in series.)

6) $C_1 = C_2 = 8.62 \times 10^{-18}$F, $C_G = 6.83 \times 10^{-18}$F and T=38.4 K

Exercise 9.8.1

1) Apply the tight-binding approximation, considering only the nearest neighbors and using a wave function of the form

$$\psi(x) = \frac{1}{\sqrt{N}} \sum_A \exp(ikx_A)\phi_{at}(x - x_A)$$

to find the energy band

$$E = \alpha + 2\beta \cos(ka)$$

where

$$\beta(<0) = \int \phi_{at}(x - a)H\phi_{at}(x)dx \quad and \quad \alpha = \int \phi_{at}(x)H\phi_{at}(x)dx.$$

The $E(k)$ relation is as represented below:

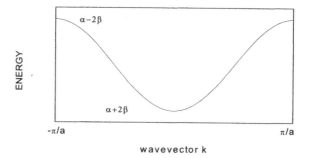

2) There is one available electron per atom to populate the p-band, which has $2N$ states when the spin degeneracy is taken into account. Therefore, this structure would be metallic, with a Fermi level located at the middle of the conduction band.

3) With a wave defined as in 1) at the bottom of the band the exponential term is equal to 1 and the corresponding wave is formed by bonding p-type orbitals:

At the top of the band $k = \pi/a$ and the exponential term takes the values 1 and -1, alternatively. The wave is formed by the sum of anti-bonding p-type orbitals.

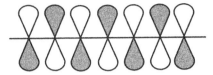

This can be compared with the lowest and highest split levels of the coupled quantum well, for which the bonding state also possesses the lowest energy.

4) By using the same tight-binding approximation as in 1), the 1D lattice, which now has a two-atom site basis, must be treated as in section 10.6 (or as the graphene sheet). The energy band is given by $E = H_{11} \pm |H_{12}|$ (equation (10.107)), which in the nearest neighbor approximation leads to a dispersion relation of the type

$$E = \alpha \pm \sqrt{\beta^2 + \gamma^2 + 2\beta\gamma \cos(ka')}$$

and therefore to the formation of two allowed energy bands.

5) Since there is one available electron per atom, the lowest band is fully occupied and the highest band is empty. The former is known as the lowest unoccupied molecular orbital (HOMO), and the latter as the highest unoccupied molecular orbital (LUMO). The conjugated polymer behaves as an intrinsic semiconductor (in practice, such polymers may be useful for low cost, light-emitting applications).

6) Conjugated polymer chains do not lead to 1D ballistic transport due to the soft character of the structure, which induces disorder and prevents them from maintaining a perfectly periodic structure. An uncompensated charge injected in the LUMO locally deforms the polymer chain and carries this deformation with it (it is called a polaron). From a general point of view, these chains behave instead like well cooked pieces of spaghetti, rather than as rigid, uncooked spaghetti.

Exercise 9.8.2

1) It is an armchair nanotube. Therefore it is metallic and is not suitable for emitting radiation, unless it is used for field emission in a vacuum.

2) We have $(m-n)/3 = 7$ integer, and thus the nanotube is metallic.

3) $(m-n)$ is not a multiple of 3, and the nanotube is semi-conductive.

4) E_G=0.585 eV. E_{m1}=0.292 eV, E_{m2}=0.585 eV and E_{m3}=1.169 eV above the energy of the K point.

5) By integrating the density of states versus energy we find the density of carriers in a given subband as a function of the Fermi energy:

$$N(E) = \frac{4}{\pi \beta_0 a \sqrt{3}} \sqrt{(E_F - E_C)^2 - (E_m - E_C)^2}$$

6) The nanotube is of the armchair type and is therefore metallic. The 1D density of states is constant and readily found to be equal to D_{1D}=8/hv_c, taking into account the valley degeneracy.

7) From Figures 9.4 and 9.11 the radius of a (n,n) armchair nanotube can be expressed as $R= 3na_0/2\pi$ and here is equal to 1.356 nm. Taking the energy origin as that of the Dirac point, the Fermi level position is E_F=$(C/L)(V_G-V_T)/eD_{1D}$=13 meV.

8) The quantized energy level splitting is ΔE=$hv_c/2L$ (equation (9.44)) and the Coulomb energy is e^2/C. The ratio between both is equal to $hv_c(C/L)/2e^2 \cong 10.4$. Thus, the level splitting prevails, but the Coulomb blockade is not negligible.

Exercise 9.8.3

1) At a very small length electrons are ballistic and there is no potential drop inside the nanotube. However, as the voltage V is increased above the optical phonon energy, there is some possibility of backscattering inside the nanotube, which induces energy loss. Thus there is an electric field which accelerates the electrons, which lose an optical phonon energy quantum as soon as their energy exceeds that threshold (see the figure below). If each electron suffers from a large number of such collisions, the average electric field ε is given by V/L, with L the nanotube length.

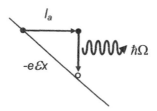

2) From the figure above it is readily found that l_a=$\hbar\Omega/e\varepsilon$= $(\hbar\Omega/eV)L$.

3) Between two collisions electrons are coherent and they lose coherence when creating an optical phonon. If we neglect the fact the electrons are accelerated between two collisions, we can apply equation (3.23) to find the transmission and making $L_0 = l_a$ in the equation gives a resistance

$$R = \frac{h}{4e^2} \frac{l_a + L}{l_a} = R_0 + \frac{h}{4e^2} \frac{l_a}{L} = R_0 + \frac{V}{I_0}$$

with

$$I_0 = \frac{4e}{h} \hbar \Omega \ .$$

4) From question 3) it is readily seen that for small voltages the resistance is that of a ballistic nanotube, and for large V the resistance is proportional to the voltage, and therefore the current saturates at I_0.

5) Taking into account the average length l_Ω traveled by an electron after its kinetic energy exceeds the optical phonon energy, the energy diagram sketched above is turned into the new one below:

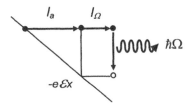

The resistance is now given by

$$R = \frac{h}{4e^2} \frac{l_a + l_\Omega + L}{l_a + l_\Omega} \ .$$

Thus, at low electric field the resistance is close to the ballistic regime, and at high electric field l_a becomes negligible compared to l_Ω and the resistance saturates at a value $R_{sat} = R_0(1 + L/l_\Omega)$. Thus, for small lengths the current no longer saturates. Resistance-voltage and current-voltage characteristics are shown below (the nanotube lengths of the R-V curves are the same as for the I-V curves).

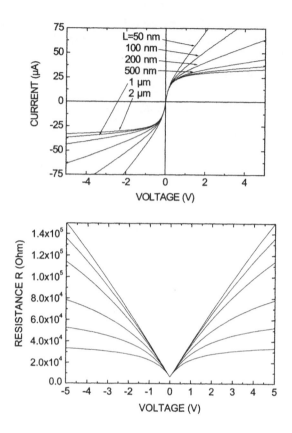

Index

Printed and bound by CPI Group (UK) Ltd, Croydon, CR0 4YY